Radosław Katarzyniak, Tzu-Fu Chiu, Chao-Fu Hong, and Ngoc Thanh Nguyen (Eds.)

Semantic Methods for Knowledge Management and Communication

T0140501

Studies in Computational Intelligence, Volume 381

Editor-in-Chief
Prof. Janusz Kacprzyk
Systems Research Institute
Polish Academy of Sciences
ul. Newelska 6
01-447 Warsaw
Poland
E-mail: kacprzyk@ibspan.waw.pl

Further volumes of this series can be found on our homepage:
springer.com

Vol. 359. Xin-She Yang, and Slawomir Koziel (Eds.)
Computational Optimization and Applications in Engineering and Industry, 2011
ISBN 978-3-642-20985-7

Vol. 360. Mikhail Moshkov and Beata Zielosko
Combinatorial Machine Learning, 2011
ISBN 978-3-642-20994-9

Vol. 361. Vincenzo Pallotta, Alessandro Soro, and
Eloisa Vargiu (Eds.)
Advances in Distributed Agent-Based Retrieval Tools, 2011
ISBN 978-3-642-21383-0

Vol. 362. Pascal Bouvry, Horacio González-Vélez, and
Joanna Kolodziej (Eds.)
Intelligent Decision Systems in Large-Scale Distributed Environments, 2011
ISBN 978-3-642-21270-3

Vol. 363. Kishan G. Mehrotra, Chilukuri Mohan, Jae C. Oh,
Pramod K. Varshney, and Moonis Ali (Eds.)
Developing Concepts in Applied Intelligence, 2011
ISBN 978-3-642-21331-1

Vol. 364. Roger Lee (Ed.)
Computer and Information Science, 2011
ISBN 978-3-642-21377-9

Vol. 365. Roger Lee (Ed.)
Computers, Networks, Systems, and Industrial Engineering 2011, 2011
ISBN 978-3-642-21374-8

Vol. 366. Mario Köppen, Gerald Schaefer, and
Ajith Abraham (Eds.)
Intelligent Computational Optimization in Engineering, 2011
ISBN 978-3-642-21704-3

Vol. 367. Gabriel Luque and Enrique Alba
Parallel Genetic Algorithms, 2011
ISBN 978-3-642-22083-8

Vol. 368. Roger Lee (Ed.)
Software Engineering, Artificial Intelligence, Networking and Parallel/Distributed Computing 2011, 2011
ISBN 978-3-642-22287-0

Vol. 369. Dominik Ryzko, Piotr Gawrysiak, Henryk Rybinski,
and Marzena Kryszkiewicz (Eds.)
Emerging Intelligent Technologies in Industry, 2011
ISBN 978-3-642-22731-8

Vol. 370. Alexander Mehler, Kai-Uwe Kühnberger,
Henning Lobin, Harald Lüngen, Angelika Storrer, and
Andreas Witt (Eds.)
Modeling, Learning, and Processing of Text Technological Data Structures, 2011
ISBN 978-3-642-22612-0

Vol. 371. Leonid Perlovsky, Ross Deming, and Roman Ilin (Eds.)
Emotional Cognitive Neural Algorithms with Engineering Applications, 2011
ISBN 978-3-642-22829-2

Vol. 372. António E. Ruano and
Annamária R. Várkonyi-Kóczy (Eds.)
New Advances in Intelligent Signal Processing, 2011
ISBN 978-3-642-11738-1

Vol. 373. Oleg Okun, Giorgio Valentini, and Matteo Re (Eds.)
Ensembles in Machine Learning Applications, 2011
ISBN 978-3-642-22909-1

Vol. 374. Dimitri Plemenos and Georgios Miaoulis (Eds.)
Intelligent Computer Graphics 2011, 2011
ISBN 978-3-642-22906-0

Vol. 375. Marenglen Biba and Fatos Xhafa (Eds.)
Learning Structure and Schemas from Documents, 2011
ISBN 978-3-642-22912-1

Vol. 376. Toyohide Watanabe and Lakhmi C. Jain (Eds.)
Innovations in Intelligent Machines – 2, 2011
ISBN 978-3-642-23189-6

Vol. 377. Roger Lee (Ed.)
Software Engineering Research, Management and Applications 2011, 2011
ISBN 978-3-642-23201-5

Vol. 378. János Fodor, Ryszard Klempous, and Carmen Paz
Suárez Araujo (Eds.)
Recent Advances in Intelligent Engineering Systems, 2011
ISBN 978-3-642-23228-2

Vol. 379. Ferrante Neri, Carlos Cotta, and Pablo Moscato (Eds.)
Handbook of Memetic Algorithms, 2011
ISBN 978-3-642-23246-6

Vol. 380. Anthony Brabazon, Michael O'Neill, and Dietmar
Maringer (Eds.)
Natural Computing in Computational Finance, 2011
ISBN 978-3-642-23335-7

Vol. 381. Radosław Katarzyniak, Tzu-Fu Chiu,
Chao-Fu Hong, and Ngoc Thanh Nguyen (Eds.)
Semantic Methods for Knowledge Management and Communication, 2011
ISBN 978-3-642-23417-0

Radosław Katarzyniak, Tzu-Fu Chiu, Chao-Fu Hong,
and Ngoc Thanh Nguyen (Eds.)

Semantic Methods for Knowledge Management and Communication

 Springer

Editors

Prof. Radosław Katarzyniak
Institute of Informatics
Wrocław University of Technology
Str. Wybrzeże Wyspiańskiego 27
50-370 Wrocław, Poland
E-mail: radoslaw.katarzyniak@pwr.wroc.pl

Prof. Tzu-Fu Chiu
Department of Industrial Management &
Enterprise Information
Aletheia University
No. 32, Chen-Li Street
Tamsui District, New Taipei City, Taiwan, R.O.C.
E-mail: chiu@mail.au.edu.tw

Prof. Chao-Fu Hong
Department of Infomation Management
Aletheia University
No. 32, Chen-Li Street
Tamsui District, New Taipei City, Taiwan, R.O.C.
E-mail: au4076@au.edu.tw

Prof. Ngoc Thanh Nguyen
Institute of Informatics
Wrocław University of Technology
Str. Wybrzeże Wyspiańskiego 27
50-370 Wrocław, Poland
E-mail: thanh@pwr.wroc.pl

ISBN 978-3-642-27055-0

DOI 10.1007/978-3-642-23418-7

Studies in Computational Intelligence

ISBN 978-3-642-23418-7 (eBook)

ISSN 1860-949X

Typeset & Cover Design: Scientific Publishing Services Pvt. Ltd., Chennai, India.

Printed on acid-free paper

9 8 7 6 5 4 3 2 1

springer.com

Preface

Knowledge management and communication have already become to be vital research and practical issues studied intensively by highly developed societies. These societies have already utilized uncountable computing techniques to create, collect, process, retrieve and distribute enormous volumes of knowledge, and created complex human activity systems involving both artificial and natural agents. In these practical contexts effective management and communication of knowledge has become badly needed to keep human activity systems ongoing. Unfortunately, the diversity of computational models applied in the knowledge management field has led to the situation in which humans (the end users of all artificial technology) find it almost impossible to utilize own products in the effective way. To cope with this problem the concept of human centered computing, strongly combined with computational collective techniques, and supported by new semantic methods has been developed and put on the current research agenda by main academic and industry centers.

In this book many interesting issues related to the above mentioned concepts are discussed in a rigorous scientific way and evaluated from practical point of view. All chapters in this book contribute directly or indirectly to the concept of human centered computing in which semantic methods are key factor of success. These chapters are extended versions of oral presentations presented during the 3rd International Conference on Computational Collective Intelligence - Technologies and Applications - ICCCI 2011 (21–23 September 2011, Gdynia, Poland) and the 1st Polish-Taiwanes Workshop on Semantic Methods for Knowledge Discovery and Communication (21–23 September 2011, Gdynia, Poland), as well as individual contributions prepared independently from these two scientific events.

September 2011

Radosław Katarzyniak
Tzu-Fu Chiu
Chao-Fu Hong
Ngoc Thanh Nguyen

Contents

Part I: Knowledge Processing in Agent and Multiagent Systems

Chapter 1: A Multiagent System for Consensus-Based Integration of Semi-hierarchical Partitions - Theoretical Foundations for the Integration Phase 3
Radosław P. Katarzyniak, Grzegorz Skorupa, Michał Adamski, Łukasz Burdka

Chapter 2: Practical Aspects of Knowledge Integration Using Attribute Tables Generated from Relational Databases 13
Stanisława Kluska-Nawarecka, Dorota Wilk-Kołodziejczyk, Krzysztof Regulski

Chapter 3: A Feature-Based Opinion Mining Model on Product Reviews in Vietnamese 23
Tien-Thanh Vu, Huyen-Trang Pham, Cong-To Luu, Quang-Thuy Ha

Chapter 4: Identification of an Assessment Model for Evaluating Performance of a Manufacturing System Based on Experts Opinions 35
Tomasz Wiśniewski, Przemysław Korytkowski

Chapter 5: The Motivation Model for the Intellectual Capital Increasing in the Knowledge-Base Organization 47
Przemysław Różewski, Oleg Zaikin, Emma Kusztina, Ryszard Tadeusiewicz

Chapter 6: Visual Design of Drools Rule Bases Using the XTT2 Method 57
Krzysztof Kaczor, Grzegorz Jacek Nalepa, Łukasz Łysik, Krzysztof Kluza

Chapter 7: New Possibilities in Using of Neural Networks Library for Material Defect Detection Diagnosis 67
Ondrej Krejcar

Chapter 8: Intransitivity in Inconsistent Judgments 81
Amir Homayoun Sarfaraz, Hamed Maleki

Part II: Computational Collective Intelligence in Knowledge Management

Chapter 9: A Double Particle Swarm Optimization for Mixed-Variable Optimization Problems 93
Chaoli Sun, Jianchao Zeng, Jengshyang Pan, Shuchuan Chu, Yunqiang Zhang

Chapter 10: Particle Swarm Optimization with Disagreements on Stagnation 103
Andrei Lihu, Ştefan Holban

Chapter 11: Classifier Committee Based on Feature Selection Method for Obstructive Nephropathy Diagnosis 115
Bartosz Krawczyk

Chapter 12: Construction of New Cubature Formula of Degree Eight in the Triangle Using Genetic Algorithm 127
Grzegorz Kusztelak, Jacek Stańdo

Chapter 13: Affymetrix Chip Definition Files Construction Based on Custom Probe Set Annotation Database 135
Michał Marczyk, Roman Jaksik, Andrzej Polański, Joanna Polańska

Part III: Models for Collectives of Intelligent Agents

Chapter 14: Advanced Methods for Computational Collective Intelligence 147
Ngoc Thanh Nguyen, Radosław P. Katarzyniak, Janusz Sobecki

Chapter 15: Identity Criterion for Living Objects Based on the Entanglement Measure 159
Mariusz Nowostawski, Andrzej Gecow

Chapter 16: Remedial English e-Learning Study in Chance Building Model 171
Chia-Ling Hsu

Chapter 17: Using IPC-Based Clustering and Link Analysis to Observe the Technological Directions 183
Tzu-Fu Chiu, Chao-Fu Hong, Yu-Ting Chiu

Chapter 18: Using the Advertisement of Early Adopters' Innovativeness to Investigate the Majority Acceptance 199
Chao-Fu Hong, Tzu-Fu Chiu, Yuh-Chang Lin, Jer-Haur Lee, Mu-Hua Lin

Chapter 19: The Chance for Crossing Chasm: Constructing the Bowling Alley 215
Chao-Fu Hong, Yuh-Chang Lin, Mu-Hua Lin, Woo-Tsong Lin, Hsiao-Fang Yang

Chapter 20: Visualization of the Technological Evolution of the DVD Business Ecosystem 231
Yan-Ru Li

Part IV: Models and Environments for Human-Centered Computing

Chapter 21: Discovering Students' Real Voice through Computer-Mediated Dialogue Journal Writing 241
Ai-Ling Wang, Dawn Michele Ruhl

Chapter 22: The $\mathcal{ALCN}$ Description Logic Concept Satisfiability as a SAT Problem 253
Adam Meissner

Chapter 23: Embedding the HEART Rule Engine into a Semantic Wiki 265
Grzegorz Jacek Nalepa, Szymon Bobek

Chapter 24: The Acceptance Model of e-Book for On-Line Learning Environment 277
Wei-Chen Tsai, Yan-Ru Li

Chapter 25: Human Computer Interface for Handicapped People Using Virtual Keyboard by Head Motion Detection 289
Ondrej Krejcar

Chapter 26: Automated Understanding of a Semi-natural Language for the Purpose of Web Pages Testing 301
Marek Zachara, Dariusz Pałka

Chapter 27: Emerging Artificial Intelligence Application: Transforming Television into Smart Television 311
Sasanka Prabhala, Subhashini Ganapathy

Chapter 28: Secure Data Access Control Scheme Using Type-Based Re-encryption in Cloud Environment 319
Namje Park

Chapter 29: A New Method for Face Identification and Determing Facial Asymmetry 329
Piotr Milczarski

**Chapter 30: 3W Scaffolding in Curriculum of Database Management
and Application – Applying the Human-Centered Computing Systems** 341
Min-Huei Lin, Ching-Fan Chen

**Chapter 31: Geoparsing of Czech RSS News and Evaluation of Its Spatial
Distribution** 353
Jiří Horák, Pavel Belaj, Igor Ivan, Peter Nemec, Jiří Ardielli, Jan Růžička

Author Index 369

Part I
Knowledge Processing in Agent
and Multiagent Systems

A Multiagent System for Consensus-Based Integration of Semi-Hierarchical Partitions - Theoretical Foundations for the Integration Phase

Radosław P. Katarzyniak, Grzegorz Skorupa, Michał Adamski, and Łukasz Burdka

Division of Knowledge Management Systems, Institute of Informatics,
Wroclaw University of Technology
Wyb.Wyspianskiego 27, 50-370 Wrocław, Poland
{radoslaw.katarzyniak,grzegorz.skorupa}@pwr.wroc.pl,
{michal.adamski,lukasz.burdka}@student.pwr.wroc.pl

Abstract. In this paper theoretical assumptions underlying design and organization of a multiagent system for knowledge integration task are presented. The input knowledge is given in form of semi-hierarchical partitions. This knowledge is distributed (produced by different agents), partial, inconsistent and requires integration phase. A central agent exists that is responsible for carrying out the integration task. A precise model for integration is defined. This model is based on the theory of consensus. An introductory discussion of computational complexity of integration step is presented in order to set up strong theoretical basis for the design of the central integrating agent. Finally, a multiagent, interactive and context sensitive strategy for integration is briefly outlined to show further design directions.

Keywords: multiagent system, semi-hierarchical partition, knowledge integration, consensus theory.

1 Introduction

Integration of knowledge is a common problem in distributed environments where multiple cooperating agents access and utilize distributed sources of knowledge about a common environment. The problem of knowledge integration has already been studied for multiple data structures and as a result various models for integration have been proposed to cope with its particular cases [6]. These models are usually idealized, contrary to actual circumstances where knowledge integration processes are hybrid, dependent on complex input knowledge profiles, interactive, based on evaluations of intermediate integration results, consisting of many steps of intensive and communication-rich communication between knowledge processors and knowledge sources [2]. However, it must be stressed that the idealized models for knowledge integration tasks are practically important because they give theoretical means for effective analysis of nature and complexity of particular knowledge integration tasks.

In this paper such idealized model for knowledge integration is presented and an initial idea of its effective utilization in the context of multiagent system is briefly

R. Katarzyniak et al. (Eds.): Semantic Methods, SCI 381, pp. 3–12.
springerlink.com   © Springer-Verlag Berlin Heidelberg 2011

outlined. It is assumed that knowledge is represented by semi-hierarchical partitions. Each semi-hierarchical partition represents an individual and usually incomplete point of view of an agent on a current state of a common environment. The target of a dedicated central agent is to integrate such incoming, incomplete and inconsistent views to produce a unified collective representation of a current state, provided that such representation has got high quality and can be computed in an effective way.

The knowledge integration task discussed in this paper is in many ways similar to knowledge integration problems defined elsewhere for the case of ordered hierarchical coverings and ordered hierarchical partitions [1,2]. In particular, it is based on the same theory of choice and consensus which provides strict computational models for socially acceptable approaches to the creation of collective opinions. However, our knowledge integration task is considered for a newly separated class of similar knowledge structures which are different to the above mentioned ordered hierarchical partitions and ordered hierarchical coverings.

The forthcoming text is organized as follows. At first, the most important details of the assumed knowledge representation method are given, as well as a related idealized consensus-based model for solving the problem of integration of semi-hierarchical partitions is proposed. At second, the chosen integration task is discussed in order to determine its computational complexity. At third, a general overview of an original strategy of knowledge integration is discussed. It is assumed that this strategy is to be realized by a multiagent system consisting of individual agents situated in a distributed processing environment. One agent is chosen for carrying out the main integration task. In the paper a few related research problems into the complexity issues are pointed in order to define future directions of design and implementation work.

2   Partial Hierarchical Coverings in Knowledge Integration Tasks

2.1   Source and Pragmatic Interpretation of Knowledge Items

Let us assume that knowledge about a world is represented by classifications of objects $O=\{o_1,o_2,...,o_N\}$. Each classification is interpreted as representation of a current state of this world produced by an individual agent from $Ag=\{ag_1,...,ag_K\}$. This knowledge can be incomplete. Agents can carry out classifications based on the attributes $A=\{a_1,a_2,..,a_M\}$ which are organized in a sequence $(a_{i1},a_{i2},...,a_{iM})$, $m,n=1...M$, where for $m\neq n$ $a_{in}\neq a_{im}$ holds. Obviously, this sequence defines M-level classification tree related to the following sequence of attribute-value atom tests: $test(a_{i1});test(a_{i2});...;test(a_{iM})$. Obviously, each attribute a_{in} refers to a particular level of the classification tree. It is further assumed that each run of classification procedure is carried out until the final test $test(a_{iM})$ is completed or the next test to be realized cannot be completed due to an unknown value of the related attribute e.g. $test(a_i)$ has been realized but due to an unknown value of a_{i+1} the test $test(a_{i+1})$ is not to be launched at all (see Example 1). Such classification procedure is an easy case of a very rich class of decision tree-based classification schemes e.g. [8].

Example 1. Let $O=\{o_1,o_2,o_3,o_4,o_5,o_6,o_7\}$, $A=\{a,b,c\}$, $V_a=\{a_1,a_2\}$, $V_b=\{b_1,b_2,b_3\}$, and $V_c=\{c_1,c_2\}$ be objects, attributes, and domains of attributes, respectively. Let the classification procedure be defined by the following sequence $test(a);test(b);test(c)$. The related decision tree produced by this sequence is presented on Fig. 1 and Fig. 2.

On Fig. 2 and Fig. 2 two classifications C_1 and C_2 are presented. The following interpretation of C_1 shows the commonsense meaning of accepted model of classification:

a) Agent ag_1 *knows* that:
- Object o_1 exhibits properties: $a=a_1$ and $b=b_2$;
- Object o_2 exhibits properties: $a=a_1$ and $b=b_2$ and $c=c_1$;
- Object o_5 exhibits property $a=a_2$.

b) Agent ag_1 *does not know* current values of attribute:
- a in object o_6;
- b in object o_5;
- c in object o_1.

c) *It does not follow from* C_1 whether agent ag_1 knows anything about actual values of the following attributes:
- b and c in object o_6;
- c in object o_5.

Object o_2 and o_3 are located in some tree leafs. Therefore they should be treated as completely classified. However, the classification of other objects in C_1 is incomplete. in this sense C_1 is treated as partial.

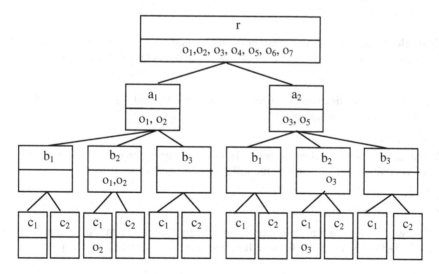

Fig. 1. Partial classification C_1

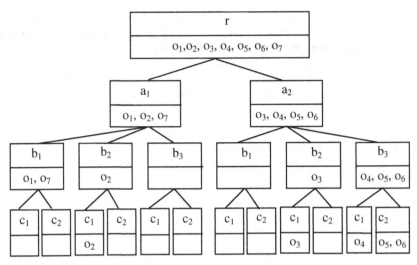

Fig. 2. Complete classification C_2

The knowledge integration task studied in the following sections will be defined for profiles of classifications produced similarly as in Example 1. It is quite obvious, that to make this integration task solvable one needs to choose a particular measure of differences between classification results. In our case we use a relatively natural measure of distance defined as the minimal number of objects' movements between the parent-child nodes, needed to transfer one classification to another. Example 2 explains this idea.

Example 2. Let us consider classifications C_1 and C_2 from Fig. 1 and Fig. 2. The following objects are located in the same nodes in both classification trees:

- o_2, o_3.

It means that the distance between classifications C_1 and C_2 results from different locations of the following objects:

- o_1, o_4, o_5, o_6, o_7.

At the same time the following holds: In order to move object o_i from its position in C_1 to its position in C_2 one needs the following numbers of movements:

- 2 for o_2;
- 3 for each of o_1, o_4, o_5 and o_6.

It follows that the minimal number of objects' movements required to transfer C_1 to C_2 is $2+3+3+3+3=14$. Such distance function is intuitive and easy to compute.

2.2 Universe of Knowledge Items

Let $UV_T(O)$ be the universe of all classifications that can ever be computed for a set O and a classification tree T. The following Corollary 1 results from the accepted classification strategy:

Corollary 1. Let T, r, W, and L denote a classification tree, the root of T, the set of all tree nodes in T different to r, and the set of all leafs of T. Each classification $C \in UV_T(O)$ is a function $C:W \cup \{r\} \to 2^O$ that fulfills the following conditions:
a) $C(r) = O$,
b) for $m, n \in W \cup \{r\}$, if n is the parent node of m, then $C(m) \subset C(n)$,
c) for $m, n \in W$, if n and m are children nodes of the same parent node, then $C(m) \cap C(n) = \varnothing$.

Definition 1. Elements of $UV_T(O)$ are called *semi-hierarchical ordered partitions*.

Definition 2. Let the distance function given in Example 2 be denoted by $\delta: UV_T(O) \times UV_T(O) \to R^+$.

The universe $UV_T(O)$ can be easily related to some classes of tree-based knowledge structures previously studied in the theory of consensus [1][2][6]. Two of them have been already mentioned and seem to be especially important for further analysis of our case. Let T, r, W, and L are interpreted as in Corollary 1. The following definitions Def. 3 and Def. 4 have been proposed elsewhere e.g. [1]:

Definition 3. Function $C:W \cup \{r\} \to 2^O$ is called a *hierarchical ordered covering* of O if and only if the following conditions are fulfilled:
a) $C(r)=O$
b) for two nodes $m, n \in W \cup \{r\}$, if n is the parent node of m, then $C(m) \subset C(n)$.

Let the *universe of hierarchical ordered coverings* of O, based on the tree T, be denoted by $V_T(O)$.

Definition 4. Function $C:W \cup \{r\} \to 2^O$ is called a *hierarchical ordered partition* of O if and only if the following conditions are fulfilled:
a) for nodes $m,n \in W \cup \{r\}$, if n is the parent node of m, then $C(m) \subset C(n)$,
b) for $m, n \in W$, if n and m are children nodes of the same parent node, then $C(m) \cap C(n) = \varnothing$,
c) $\bigcup_{x \in L} C(x) = O$.

Let the *universe of hierarchical ordered partitions* of O, based on tree T, be denoted by $U_T(O)$.
    It is easy to see that Corollary 2 holds:

Corollary 2. For a given tree T:

$$U_T(O) \subset UV_T(O) \subset V_T(O).$$

This fact can be used to expect some undesirable computational problems related to knowledge integration tasks defined for items form $UV_T(O)$.

2.3 Idealized Model for Knowledge Integration Task

The theory of choice and consensus makes it possible to define our idealized model for knowledge integration task. Namely, this task can be treated as equivalent to the following problem of consensus choice:

Definition 5. Let $UV_T(O)$ be given. Let $\Pi(UV_T(O))$ and $\Pi^*(UV_T(O))$ be the set of all subsets without and with repetitions of $UV_T(O)$, respectively. Elements of $\Pi^*(UV_T(O))$ are called knowledge profiles. Let $C=\{C_1,C_2,...,C_K\}$, $C\in\Pi^*(UV_T(O))$ be given. The knowledge integration task is defined by the following elements:

a) a distance function d: $UV_T(O)\times UV_T(O)\rightarrow R^+$
b) a choice functions R_n: $\Pi^*(UV_T(O)) \rightarrow \Pi(UV_T(O))$, such that for n=1,2,... and $C\in\Pi^*(UV_T(O))$, profile $C^*\in R_n(C)$ if and only if

$$\sum_{X\in C}\left[d(C^*,X)\right]^n = \min_{Y\in UV_T(O)} \left(\sum_{X\in C}\left[d(Y,X)\right]^n\right).$$

Elements of $R_n(C)$ can be treated as alternative structures representing the result of knowledge integration step realized for a particular input profile $C\in\Pi^*(UV_T(O))$.

Def. 5 sets up a general scheme of our knowledge integration task. There are two practical problems that need to be solved when Def. 5 is used. The first problem refers to computational complexity of actual implementations of Def. 5. Namely, it has already been proven that for the similar task defined for profiles from $\Pi^*(U_T(O))$ and $\Pi^*(V_T(O))$ this complexity is strongly influenced by distance function d and a choice function R_n. The second problem refers to the quality of knowledge which can be characterized by different levels of consistency and completeness. In the idealized strategy of integration task, the consistency of input profiles is not considered although it determines the quality of the final result. In the forthcoming sections some hints are given to solving both problems in an effective way.

3 Implementing of Knowledge Integration Phase

3.1 Computational Complexity of Integration Step

The idealized approach to knowledge integration task proposed in Def. 5 has already been applied and studied for other popular knowledge structures, in particular for the already mentioned profiles of elements from $U_T(O)$ and $V_T(O)$ [1]. It has already been proven that the integration of profiles from $U_T(O)$ and $V_T(O)$ leads to unacceptable computational complexity of integration task. Example 3 explains it in more detail (see also [1]).

Example 3. Let $U_T(O)$ and $C=\{C_1,C_2,...,C_K\}$, $C\in\Pi^*(U_T(O))$ be given. Let $\eta:U_T(O)\times U_T(O)\rightarrow R^+$ be a distance function such that d(C',C'') is the minimal number of objects that have to be moved from one tree leaf to another in order to transfer C' into C'' e.g. $\eta(C_1,C_2) = 5$ (see Fig. 1 and Fig. 2) Note: The distance function η differs to δ (see Def. 3). In [1] the following theorems were proved:

a) If functions R_1 and η are applied to implement Def. 5, then the choice of consensus is computationally tractable and can be realized in a polynomial time. An example is given in [1].

b) If functions R_n, n≥2, and η are applied to implement Def. 5, then the integration task becomes computationally difficult and is equivalent to some NP-complete problem.

Similar results were obtained for the universe $V_T(O)$ and another distance functions [1] [2]. Due to Corollary 2 it is reasonable to expect that the same situation can take place for profiles from $UV_T(O)$.

Let us assume that the distance function δ (see Def. 2) and the choice function R_1 are used to implement the strategy described in Def. 5. The following theorem can be proved:

Theorem 1. If $C=\{C_1,C_2,...,C_K\}$, $C \in \Pi^*(UV_T(O))$, then $R_1(C)$ can be computed in polynomial time.

Proof. Let M, pos_T and $\underline{\delta}$ be given as follows:

$M=card(W \cup \{r\})$ is the number of nodes of the tree T (Note: numbers 1...M identify in unique way the position of all tree nodes in the given tree structure),

pos_T: $O \times UV_T(O) \rightarrow \{1..M\}$ where and $pos_T(o,X)$ is the node number in which object $o \in O$ is located in hierarchical partition $X \in UV_T(O)$,

$\underline{\delta}$: $K \times K \rightarrow R^+$ where $\underline{\delta}(p,q)$ is the length of the shortest path between nodes p and q in the tree T.

Let us consider the following algorithm:

Algorithm P

Create an empty partition $C_c \in UV_T(O)$.
For each object $o_i \in \{o_1,o_2,...,o_N\}$ **do begin**
  Step 1: compute the sums Σ_1 .... Σ_K:
  $\Sigma_1 := \underline{\delta}(1,pos_T(o_i,C_1))+...+\underline{\delta}(1,pos_T(o_i,C_K))$
  ....
  $\Sigma_M := \underline{\delta}(M,pos_T(o_i,C_1))+...+\underline{\delta}(M,pos_T(o_i,C_K))$.
  Step 2: choose $s \in \{1...M\}$ such that $\Sigma_s = \min_{p=1..M}\left(\Sigma_p\right)$.
  Step 3: include o_i to node s in partition C_c.
end.

At first, it can be proven that C_c fulfills the requirement:

$$\sum_{X \in C} d(C_c,X) = \min_{Y \in UV_T(O)}\left(\sum_{X \in C} d(Y,X)\right).$$

In consequence $C_c=R_1(C)$ (see Def. 5).

At second, it can be proven that Algorithm P computes $R_1(C)$ in polynomial time. It is easy to notice that:

1. Due to polynomial complexity of $pos_T(p,q)$), Step 1 is realized in polynomial time.
2. Step 2 and Step 3 are realized in polynomial time.

It follows from 1 and 2 that Algorithm P is polynomial.

Unfortunately, this desirable feature does not hold in applications where the same distance function δ is combined with the choice functions $R_n(C)$, $n \geq 2$. Namely, in these cases the knowledge integration task becomes NP-complete. The proof is to be published elsewhere. At this stage it is enough to mention that similar results for ordered hierarchical partitions and ordered hierarchical coverings are given in [1].

It is quite obvious that NP-completeness exhibited by some implementations of Def. 5 forces systems designers to develop and implement heuristic solutions.

3.2 Coping with Inconsistency of Knowledge Profiles

The second problem that can influence the quality of final integration results originates from low consistency of incoming input knowledge profiles. It often happens in real circumstances that input profiles are highly inconsistent and/or can be apparently divided into disjoint classes of knowledge items. Humans have developed multiple cognitive strategies to cope with profile's inconsistency. In particular they can use appropriate methods to remove from input knowledge profiles items that decrease the knowledge consistency. It is also possible for them to pre-processed the input knowledge profile by evaluation of particular profile's items on the base of knowledge source value. In such case rich communication and cooperation between agents is required.

In another approach the integrating agent can be forced to accept the low consistency of input knowledge in order to reflect this feature by computing multiple alternatives for the knowledge. To achieve this or similar target the agent can start the knowledge integration phase by computing separate clusters of collected input knowledge items and then deriving separate representatives for each of these clusters. in this case the natural inconsistency of knowledge collected in a particular context is accepted and treated as important feature of a problem domain. it easily follows that in order to implement this strategy for improving the quality of input profiles various data mining techniques, including clustering methods, can be effectively applied.

4   Outline of Multiagent Strategy for Knowledge Integration Task

Let us now summarize the above discussion by outlining at a very general level the following multiagent strategy for our knowledge integration task. This way further directions of necessary research and development related to our knowledge integration task can be better defined as well as possible implementation methodologies determined.

Central agent is responsible for gathering and integrating knowledge. She should also measure and perform tasks to increase quality of obtained results. The integration process can be divided into a few phases: gathering knowledge, measuring consistency of obtained knowledge, if necessary choosing strategies to increase result consistency and finally integrating knowledge according to chosen strategies.

4.1 Choosing Trustworthy Sources When Gathering Knowledge

Central agent can try to increase knowledge consistency at first stage by choosing only most trustworthy sources. In such case it is assumed agent knows how trustful each source is. According to this knowledge agent chooses data from only K most trustful sources. Later agent checks if gathered data contains enough information about partial classifications of objects. If some information is missing, agent asks further sources only about the missing data. Data gathered in such manner constructs set C from Def. 5.

4.2 Knowledge Integration Process

Knowledge integration is run according to algorithm P described within the proof of Theorem 1. This algorithm has polynomial complexity. Various distance functions $\underline{\delta}$ can be proposed to obtain a desired result properties. Functions δ and η were proposed in previous paragraphs as examples. There are many other functions that need testing.

4.3 Clustering Knowledge in Case of a Low Consistency

Later agent has to measure the consistency of obtained result C*. Result consistency measure p(C*,C) = card(C) / (card(C)+d(C*,C)) may be used. Intuitively low value of p means that result if inconsistent. If the measure is less than defined threshold agent must accept poor consistency of input knowledge. Proposing a few more consistent integration results is required. In such case agent can divide input data into separate more consistent sets. This can be achieved using clustering algorithms. K-Means [4] method with a distance function $\underline{\delta}$ may be used. This well known algorithm finds K clusters among data. When facing inconsistent integration result agent runs K-Means algorithm for K=2 and proposes 2 separate results. If the consistency is still below required threshold number of clusters (K) is increased and the process is repeated.

Another strategy to obtain a few alternative but consistent integrated knowledge representations is to find out the right number of clusters in data and run the integration for each found cluster. This approach is similar to previous one but does not measure consistency directly. Finding out the right number of clusters is a well known problem and can be solved using various methods (see [5] for many examples). Silhouette Validation Method used with a measure called Overall Average Silhouette Width [7] is one example.

5 Conclusions

In this paper detailed theoretical basis for effective implementation of knowledge integration task has been presented and discussed. The knowledge integration task was defined for profiles of semi-hierarchical ordered partitions, which are a subclass of hierarchical ordered coverings and a super-class of hierarchical ordered partitions. Due to the fact that many consensus problems for profiles consisting of hierarchical ordered partitions and coverings are equivalent to NP-complete decision problems, it can be expected that the same situation will take place for profiles of semi-hierarchical ordered partitions.

Acknowledgements. This paper was partially supported by Grant no. N N519 407437 funded by Polish Ministry of Science and Higher Education (2009-2012).

References

1. Daniłowicz, C., Nguyen, N.T.: Methods of consensus choice for profiles of ordered coverings and ordered partitions. Wrocław University of Technology, Wrocław (1992) (in Polish)
2. Daniłowicz, C., Nguyen, N.T., Jankowski, Ł.: Methods of representation choice for knowledge state of multiagent systems. Oficyna wydawnicza Politechniki Wrocławskiej, Wrocław (2002) (in Polish)
3. Duong, T.H., Nguyen, N.T., Jo, G.S.: A Method for Integration of WordNet-based Ontologies Using Distance Measures. In: Lovrek, I., Howlett, R.J., Jain, L.C. (eds.) KES 2008, Part I. LNCS (LNAI), vol. 5177, pp. 210–219. Springer, Heidelberg (2008)
4. Lloyd, S.P.: Least squares Quantization in PCM. IEEE Transactions on Information Theory 28(2), 129–137 (1982)
5. Milligan, G.W., Cooper, M.C.: An Examination of Procedures for Determining the Number of Clusters in a Data Set. Psychometrika 50(2), 159–179 (1985)
6. Nguyen, N.T.: Advanced Methods for Inconsistent Knowledge Management. Springer, London (2007)
7. Rousseeuw, P.J.: Silhouettes: a Graphical Aid to the Interpretation and Validation of Cluster Analysis. Computational and Applied Mathematics 20, 53–65 (1987)
8. Silla, C.N., Freitas, A.A.: A survey of hierarchical classification across different application domains. Data Mining and Knowledge Discovery 22, 31–72 (2011)

Practical Aspects of Knowledge Integration Using Attribute Tables Generated from Relational Databases

Stanisława Kluska-Nawarecka[1,2], Dorota Wilk-Kołodziejczyk[3], and Krzysztof Regulski[4]

[1] Foundry Research Institute, Cracow, Poland
[2] Academy of Information Technology (WSInf), Łódź, Poland
[3] Andrzej Frycz Modrzewski University, Cracow, Poland
[4] AGH University of Science and Technology, Cracow, Poland
nawar@iod.krakow.pl, wilk.kolodziejczyk@gmail.com,
regulski@tempus.metal.agh.edu.pl

Abstract. Until now, the use of attribute tables, which enable approximate reasoning in tasks such as knowledge integration, has been posing some difficulties resulting from the difficult process of constructing such tables. Using for this purpose the data comprised in relational databases should significantly speed up the process of creating the attribute arrays and enable getting involved in this process the individual users who are not knowledge engineers. This article illustrates how attribute tables can be generated from the relational databases, to enable the use of approximate reasoning in decision-making process. This solution allows transferring the burden of the knowledge integration task to the level of databases, thus providing convenient instrumentation and the possibility of using the knowledge sources already existing in the industry. Practical aspects of this solution have been studied on the background of the technological knowledge of metalcasting.

Keywords: attribute table, knowledge integration, databases, rough sets, methods of reasoning.

1 Introduction

The rough logic based on rough sets developed in the early '80s by Prof. Zdzislaw Pawlak [1] is used in the analysis of incomplete and inconsistent data. Rough logic enables modelling the uncertainty arising from incomplete knowledge which, in turn, is the result of the granularity of information. The main application of rough logic is classification, as logic of this type allows building models of approximation for a family of the sets of elements, for which the membership in sets is determined by attributes. In classical set theory, the set is defined by its elements, but no additional knowledge is needed about the elements of the universe, which are used to compose the set. The rough set theory assumes that there are some data about the elements of the universe, and these data are used in creation of the sets. The elements that have the same information are indistinguishable and form the, so-called, elementary sets.

R. Katarzyniak et al. (Eds.): Semantic Methods, SCI 381, pp. 13–22.
springerlink.com   © Springer-Verlag Berlin Heidelberg 2011

The set approximation in a rough set theory is achieved through two definable sets, which are the upper and lower approximations. The reasoning is based on the attribute tables, i.e. on the information systems, where the disjoint sets of conditional attributes C and decision attributes D are distinguished (where A is the total set of attributes and $A = C \cup D$.).

So far, attribute tables used for approximate reasoning at the Foundry Research Institute in Cracow were generated by knowledge engineers and based on data provided by experts, using their specialist knowledge [2]. The reasoning was related with the classification of defects in steel castings. Yet, this process was both time-consuming and expensive. Automatic or at least semi-automatic creation of attribute arrays would allow the integration of technological knowledge, already acquired in the form of e.g. relational databases, with the methods of approximate reasoning, the results of which are becoming increasingly popular in processes where it is necessary to operate on knowledge incomplete and uncertain [3, 4, 5].

2 Relational Data Model

2.1 Set Theory vs. Relational Databases

The relational databases are derived in a straight line from the set theory, which is one of the main branches of mathematical logic. Wherever we are dealing with relational databases, we *de facto* operate on sets of elements. The database is presented in the form of arrays for entities, relationships and their attributes. The arrays are structured in the following way: entities – rows, attributes - columns, and relationships - attributes. The arrays, and thus the entire database, can be interpreted as relations in a mathematical meaning of this word. Also operations performed in the database are to be understood as operations on relations. The basis of such model is the relational algebra that describes these operations and their properties. If sets A_1, A_2, ... A_n are given, the term "relation r" will refer to any arbitrary subset of the Cartesian product $A_1 A_2 ... A_n$. A relation of this type gives a set of tuples $(a_1, a_2, ..., a_n)$, where each $a_i \in A_i$. In the case of data on casting defects, the following example can be given:

damage-name= {cold laps, cold shots}
damage-type= {wrinkles, scratch, fissure, metal beads}
distribution = {local, widespread}
where relation:
   r= {(cold laps, wrinkles, local), (cold shots, metal beads, NULL), (cold laps, fissure, widespread)}

is determined on a damage-name $\times$ damage-type $\times$ distribution basis. In a relational data model A_1, A_2, ..., A_n mean attributes, while $R = (A_1, A_2, ..., A_n)$ is schema of relation R e.g. damage-schema = (damage-name, damage-type, distribution), r(R) means instance r of relation with schema R (e.g. damage (damage-schema). On the other hand, $t = (a_1, a_2, ..., a_n)$, $t \in$ r, denotes a tuple of relation r(R). The current value of the relationship (relationship instance) can be presented in a tabular form, where:

- columns correspond to attributes,
- header corresponds to the scheme of relation,
- elements of the relationship - tuples are represented by rows.

It is customary to present a model of a database - schema of relationship with ER (entity relationship) models to facilitate the visualisation. The simplest model of a database on defects in steel castings can take the form shown in Figure 1.

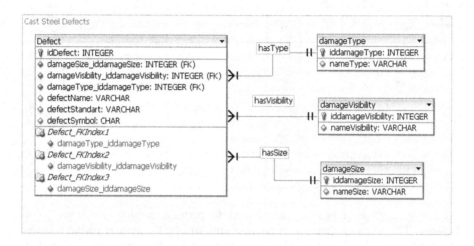

Fig. 1. A fragment of ER database model for defects in steel castings

2.2 Generating Attribute Tables Based on Relational Databases

As can be concluded from this brief characterisation of the relational databases, even their structure, as well as possible set theory operations (union, intersection, set difference, and Cartesian product) serve as a basis on which the attribute tables are next constructed, taking also the form of relationships. Rows in an attribute array define the decision rules, which can be expressed as:

$$IF \dots THEN \dots: X \rightarrow Y \qquad (1)$$

where prerequisite $X = x_1 \wedge x_2 \wedge .. \wedge x_n$ is the conditional part of a rule, and Y (conclusion) is its decision part. Each decision rule sets decisions to be taken if conditions given in the table are satisfied. Decision rules are closely related with approximations. Lower approximations of decision classes designate deterministic decision rules clearly defining decisions based on conditions, while upper approximations appoint the non-deterministic decision rules.

The attributes with an area ordered by preference are called criteria because they refer to assessment in a specific scale of preference. An example is the row in a decision table, i.e. an object with a description and class assignment.

It is possible, therefore, to generate an attribute table using a relational database. The only requirement is to select from the schema of relations, basing on the expert knowledge, the attributes that should (and can) play the role of decision attributes in the table, and also a set of conditional attributes, which will serve as a basis for the classification process.

In the case of Table 1, the conditional attributes will be attributes a_4-a_{12}, and the decision attributes will be a_1-a_3, since the decision is proper classification of defect.

Table 1. Fragment of attribute table for defects in steel castings

Attribute symbol	a_1	a_2	a_3	a_4	a_5	a_6	a_7	a_8	a_9
SYMBOL OF OBJECT IN ARRAY	DAMAGE SYMBOL	DAMAGE NAME	STANDARD	DAMAGE TYPE	DISTRIBUTION	LOCATION	OCCURRENCE	DAMAGE SHAPE	TECHNOLOGICAL OPERATION
x_1	341	COLD LAPS	CZ	WRINKLES SCRATCH, EROSION SCAB	LOCAL	INSERT WALL, CHAPLET, SURFACE	NUMEROUS	NARROW, ROUNDED EDGES	CASTING DESIGN, POURING, COOLING
x_2	w207	COLD LAP	PL	FISSURE, SCRATCH	LOCAL	SURFACE	SINGLE	NARROW, ROUNDED EDGES	GATING SYSTEM DESIGN, POURING
x_3	w407	COLD SHOTS	PL	METAL BEADS		INTERIOR		SPHER-ICAL	GATING SYSTEM DESIGN, POURING
x_4	c311	COLD LAP, COLD SHOTS	FR	DIS-CONTINUITY, FISSURE	WIDESPREAD	SURFACE, SUB-SURFACE AREA	NUMEROUS	ROUNDED EDGES, NARROW	FEEDING SYSTEM, DESIGN, POURING
x_5	c331	COLD LAP NEAR CORE OR OTHER METALLIC PART	FR	DIS-CONTINUITY	LOCAL	NEAR INSERTS	DATA NOT AVAILABLE	CURVED WALLS	POURING, SOLID-IFICATION

Creating attribute tables we are forced to perform certain operations on the database. The resulting diagram of relationships will be the sum of partial relations, and merging will be done through one common attribute. In the case of an attribute table, the most appropriate type of merging will be external merging, since in the result we expect to find all tuples with the decision attributes and all tuples with the conditional attributes, and additionally also those tuples that do not have certain conditional attributes, and as such will be completed with NULL values.

3 Classification Using Rough Set Theory

The basic operations performed on rough sets are the same as those performed on classical sets. Additionally, several new concepts not used in classical sets are introduced.

3.1 Indiscernibility Relation

For each subset of attributes, the pairs of objects are in the relation of indiscernibility if they have the same values for all attributes from the set B, which can be written as:

$$IND(B) = \{x_i, x_j \in U : \forall b \in B, f(x_i, b) = f(x_j, b)\} \qquad (2)$$

The relation of indiscernibility of elements x_i and x_j is written as $x_i \; IND(B) \; x_j$. Each indiscernibility relation divides the set into a family of disjoint subsets, also called abstract classes (equivalence) or elementary sets. Different abstract classes of the indiscernibility relation are called elementary sets and are denoted by $U \; / \; IND \; (B)$. Classes of this relation containing object x_i are denoted by $[x_i]_{IND(B)}$. So, the set

$[x_i]_{IND(B)}$ contains all these objects of the system S, which are indistinguishable from object x_i in respect of the set of attributes B [6]. The abstract class is often called an elementary or atomic concept, because it is the smallest subset in the universe U we can classify, that is, distinguish from other elements by means of attributes ascribing objects to individual basic concepts.

The indiscernibility relationship indicates that the information system is not able to identify as an individual the object that meets the values of these attributes under the conditions of uncertainty (the indeterminacy of certain attributes which are not included in the system). The system returns a set of attribute values that match with certain approximation the identified object.

Rough set theory is the basis for determining the most important attributes of an information system such as attribute tables, without losing its classificability as compared with the original set of attributes. Objects having identical (or similar) names, but placed in different terms, make clear definition of these concepts impossible. Inconsistencies should not be treated as a result of error or information noise only. They may also result from the unavailability of information, or from the natural granularity and ambiguity of language representation.

To limit the number of redundant rules, such subsets of attributes are sought which will retain the division of objects into decision classes the same as all the attributes. For this purpose, a concept of the reduct is used, which is an independent minimal subset of attributes capable of maintaining the previous classification (distinguishability) of objects. The set of all reducts is denoted by RED (A).

With the notion of reduct is associated the notion of core (kernel) and the interdependencies of sets. The set of all the necessary attributes in B is called kernel (core) and is denoted by core (B). Let $B \subseteq A$ and $a \in B$. We say that attribute a is superfluous in B when:

$$IND(B) = IND(B - \{a\}) \tag{3}$$

Otherwise, the attribute a is indispensable in B. The set of attributes B is independent if for every $a \in B$ attribute a is indispensable. Otherwise the set is dependent.

The kernel of an information system considered for the subset of attributes $B \subseteq A$ is the intersection of all reducts of the system

$$core(B) = \cap \, RED(A) \tag{4}$$

Checking the dependency of attributes, and searching for the kernel and reducts is done to circumvent unnecessary attributes, which can be of crucial importance in optimising the decision-making process. A smaller number of attributes means shorter dialogue with the user and quicker searching of the base of rules to find an adequate procedure for reasoning. In the case of attribute tables containing numerous sets of unnecessary attributes (created during the operations associated with data mining), the problem of reducts can become a critical element in building a knowledge base. A completely different situation occurs when the attribute table is created in a controlled manner by knowledge engineers, e.g. basing on literature, expert knowledge and/or standards, when the set of attributes is authoritatively created basing on the available knowledge of the phenomena. In this case, the reduction of attributes is not necessary, as it can be assumed that the number of unnecessary attributes (if any) does not affect the deterioration of the model classificability.

3.2 Query Language

Query language in information systems involves rules to design questions that allow the extraction of information contained in the system. If the system represents information which is a generalisation of the database in which each tuple is the realisation of the relationship which is a subset of the Cartesian product (data patterns or templates), the semantics of each record is defined by a logical formula assuming the form of [8]:

$$\varphi_i=[A_1=a_{i,1}] \wedge [A_2=a_{i,2}] \wedge \ldots \wedge [A_n=a_{i,n}] \qquad (5)$$

The notation $A_j=a_{i,j}$ means that the formula $_i$ is true for all values that belong to the set $a_{i,j}$. Hence, if $a_{i,j}=\{a_1, a_2, a_3\}$, $A_j=a_{i,j}$ means that $A_i=a_1 \vee A_i=a_2 \vee A_i=a_3$, while the array has a corresponding counterpart in the formula:

$$\Psi= \varphi_1 \vee \varphi_2 \vee .. \vee \varphi_m \qquad (6)$$

If the array represents some rules, the semantics of each row is defined as a formula:

$$\rho_i=[A_1=a_{i,1}] \wedge [A_2=a_{i,2}] \wedge \ldots \wedge [A_n=a_{i,n}] \Rightarrow [H=h_i] \qquad (7)$$

On the other hand, to the array of rules is corresponding a conjunction of formulas describing the rows. The decision table (Table 1.) comprises a set of conditional attributes $C=\{a_4, a_5, a_6, a_7, a_8, a_9\}$ and a set of decision attributes $D= \{a_1, a_2, a_3\}$. Their sum forms a complete set of attributes $A = C \cup D$. Applying the rough set theory, it is possible to determine the elementary sets in this table. For example, for attribute a_4 (*damage type*), the elementary sets will assume the form:

− $E_{wrinkles} = \{\emptyset\}$; $E_{scratch} = \{ \emptyset \}$; $E_{erosion\ scab} = \{ \emptyset \}$; $E_{fissure} = \{ \emptyset \}$;
− $E_{wrinkles,\ scratch,\ erosion\ scab} = \{ x_1 \}$; $E_{cold\ shots} = \{x_3\}$; $E_{fissure,\ scratch} = \{x_2\}$;
− $E_{discontinuity} = \{ x_5 \}$; $E_{discontinuity,\ fissure} = \{ x_4 \}$;
− $E_{wrinkles,\ scratch,\ erosion\ scab,\ fissure} = \{\emptyset\}$; $E_{wrinkles,\ scratch,\ erosion\ scab,\ fissure,\ cold\ shots} = \{\emptyset\}$;
− $E_{wrinkles,\ scratch,\ erosion\ scab,\ cold\ shots} = \{\emptyset\}$; $E_{discontinuity,\ fissure,\ cold\ shots} = \{\emptyset\}$;
− $E_{discontinuity,\ fissure,\ wrinkles,\ scratch,\ erosion\ scab} = \{\emptyset\}$;

Thus determined sets represent a partition of the universe done in respect of the relationship of indistinguishability for an attribute "*distribution*". This example shows one of the steps in the mechanism of reasoning with application of approximate logic. Further step is determination of the upper and lower approximations in the form of a pair of precise sets. Abstract class is the smallest unit in the calculation of rough sets. Depending on the query, the upper and lower approximations are calculated by summing up the appropriate elementary sets.

The sets obtained from the Cartesian product can be reduced to the existing elementary sets.

Query example: $t_1=$ (damage type, {discontinuity, fissure}) · (distribution, {local})

When calculating the lower approximation, it is necessary to sum up all the elementary sets for the sets of attribute values which form possible subsets of sets in a query:

$\underline{S}$ (t₁) = (damage type, {discontinuity, fissure}) · (distribution, {local}) + (damage type, {discontinuity})] · (distribution, { local})

The result is a sum of elementary sets forming the lower approximation:

E $_{discontinuity, local}$ ∪ E $_{discontinuity, fissure, local}$ = {x₅}

The upper approximation for the submitted query is:

$\underline{S}$ (t₁) = (damage type, {discontinuity, fissure}) · (distribution, {local}) + (damage type, {discontinuity})] · (distribution, {local}) + (damage type, {discontinuity, scratch})] · (distribution, {local})

The result is a sum of elementary sets forming the upper approximation:

E$_{discontinuity,local}$ ∪ E$_{fissure,scratch,local}$ ∪ E$_{discontinuity,fissure,local}$ = { x₂, x₅, }

3.3 Reasoning Using RoughCast System

The upper and lower approximations describe a rough set inside which there is the object searched for and defined with attributes. The practical aspect of the formation of queries is to provide the user with an interface such that its use does not require knowledge of the methods of approximate reasoning, or semantics of the query language.

It was decided to implement the interface of a RoughCast reasoning system in the form of an on-line application leading dialogue with the user applying the interactive forms (Fig. 2a). The system offers functionality in the form of an ability to classify objects basing on their attributes [7]. The attributes are retrieved from the database, to be presented next in the form of successive lists of the values to a user who selects appropriate boxes. In this way, quite transparently for the user, a query is created and sent to the reasoning engine that builds a rough set. However, to make such a dialogue possible without the need for the user to build a query in the language of logic, the system was equipped with an interpreter of queries in a semantics much more narrow than the original Pawlak semantics. This approach is consistent with the daily cases of use when the user has to deal with specific defects, and not with hypothetical tuples. Thus set up inquiries are limited to conjunctions of attributes, and therefore the query interpreter has been equipped with one logical operator only. The upper and lower approximations are presented to the user in response.

The RoughCast system enables the exchange of knowledge bases. When working with the system, the user has the ability to download the current knowledge base in a spreadsheet form, edit it locally on his terminal, and update in his system. The way the dialogue is carried out depends directly on the structure of decision-making table and, consequently, the system allows reasoning using arrays containing any knowledge, not just foundry knowledge.

The issue determining to what extent the system will be used is how the user can acquire a knowledge base necessary to operate the system. So far, as has already been mentioned, this type of a database constructed in the form of an array of attributes was compiled by a knowledge engineer from scratch. However, the authors suggest to develop a system that would enable acquiring such an array of attributes in a semi-atomatic mode through, supervised by an expert, the initial round of queries addressed to a relational database in a SQL language (see 2.2).

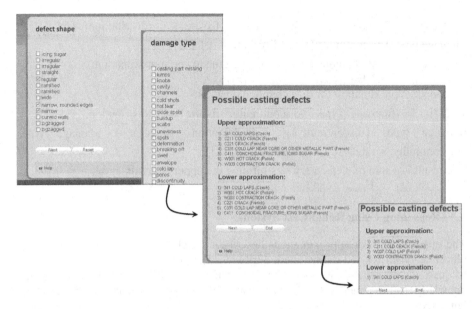

Fig. 2. Forms selecting attribute values in the RoughCast system, upper and lower approximations calculated in a single step of reasoning and the final result of dialogue for the example of "cold lap" defect according to the Czech classification system

4   Knowledge Integration for Rough Logic-Based Reasoning

The problems of knowledge integration have long been the subject of ongoing work carried out by the Foundry Research Institute, Cracow, jointly with a team from the Faculty of Metals Engineering and Industrial Computer Science, AGH University of Science and Technology, Cracow [9, 10].

Various algorithms of knowledge integration were developed using a variety of knowledge representation formalisms. Today, the most popular technology satisfying the functions of knowledge integration includes various ontologies and the Semantic Web, but it does not change the fact that the relational databases remain the technique most commonly used in industrial practice for the data storage. On the one hand, to thus stored data the users get access most frequently, while - on the other - the databases are the easiest and simplest tool for quick data mining in a given field of knowledge. Therefore, the most effective, in terms of the duration of the process of knowledge acquisition, would be creating the knowledge bases from the ready databases. Studies are continued to create a coherent ontological model for the area which is metals processing, including also the industrial databases.

One of the stages in this iterative process is accurate modelling of the cases of the use of an integrated knowledge management system. A contribution to this model can be the possibility of using attribute tables for reasoning and classification. The process of classification is performed using a RoughCast engine, based on the generated attribute table. The database from which the array is generated does not necessarily have to be dedicated to the system. This gives the possibility of using nearly any industrial database. The only requirement is to select from among the attributes present in the base the sets of conditional and decision attributes. If there are such sets, we can generate the attribute table using an appropriate query.

An example might be a database of manufacturers of different cast steel grades (Fig. 3).

cast steel symbol acc. to PN-EN	design ation no.	cast steel symbol	symbol used in foundry	standard requirements	application	foundry	foundry's website	casting weight	arbitrary keywords
LH14 (GX12Cr12)			LH14	PN-86/H83158		Magnus-Nord	www.magnus-nord.pl	from 0,5 to 1500 kg.	corrosion-resistant cast steel, LH14,
			LOH18N10M2	PN-86/H83158		Magnus-Nord	www.magnus-nord.pl		corrosion-resistant cast steel
			LOH18N10M2	PN			http://www.hardkop.pl/		corrosion-resistant cast steel, staliwo wysokostopowe
LH14			LH14	PN-86/H-83158		PIOMA-ODLEWNIA	http://pioma-odlewnia.com.pl	5 - 5000kg	castings + acid-resistant cast steel
	1.4027		LH14	wg DIN G-X20Cr14		MOD-Guss	http://www.mod-guss.com.pl	up to 800 kg	acid-resistant cast steel
LH14				wg PN			http://www.hardkop.pl/		LH14+ + alloyed cast steel
LH25N19S2 (G-X40CrNiSi 25 20)	1.4848			DIN 17465		Magnus-Nord	www.magnus-nord.pl	0,5 - 1500 kg.	
LH25N19S2 (G-X40CrNiSi 25 20)	1.4848					MOD-Guss	http://www.mod-guss.com.pl	30 kg	heat-resistant cast steel
GP-240GH (L20)	1.0619					Odlewnia Polna S.A.	http://www.odlewniapolna.pl/	0,2- 50 kg	carbon cast steel
GP-240GH	1.0619			PN-EN 10213-2		odlewnia Rawicz	http://www.odlewnia-rawicz.pl	1,5 - 1800 kg	carbon cast steel
GP-240GH	1.0619			PN-EN 10213-2			http://www.mod-guss.com.pl/pol/materialy.htm		carbon cast steel

Fig. 3. Fragment of cast steel manufacturers database

Using such a database, the user can get to answer to the question which foundries produce the cast steel of the required mechanical properties, chemical composition or casting characteristics.

The decision attributes will be here the parameters that describe the manufacturer (the name of foundry) as well as a specific grade of material (the symbol of the alloy), while the conditional attributes will be user requirements concerning the cast steel properties. Using thus prepared attribute table, one can easily perform the reasoning.

5 Summary

The proposed by the authors procedure to create attribute tables and, basing on these tables, conduct the process of reasoning using the rough set theory enables a significant reduction in time necessary to build the models of reasoning. Thus, the expert contribution has been limited to finding out in the database the conditional and decision attributes – other steps of the process can be performed by the system administrator. This solution allows a new use of the existing databases in reasoning about quite different problems, and thus - the knowledge reintegration. Reusing of knowledge is one of the most important demands of the Semantic Web, meeting of which should increase the usefulness of industrial systems.

Commissioned International Research Project financed from the funds for science decision No. 820/N-Czechy/2010/0.

References

1. Pawlak, Z.: Rough sets. Int. J. of Inf. and Comp. Sci. 11(341) (1982)
2. Kluska-Nawarecka, S., Wilk-Kołodziejczyk, D., Górny, Z.: Attribute-based knowledge representation in the process of defect diagnosis. Archives of Metallurgy and Materials 55(3) (2010)
3. Wilk-Kołodziejczyk D.: The structure of algorithms and knowledge modules for the diagnosis of defects in metal objects, Doctor's Thesis, AGH, Kraków 2009 (in Polish)
4. Kluska-Nawarecka, S., Wilk-Kołodziejczyk, D., Dobrowolski, G., Nawarecki, E.: Structuralization of knowledge about casting defects diagnosis based on rough set theory. Computer Methods In Materials Science 9(2) (2009)
5. Regulski, K.: Improvement of the production processes of cast-steel castings by organizing the information flow and integration of knowledge, Doctor's Thesis, AGH, Kraków (2011)
6. Szydłowska, E.: Attribute selection algorithms for data mining. In: XIII PLOUG Conference, Kościelisko (2007) (in Polish)
7. Walewska, E.: Application of rough set theory in diagnosis of casting defects, MSc. Thesis, WEAIIE AGH, Kraków (2010) (in Polish)
8. Ligęza, A., Szpyrka, M., Klimek, R., Szmuc, T.: Verification of selected qualitative properties of array systems with the knowledge base (in Polish). In: Bubnicki, Z., Grzech, A. (eds.) Knowledge Engineering and Expert Systems, pp. s. 103–s. 110. Oficyna Wydawnicza Politechniki Wrocławskiej, Wrocław (2000)
9. Kluska-Nawarecka, S., Górny, Z., Pysz, S., Regulski, K.: An accessible through network, adapted to new technologies, expert support system for foundry processes, operating in the field of diagnosis and decision-making, Innovations in foundry, Part 3. In: Sobczak, J. (ed.) Instytut Odlewnictwa, Kraków, pp. s. 249–s. 261 (2009) (in Polish)
10. Dobrowolski, G., Marcjan, R., Nawarecki, E., Kluska-Nawarecka, S., Dziadus, J.: Development of INFOCAST: Information system for foundry industry. TASK Quarterly 7(2), 283–289 (2003)

A Feature-Based Opinion Mining Model on Product Reviews in Vietnamese

Tien-Thanh Vu, Huyen-Trang Pham, Cong-To Luu, and Quang-Thuy Ha

Vietnam National University, Hanoi (VNU), College of Technology,
144, Xuan Thuy, Cau Giay, Hanoi, Vietnam
{thanhvt,trangph,tolc,thuyhq}@vnu.edu.vn

Abstract. Feature-based opinion mining and summarizing (FOMS) of reviews is an interesting issue in opinion mining field. In this paper, we propose an opinion mining model on Vietnamese reviews on mobile phone products. Explicit/Implicit feature-words and opinion-words were extracted by using Vietnamese syntax rules as same as synonym feature words were grouped into a feature, which belongs to the feature dictionary. Customers' opinion orientations and summarization on features were determined by using VietSentiWordNet and suitable formulas.

Keywords: feature-word, feature-based opinion mining system, opinion summarization, opinion-word, reviews, syntactic rules, VietSentiWordnet dictionary.

1 Introduction

Feature-based opinion mining and summarizing (FOMS) on multiple reviews is an important problem in the opinion mining field [5,7,8,11,13,14]. This problem involves three main tasks [5]: (1) extracting features of the product that customers have expressed their opinions on; (2) for each feature, determining whether the opinion of each customer is positive, negative or neutral; and (3) producing a summary for all of customers on all of features.

There are many researches have done for improvement FOMS systems [2,6,8,9,11,13,14]. Two very important tasks to improve FOMS systems are finding rules to extract feature words and opinion words as same as grouping synonym feature phrases.

In this work, we proposed a feature-based opinion mining model on Vietnamese customer reviews in the domain of mobile phones products. Explicit/Implicit feature words and opinion words were extracted by using Vietnamese syntax rules as same as synonym feature words were grouped into a feature, which belongs to the feature dictionary. Customer's opinion orientation and summarization on features was determined by using VietSentiWordNet and suitable formulas.

The rest of this article is organized as following. In the second section, related works on solutions to extract features and opinions are shown. In next section, we focus on our model with four phases. Experiments and remarks are described in the fourth section. Conclusions are shown in the last section.

R. Katarzyniak et al. (Eds.): Semantic Methods, SCI 381, pp. 23–33.
springerlink.com   © Springer-Verlag Berlin Heidelberg 2011

2 Related Work

2.1 Feature Extraction

Feature extraction is one of main tasks for feature-based opinion mining. M. Hu and B. Liu, 2004 [6] proposed a technique based on associated rules mining to extract product features. The main idea of this approach was reviewers usually use the synonym words to review about the same product features, then sets of N/NP which frequently occur in reviews could be considered as product features. D.Marcu and A. Popescu, 2005 [7] proposed an algorithm to determine an N or NP be a feature or not by its PMI weight. With hypothesis that product features were mentioned in product reviews more frequently than normal documents, S. Christopher et al, 2007 [2] introduced a language model for extracting product features. S. Veselin and C. Cardie, 2008 [11] considered extracting features as solving related topics, then authors gave a classification model to examine if the two opinions were the same features. L. Zhang and B. Liu, 2010 [13] used the double propagation method [9] for two innovations for feature extraction, the former based on part-whole relation and the later based on "No" pattern.

By using the double propagation approach for mining semantic relations between features and opinion words, G. Qiu et al, 2011 [8] considered to find rules which extracting feature words and opinion words. The method showed some effective results but for only small size data set. Z. Zhai et al, 2010 [14] proposed a constrained semi-supervised learning method to group synonym feature words for summary of product features-based opinions. The method outperformed the original EM and the state-of-the-art existing methods by a large margin.

In this work, we propose some explicit/implicit feature extracting rules and a solution grouping synonym feature words in Vietnamese reviews not only in a sentence but also in sequences of sentences.

2.2 Opinion Words Extraction

In 1997, V. Hatzivassiloglou and K. McKeown [4] proposed a method for identifying orientation of opinion adjectives (positive, negative or neutral) by detecting a pair of words connected by the conjunction of large data sets. P. D. Turney and M. L. Littman, 2003 [10] determined PMI information of terms with both positive and negative sets as a measure of semantic combining.

M. Hu and B. Liu, 2004 [6], S. Kim and E. Hovy, 2006 [6] considered the strategy based on dictionary by using a small set of boost opinion words and an online dictionary. The strategy, first, created small seeds of opinion word with known directions by hand, then enriched this seeds by searching in the synonyms and antonyms WordNet.

Recently, G. Qiu et al, 2011 [8] used double propagation rules to extract not only feature words but also opinion words because of semantic relation between feature words and opinion words.

2.3 Feature-Based Opinion Mining System on Vietnamese Product Reviews

Binh Thanh Kieu and Son Bao Pham, 2010 [1] proposed opinion analysis system in "computer" product in Vietnamese reviews using rule-based method for constructing automatic evaluation of users' opinion at sentence level. But this system could not detect implicit features which occurred in sentences without feature words as same as considered for feature words in only one sentence.

3 Our Approach

Fig 1 describes proposed model for feature-based opinion mining and summarizing on reviews in Vietnamese. The input was a Vietnamese product name. The output was a summary, which showed the numbers of positive, negative or neutral reviews for all of features.

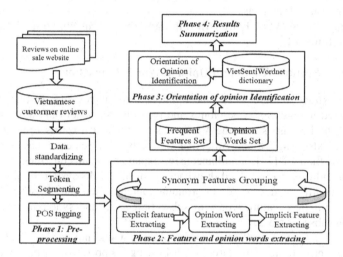

Fig. 1. Model for Feature-based Opinion Mining and Summarizing in Vietnamese Reviews

Firstly, the system crawled all reviews on the product from the online sale website, then entered pre-processing phase to standardize data, to segment token and to tag Part-of-Speech. After that, it extracted all of explicit feature words and opinion words, respectively. From the extracted opinion words, it then identified the implicit feature words. From the set of all extracted explicit words and implicit feature words, we built a synonym feature dictionary. Based on the dictionary, the system changed all of the extracted explicit feature words and implicit feature words into features. Then, all of the infrequent features were removed and the remaining features became opinion features for opinion mining. Opinion orientations based on opinion features and opinion words were determined. Finally, the system summarized discovered information.

The model includes four main phases: (1) Pre-processing; (2) Extracting for feature words and opinion words; (3) Orientation of opinion identification; (4) Summarizing.

3.1 Phase 1: Pre-processing

- Data Standardizing: We adopt combining N-gram statistic and HMM model method for the purpose of switching from unsigned to sign Vietnamese, such as "**hay qua**" switched into "**hay quá**" (great).
- Token Segmenting: We use WordSeg tool [3] to practice this task. The following shows a review sentence: "**Các tính năng nói chung là tốt**" (Features are generally good.). After token segmenting, we have the follow result: **Các | tính năng | nói chung | là | tốt.**
- Pos Tagging: WordSeg tool used again for this task. The obtained result from above example is: **Các /Nn tính năng /Na nói chung /X là /Cc tốt /Aa**, in which /N is a Noun, /A is an adjective.

Example 1: There is a customer review
"**Con này có đầy đủ tính năng. Nó cũng khá là dễ dùng**"
("This mobile has full of functions. It is also quite easy to use").
There is the result of the phase 01
Con /Nc này /Pp có /Vts đầy đủ /An tính năng /Na . /. Nó /Pp cũng /Jr khá /Aa là /Cc dễ /Aa dùng /Vt.

3.2 Phase 2: Feature Words and Opinion Words Extraction

This phase extracts feature words and opinion words in reviews. In this subsection, we considered feature words be Nouns and opinion words not only adjectives as [5] but also verbs because sometime Vietnamese verbs also express opinions. So, we focused on extracting Noun, Adjective or Verb in a sentence based on feature extraction method of [14], simultaneously, expanding syntactic rules to match the domain. In addition, we resolved drawback point of FOMS system by proposing the method to identify feature words in pronoun-contained sentence in subsection 3.2.2, determining implicit feature words in subsection 3.2.3 and synonym feature words grouping in subsection 3.2.4.

A limit when using WordSeg was noun phrases would not be identified. An NP in Vietnamese has basic structure as following: <previous adjunct><center N (CN)><next adjunct>. Here, we defined:

-   <Previous adjunct> may be Classification N – NT, such: **con, cái, chiếc, quả**, etc; or number N – Nn, such: **các** (all), **mỗi** (any), etc.
-   <Next adjunct> may be pronoun – P, such: **này** (this), **đó** (that), etc.

An NP can lack of the previous or next adjunct, but cannot do that with center N.

3.2.1 Explicit Feature Words Extraction
Explicit feature words are feature words which appear directly in the sentence. This step extracted those feature words relying on three syntactic rules, they are: part-whole relation, "No" patterns and double propagation rule.

a) First rule: Part-whole relation. Because a feature word is a part of an object - O (probably a product name, or presented by words like: "**máy**", "**em**", ("mobile"),etc.), it can be based on this rule to extract feature words. The following cases demonstrate this rule:

- N/NP + prep + O. We added "**từ**" (from) to preposition list compare with [14]. For example, the following phrase: "**Màn hình\<N\> từ điện thoại\<O\>**" ("The screen\<N\> from this mobile\<O\>"), so "**màn hình**" (screen) is a feature word.

- O + với (with) + N/NP. For example: "**Samsung Galaxy Tab\<O\> với những tính năng\<NP\> hấp dẫn**" ("Samsung Galaxy Tab\<O\> with attractive functions\<NP\>"), in which "**những tính năng**" (functions) is a NP, thus "**tính năng**" (function) is a feature word.

- N + O or O + N. Here, N is a product feature word, such as "**Màn hình\<N\> Nokia E63\<O\>**" or "**Nokia E63\<O\> màn hình\<N\>**" ("The Nokia E63 screen"), so "**màn hình**" (screen) is a feature word.

- O + V + N/NP. For example, "**Iphone\<O\> có những tiện ích\<NP\>**" ("Iphone\<O\> has facilities\<NP\>"), "**tiện ích**" (facility) is a feature word.

b) Second rule: "No" patterns. This rule has following base form:

Không (Not)/không có (Not)/ thiếu (Lack of)/ (No)/etc + N/NP, such as "**không có GPRS\<N\>**" ("have no GPRS\<N\>"), GPRS is considered as a feature word.

c) Third rule: Double Propagation. This rule based on the interaction between feature word and opinion word, or feature words, or opinion words each other in the sentences because they usually have a certain relationship.

- Using opinion words to extract feature word:

+ N/NP → {MR} →A. For example, "**Màn hình này tốt**" ("This display is good"), is parsed be "**màn hình này**" (this display)→{sub-pre} → **tốt** (good), so the feature word is "**màn hình**".

+ A→ {MR} → N/NP. For example, "**đầy đủ tính năng**" ("full of functions"), is parsed be **đầy đủ** (full) → {determine} → **tính năng** (functions). The feature word is "**tính năng**" (function).

+ V ← {MR} ← N/NP. For example, "**tôi rất thích chiếc camera này**" ("I like this camera so much"), is parsed be **thích** (like) ←{add} ←**chiếc camera này** (this camera). The feature word is "camera".

+ N/NP →{MR}$_1$→V← {MR}$_2$←A. For example, "**Màn hình hiển thị rõ nét**" ("the screen display clearly"), is parsed be **màn hình** (screen) → {sub-pre} → **hiển thị** (display) ← {add} ← **rõ nét** (clearly). The feature word is "**màn hình**" (screen).

In particular, there is a relation {MR} between these feature words and opinion words. The {MR} includes three types of base Vietnamese syntactic relation, they are **Determine**d to demonstrate the location of predicate, **Add** to illustrate the location of complement and **Sub-pre** to shows subject-predicative in the sentence.

- Using extracted feature to extract feature word:

N/NP1 →{conj} → N/NP2, in which, ether N1/CN in NP1 or N2/CN in NP2 is an extracted feature word. {Conj} refers to a conjunction or a comma.

3.2.2 Opinion Word Extraction

In general, this task extracted Adjectives/Verbs in the sentences which contain discovered feature word. Along with them were sentiment strengths and negative words. If the adjectives are connected to each other by commas or semicolons or conjunctions, we will extract all of these adjectives and considering them as opinion words.

In the case of extracting opinion word in pronoun sentence-such as: "**Tôi cảm thấy thích thú với những tính năng của chiếc điện thoại này. Tuy nhiên, nó hơi rắc rối.**" ("I like functions of this mobile. However, they are quie complicated."). How to understand the word "**nó**" (it) refers to "**tính năng**" (function) feature? We proposed a solution for this problem based on the observation of the adjacent pronoun sentences with the sentence which contains extracted feature word.

Suppose s_i be an extracted feature-contained sentence, s_{i+1} be the next sentence, we have an **if-then rule**:

```
If

- s_{i+1} doesn't appear a new feature word, and it begins
with a pronoun like "nó" (it) or "chúng" (they),etc.

- s_{i+1} doesn't appear a new feature word, and it begins
with an opposite word, and be followed by a pronoun
like "nó" (it) or "chúng" (they), etc.

then the opinion words in s_{i+1} is shown on the feature
word which appeared in s_i.
```

In above example, "**tính năng**" (function) feature word has corresponding opinion words: "**đầy đủ**" (full) and "**rắc rối**" (complicated).

Table 1. Some examples of using opinion words to extract implicit feature words.

Opinion word	Feature word
To (big), **nhỏ** (small), **cồng kềnh** (bulky), ...	**Kích cỡ** (size)
Cao (high), **rẻ** (cheap), **đắt** (expensive), ...	**Giá thành** (price)
Đẹp (nice), **xấu** (bad), **sang trọng** (luxury), ...	**Kiểu cách** (style)
Chậm (slow), **nhanh** (fast), **nhạy** (sensitive), ...	**Bộ xử lý** (processor)

3.2.3 Implicit Features Identification

Implicit feature words are feature words which do not appear directly in sentence but via opinion words in the sentence. For the domain of "mobile phone" products, an adjective dictionary is pre-constructed to identify the implicit feature words with opinion word. Table 1 shows some examples of using opinion words (in the left column) to identify relative implicit feature word (in the right column).

3.2.4 Grouping Synonym Feature Words

Because a opinion feature may be expressed by some feature words then synonym feature words should be grouping. To make sense summarization phase, it need to group feature words which express same opinion feature to a cluster. A solution for grouping near-synonym feature words showed in [12].

For example, a group of feature words with name "**kiểu dáng**" (appearance), can have many feature words such as: "**thiết kế**" (design) or even "**kiểu dáng**" (appearance)...

3.2.5   Frequent Features Identification

Target of this step was to define frequent features in reviews, to reject redundant features. To find frequent features, we computed frequency of features and rejected features which had frequency lower than threshold to exclude redundant features.

Let t_i be the number impression of feature f_i and h be the number of reviewers. So impression rating of feature f_i is: $tf_i = t_i$ / h. Here, we chose the value of 0.005 to the threshold.

Example 2. (Using the sentence in Example 1): After step 2, feature word "**tính năng**" (function) and opinion words "**đầy đủ**" (full) and "**dễ**" (easy) have been extracted.

3.3   Phase 3: Determining the Opinion Orientation

Opinion orientation of each customer on each opinion feature will be determined in this phase through two steps. Firstly, the opinion weight of the customer on each feature, which is considered by the customer, will be determined. Secondly, opinion orientation on the feature is classified into one of three classes of positive, negative or neutral.

In the first step, the VietSentiWordNet was used, which expresses positive and negative weights at the word level. Some modifies to the dictionary had been made for according with the domain of "mobile phone" reviews. Firstly, the weight of some opinion words is modified. Secondly, the nature weights of Verbs, which could not have opinion meaning, will be assigned to 1. Finally, some opinion words which are usually used in "mobile phone" reviews was added.

Let's consider a customer's reviews. Denote *ts* be the opinion weight on the feature in the review, ts_i be the weight of the ith opinion words on the feature in the review (denoted by **word**$_i$), w_i be opinion weight of **word**$_i$ in dictionary (w_i be selected as the positive degree if **word**$_i$ was a positive and as the negative degree if **word**$_i$ was a negative) then *ts* will be determined as:

$$ts = \sum_{1}^{m} ts_i ,$$ where m be the number of opinion words on the feature in the

review. In without"No" rule cases, ts_i was determined as w_i if there was no hedge word, and t_i was determined as $h*w_i$ if there was a hedge weight h. The "No" rule reversed the value of ts_i.

In the second step, opinion orientation for the feature will be classified into one of three classes of positive/ negative or neutral based on the ts weight.

Example 3. Using Example 2, the weights on "**tính năng**" (function) feature will be determined.. Opinion weights of "**đầy đủ**" (full) and "**dễ**" (easy) are 0.625 and 0.625 respectively. That the opinion weight of the customer on "**tính năng**" (function) feature is 1.25 be the sum of 0.625 and 0.625. The weight is greater than threshold value of 0.2 then the opinion orientation of the customer on "**tính năng**" is positive.

3.4   Phase 4: Summarization

The summarization will be determined by enumeration on all of customer's opinion orientation on all of features.

4 Experiments

This work only focuses on "mobile phone" product. There were 669 reviews on ten most popular products were crawled from http://www.thegioididong.com then. Table 2 shows the number of crawled and standardized reviews for each product. It is clear from the table that the small size of data may be the causes of limits of this work.

Table 2. Total of crawled and standardized reviews

Product names	Number of reviews
LG GS290 Cookie Fresh.	77
LG Optimums One P500	45
LG Wink Touch T300	41
Nokia c5-03	89
Nokia e63	61
Nokia E72	68
Nokia N8	88
Nokia X2-01	79
Samsung galaxy tab	42
Samsung star s5233w	79

4.1 Feature Extraction Evaluation

Table 2 showed 669 standardized reviews on ten products. Subsequently, we evaluated the result after feature extracting phase using Vietnamese syntactic rules. Table 3 illustrates the effectiveness of the feature extraction. For each product, we read all of those reviews and list to features from them. Then, we counted true features in the list which the system discovered. The precision, recall and F1 are illustrated in Columns 2, 3 and 4 respectively. It can be seen that results of frequent features extraction step are good with all values of F1 above 85%.

Furthermore, to illustrate the effectiveness of our feature extraction step, we compared the features generated using base method of [14] in which we adopted it for Vietnamese reviews. In the baseline [14], the F1 is just under 67%, the average recall is 60.68%, and average precision is 70.13% which are significantly lower than those of our innovation. We saw that there were three major reasons that lead to its poor results: Firstly, Vietnamese syntax rules have many differences in comparision with English syntax rules, for example, in Vietnamese, N comes before adjective, whereas English is opposite. Secondly, in baseline, the authors do not process grouping synonym features case, so the result is not really high. Finally, the authors do not process implicit features case, which led to recall in baseline is quite low. Comparing the average result in Table 3, we can clearly see that the proposed method is much more effective.

4.2 Whole System Evaluation

For each feature, the system extract opinion word from reviews which mention to this feature in 669 crawled reviews, calculating opinion weight, identification orientation

Table 3. Results of frequent features extraction (MF: Number of manual features; SF: Number of features found by the system)

Product names	MF	SF	Precision (%)	Recall (%)	F1 (%)
LG GS290 Cookie Fresh	18	19	94.74	100.00	97.37
LG Optimums One P500	17	18	88.89	94.12	91.50
LG Wink Touch T300	11	11	100.00	100.00	100.00
Nokia C5-03	22	23	86.96	90.91	88.93
Nokia E63	23	23	91.30	91.30	91.30
Nokia E72	26	28	82.14	88.46	85.30
Nokia N8	22	24	87.50	95.45	91.48
Nokia X2-01	15	19	73.68	93.33	83.51
Samsung Galaxy Tab	18	16	87.50	94.44	89.72
Samsung Star s5233w	15	20	85.00	93.33	90.42
Average			87.06	93.58	90.32

of opinion, and putting into positive, negative or neutral categories. After that, we obtain positive, negative and neutral reviews for all features of each product and then we evaluate performance of whole system by precision, recall and F1 measures for each product. According to the Table 4, the precision and recall of our system are quite satisfactory with approximate 65% and 62% respectively.

Table 4. Precision, Recall and F1 of Feature-based Opinion Mining Model on Vietnamese mobile phones Reviews

Product names	Precision (%)	Recall (%)	F1 (%)
LG GS290 Cookie Fresh	72.81	70.94	71.87
LG Optimums One P500	56.45	42.17	49.31
LG Wink Touch T300	65.31	55.17	60.24
Nokia C5-03	61.62	48.80	55.21
Nokia E63	68.66	62.16	65.41
Nokia E72	62.34	64.86	63.60
Nokia N8	64.84	66.94	65.89
Nokia X2-01	64.06	68.33	66.20
Samsung Star s5233w	66.05	68.15	67.10
Samsung Galaxy Tab	62.30	63.33	62.81
Average	64.99	62.19	63.59

Finally, the system generates a chart which summarizing the extracted information. Figure 2 shows an example about a summary of the reviews of customers on each LG Wink Touch T300's feature.

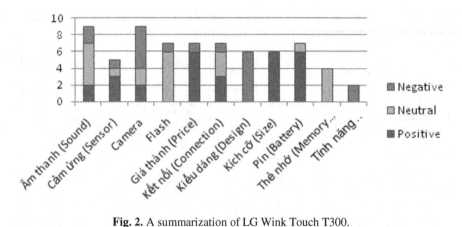

Fig. 2. A summarization of LG Wink Touch T300.

5 Conclusion

In this paper, we present an approach to build an opinion mining model of customer reviews according to features based on Vietnamese syntax rules and VietSentiWordNet dictionary. Our approach has been handing the limitations that the current FOMS systems have not yet resolved, such as: our model has identified implicit features, grouping synonym features and determining features which appear in pronoun-contained sentences. We also apply our model to implement FOMS system on "mobile phone" reviews in Vietnamese and achieved good results of approximately 90% on feature extraction step, about 68% on opinion words extraction and nearly 64% on general system, that results confirmed the correctness of our approach.

Methods to determine automatically the map of opinion words to implicit words as same as grouping feature words will be take into consideration.

Acknowledgments. This work was supported in part by the VNU-Project QG.10.38.

References

1. Kieu, B.T., Pham, S.B.: Sentiment Analysis for Vietnamese. In: 2010 Second International Conference on Knowledge and Systems Engineering, pp. 152–157 (2010)
2. Christopher, S., Bierhoff, K., Chang, E., Felker, M., Ng, H., Jin, C.: Product-feature Scoring from Reviews. In: ACM Conference on Electronic Commerce, pp. 182–191 (2007)
3. Pham, D.D., Tran, G.B., Pham, S.B.: A Hybrid Approach to Vietnamese Word Segmentation using Part of Speech tags. In: 2009 First International Conference on Knowledge and Systems Engineering, pp. 154–161 (2009)
4. Hatzivassiloglou, V., McKeown, K.: Predicting the semantic orientation of adjectives. In: ACL 1997, pp. 174–181 (1997)
5. Hu, M., Liu, B.: Mining and Summarizing in Customer Reviews. In: KDD 2004, pp. 168–177 (2004)

6. Kim, S., Hovy, E.: Automatic identification of pro and con reasons in online reviews. In: COLING (Posters), pp. 483–490 (2006)
7. Marcu, D., Popescu, A.-M.: Towards developing probabilistic generative models for reasoning with natural language representations. In: Gelbukh, A. (ed.) CICLing 2005. LNCS, vol. 3406, pp. 88–99. Springer, Heidelberg (2005)
8. Qiu, G., Liu, B., Bu, J., Chen, C.: Opinion Word Expansion and Target Extraction through Double Propagation. Computational Linguistics 37(1), 9–27 (2011)
9. Qiu, G., Liu, B., Bu, J., Chen, C.: Expanding Domain Sentiment Lexicon through Double Propagation. In: IJCAI 2009, pp. 1199–1204 (2009)
10. Turney, P.D., Littman, M.L.: Measuring praise and criticism: Inference of semantic orientation from association. ACM Trans. Inf. Syst. 21(4), 315–346 (2003)
11. Veselin, S., Cardie, C.: Topic identification for Fine-grained Opinion Analysis. In: COLING 2008, pp. 817–824 (2008)
12. Pham, H.-T., Vu, T.-T., Luu, C.-T., Ha, Q.-T.: A solution for grouping Vietnamese synonym features in product reviews. In: KSE 2011, Hanoi, Vietnam (will be submitted, 2011)
13. Zhai, Z., Liu, B., Xu, H., Jia, P.: Grouping Product Features Using Semi-Supervised Learning with Soft-Constraints. In: COLING 2010, pp. 1272–1280 (2010)
14. Zhang, L., Liu, B.: Extracting and Ranking Product Features in Opinion Documents. In: COLING (Posters), pp. 1462–1470 (2010)

Identification of an Assessment Model for Evaluating Performance of a Manufacturing System Based on Experts Opinions

Tomasz Wiśniewski and Przemysław Korytkowski

West Pomeranian University of Technology in Szczecin, ul. Zolnierska 49,
71-210 Szczecin, Poland
{twisniewski,pkorytkowski}@wi.zut.edu.pl

Abstract. In the paper a procedure for identification of an assessment model for evaluating performance of a manufacturing system is presented. Methodology applied here allows incorporating both explicit and tacit knowledge of the decision-makers. The methodology is based on aggregation of expert opinion using linguistic values and pair-wise comparison at distinctive points.

Keywords: assessment model, linguistic values, manufacturing system.

1 Introduction

In the paper a procedure for identification of an assessment model for evaluating performance of a manufacturing system is presented. Manufacturing systems are usually complicated organisms from both technological and organizational point of view. They might comprise several production cells or departments, and be organized as a hierarchical structure. Decisions that have to be taken are ranging from a short-term with a limited influence on the system to a long-term that might completely rebuild a production process. Methodology presented below applied to a manufacturing field would give an aid to a decision-maker in making an informed decision. In case of manufacturing systems decisions depend on both quantitative and qualitative factors. Quantitative factors are usually as follows: throughput time, output capacity, work-in-progress, waiting time before some equipment of a production cell, machines utilization levels etc. and qualitative factors: workforce attitude, customers' satisfaction. Although it is relatively easy to list all important factors that have to be taken into account during decision-making process, it is much harder to build a model incorporating all interrelated decision variables.

Moreover, in case of complicated systems such as manufacturing ones decision-maker usually unwittingly takes into account also the tacit knowledge and judgments about certain parameters that could not be even taken into account explicitly during decision-making process.

Methodology applied here allows incorporating both explicit and tacit knowledge of the decision-maker. The methodology is based on aggregation of expert opinion using linguistic values and pair-wise comparison at distinctive points [8]. Linguistic values and use of experts' (decision-makers) opinions in pair-wise comparison of a

R. Katarzyniak et al. (Eds.): Semantic Methods, SCI 381, pp. 35–45.
springerlink.com

selected decision variants leads to higher adequacy of the assessment model (checked by a verification procedure). In order to enable pair-wise comparison of variants the decision problem is decomposed into a set of sub problems. The decomposition is carried out taking into account relationships between input parameters.

The outline of the methodology is as follows:

Step 1: Identification of input parameters to a model with theirs domains
Step 2: Linguistic values definition for input parameters
Step 3: Decomposition into sub problems
Step 4: Distinctive points' identification (decision variants in sub problems)
Step 5: Pair-wise comparison of distinctive points' and their ranking classification

The procedure is repeated for each sub problem. Obtained results are then gradually aggregated until a whole problem is solved.

2 Literature Review

Performance evaluation is one of typical tasks in case of manufacturing systems. The performance evaluation is performed when for example some changes or a completely new manufacturing system is planned and some alternatives configurations have to be considered and assessed. As mentioned in above manufacturing systems are complicated entities with certain number of deterministic and stochastic, quantitative and qualitative factors. Literature on the topic is broad thus only the most significant positions are mentioned. Stochastic manufacturing systems were analyzed in [2], [3], and [7]. In this approach Markov models and queuing networks are applied. Lately discrete-event simulation methodology becomes more and more popular as a tool for manufacturing systems modeling [1]. Although accuracy of obtained models is satisfying, the performance of the manufacturing system there is considered usually in a form of a one parameter, for ex. output capacity or work in progress.

This is certainly not enough to embrace a real nature of a manufacturing system. Thus a multi criteria decision analysis approaches were employed [4] for assessing the performance. Several methods are used for this task; let's point out the most widely used: preference modeling, multi-attribute utility functions, decision rules, verbal decision analysis. Some of them are capable of incorporating decision parameters in a fuzzy form. The proposed in the paper methodology is a combination of some of the above approaches, i.e. discrete-event simulation model serves as a parameters source, decision parameters are fuzzy, pair-wise comparison is used. This better reveals both explicit and tacit decision-maker knowledge.

3 Model Description

Analyzed object in our research is a model of multiproduct manufacturing system with dynamic priority rules on the basic of commercial offset printing. The problem of scheduling in job shops has been extensively studied for many years and it attracts the attention of the researchers and practitioners equally. The problem is usually characterized as one in which a set of jobs, each consisting of one or more operations to be performed in a specified sequence on specified machines and requiring some process times, is to be processed over a period of time. The objective is to determine the job schedules that minimize a measure (or multiple measures) of performance [9]. In situation when capital investments are hard to finance opportunities for manufacturing system performance improvement could be attained by better management including priority rules assignment. We have studied mutual impact of priority rule (FIFO – First-In-First-Out, EDD – Early Due Date, LOR – Least Operations Remaining, MOR – Most Operations Remaining, SPT – Shortest Processing Time, LPT – Longest Processing Time), system workload (by means of machines utilization) and input buffer capacities. Optimal input buffers' capacities are taken out from out recent paper [6].

Since testing priority rules in real-world production is absolutely impossible, discrete-event simulation has often been adopted to evaluate the performance of dispatching rules for rule selection. Analysis of large and complex stochastic systems is a difficult task due to the complexities that arise when randomness is embedded within a system. ARENA simulation package from Rockwell Software [5] was used for modeling the manufacturing system. A typical commercial printing facility (fig. 1) will serve as a test bench for the analysis.Simulation runs parameters which were selected in order to assure reliable results: warm-up time is 1 day, replication length is 100 days, number of replications is 5.

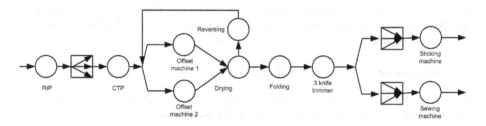

Fig. 1. Manufacturing process structure of a commercial offset printing

All 9 workstations in the manufacturing system have fixed input buffer capacities (fig. 1). Transport times between workstations are omitted since there is no batching and semi products are transferred immediately after technological operation competition. Customers orders' arrival patterns are exponential (servicing times and standard deviations are real times from machines specification. Five priority rules are analyzed (to compare also FIFO rule is analyzed): $pr_i \in \{FIFO, EDD, LOR, MOR, SPT, LPT\}$. Equipment utilization is examined on the following levels: *50%, 55%, 60%, 65%, 70%, 75%, 80%, 85%, 90%, 95%*. Three types of product exist in system: leaflet, brochure, and book.

Assessment model of manufacturing system is based on expert's opinion. It will depend on seven attributes related to waiting times and queues length. These parameters are taken from discrete-event simulation experiments on the model. Outcome of the assessment model is system efficiency presented in points. This evaluation model reflects expert inference about which system is better in given configuration. System assessment will be run as in figure 2.

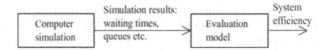

Fig. 2. Schema of system efficiency assessment

3.1 Identification of Input Parameters to a Model with Theirs Domains

x_1 – waiting time (leaflet) $x_1 \in [0,1110]$ h
x_2 – waiting time (brochure) $x_2 \in [0,500]$ h
x_3 – waiting time (book) $x_3 \in [0,2000]$ h
x_4 – WIP – work in progress (leaflet) $x_4 \in [0,1110]$ item
x_5 – WIP – work in progress (brochure) $x_5 \in [0,500]$ item
x_6 – WIP – work in progress (book) $x_6 \in [0,2000]$ item
x_7 – number waiting in queue before offset machine $x_7 \in [0,80]$ item
y – system efficiency $y \in [0,100]$ point

All consideration areas are determined on the basis of earlier results from simulation model. The longer waiting times in queues (x_1, x_2, x_3) for each type of product, the worse the system efficiency is. Similarly attributes WIP for each type of product (x_4, x_5, x_6) and number waiting in queue before offset machine (x_7) have negative impact on system efficiency. Products that spend a long time in queues cause loss because it lengthens throughput time. Bigger costs are caused also by bigger WIP and queues in system. It is related to a physical place restriction.

3.2 Decomposition into Sub Problems

Because of the fact that system output y depends on seven attributes x_i it was made decomposition of linguistic global knowledge base on two 3-dimension and two 2-dimension bases. Therefore it is necessary to aggregate x_i in threes and then in couples. Aggregation will be look like in figure 3.

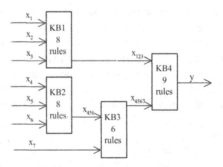

Fig. 3. Decomposition into sub problems

Thanks to decomposition we use only 31 rules (instead 2^7 in case with one knowledge base). Rules in decomposed model are also easier and possible to get from expert. First partial knowledge base (KB1) aggregates three connected attributes: x_1 – waiting time (leaflet), x_2 – waiting time (brochure), x_3 – waiting time (book). Output value is x_{123} total evaluation of waiting time for all kinds of product in points [0,100pts.] 0 – the best, 100pt – the worst. Linguistic values definitions for input parameters for KB1 are in figure 4.

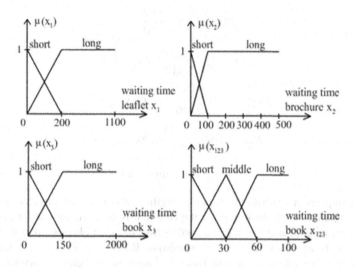

Fig. 4. Linguistic areas for KB1

x_1 – waiting time (leaflet): {short (S), long (L)}

$$\mu(x_1)_S = \frac{200 - x_1}{200} \tag{1}$$

$$\mu(x_1)_L^L = \frac{x_1}{200}, \text{for } x_1 \leq 200 \tag{2}$$

$$\mu(x_1)_L^R = 1, \text{for } x_1 > 200 \tag{3}$$

x_2 – waiting time (brochure): {short (S), long (L)}

$$\mu(x_2)_S = \frac{100 - x_2}{100} \tag{4}$$

$$\mu(x_2)_L^L = \frac{x_2}{100}, \text{for } x_2 \leq 100 \tag{5}$$

$$\mu(x_2)_L^R = 1, \text{for } x_2 > 100 \tag{6}$$

x_3 – waiting time (book): {short (S), long (L)}

$$\mu(x_3)_S = \frac{150 - x_3}{150} \tag{7}$$

$$\mu(x_3)_L^L = \frac{x_3 - 150}{100}, \text{for } x_3 \leq 150 \tag{8}$$

$$\mu(x_3)_L^R = 1, \text{for } x_3 > 150 \tag{9}$$

x_{123} – total evaluation of waiting time: {short (S), middle (M), long (L)}

$$\mu(x_{123})_S = \frac{30 - x_{123}}{30} \tag{10}$$

$$\mu(x_{123})_M^L = \frac{x_{123}}{30}, \text{for } x_{123} \leq 30 \tag{11}$$

$$\mu(x_{123})_M^R = \frac{60 - x_{123}}{30}, \text{for } x_{123} > 30 \tag{12}$$

$$\mu(x_{123})_L^L = \frac{x_{123} - 30}{30}, \text{for } x_{123} \leq 60 \tag{13}$$

$$\mu(x_{123})_L^R = 1, \text{for } x_{123} > 60 \tag{14}$$

Pair-wise comparison of distinctive points method was used for develop KB1. All rules are compared by expert in pairs – the rule that is more important for the expert wins 1 pts., in case of a draw 1/2 pts., and if lose the rule obtains 0pts. Ranking is created on the basis of all pair-wise comparisons. Rules – attributes with distinctive points in consideration space are in table 1. Pair-wise comparison with rank and criterion are in table 2.

Table 1. Distinctive points (rules) for KB1

Rule no.	1	2	3	4	5	6	7	8
x_1	S	S	S	S	L	L	L	L
x_2	S	S	L	L	S	S	L	L
x_3	S	L	S	L	S	L	S	L

Table 2. Pair-wise comparison of distinctive points and their ranking classification for KB1

Rule no.	1	2	3	4	5	6	7	8	Sum	Rank place	Criterion [pts.]
1	x	1	1	1	1	1	1	1	7	1	0
2	0	x	0	1	1	1	1	1	6	2	20
3	0	1	x	1	1	1	1	1	6	2	20
4	0	0	0	x	0	1	1	1	3	4	60
5	0	0	0	1	x	1	1	1	4	3	40
6	0	0	0	0	0	x	0	1	4	3	40
7	0	0	0	0	0	1	x	1	2	5	80
8	0	0	0	0	0	0	0	x	0	6	100

Criterion value of the evaluation of waiting time assigned on the basis of indifference Laplace's principle. Evaluation range (Δ) between places in rank was obtained from equation (*). In the figure 5 was presented rules with place in rank.

$$\Delta_{pts} = \frac{100pts}{6-1} = 20pts \qquad (15)$$

Eight rules with numerical values of conclusion were obtained:

R1(KB1): IF (x_1~0) AND (x_2~0) AND (x_3~0) THEN (Criterion ~ 0)
R2(KB1): IF (x_1~0) AND (x_2~0) AND (x_3~100) THEN (Criterion ~ 20)
R3(KB1): IF (x_1~0) AND (x_2~100) AND (x_3~0) THEN (Criterion ~ 20)
R4(KB1): IF (x_1~0) AND (x_2~100) AND (x_3~100) THEN (Criterion ~ 60)
R5(KB1): IF (x_1~100) AND (x_2~0) AND (x_3~0) THEN (Criterion ~ 40)
R6(KB1): IF (x_1~100) AND (x_2~0) AND (x_3~100) THEN (Criterion ~ 40)
R7(KB1): IF (x_1~100) AND (x_2~100) AND (x_3~0) THEN (Criterion ~ 80)
R8(KB1): IF (x_1~100) AND (x_2~100) AND (x_3~100) THEN (Criterion ~ 100)

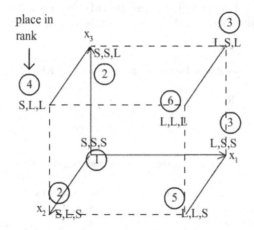

Fig. 5. Distinctive points and place in rank for KB1

Due to lack of space there will be presented only tables with ranks and criterion evaluation for other knowledge bases. Second partial knowledge base KB2 aggregate three connected attributes: x_4 – WIP – work in progress (leaflet) {small (S), big (B)}, x_5 – WIP – work in progress (brochure), {small (S), big (B)}, x_6 – WIP – work in progress (book), {small (S), big (B)}). Output value is total evaluation of work in progress in whole system in points [0,100pts.] - x_{456} {small (S), middle (M), big (B)}.

Table 3. Distinctive points (rules) for KB2

Rule no.	1	2	3	4	5	6	7	8
x_4	S	S	S	S	B	B	B	B
x_5	S	S	B	B	S	S	B	B
x_6	S	B	S	B	S	B	S	B

Table 4. Pair-wise comparison of distinctive points and their ranking classification for KB2

Rule no.	1	2	3	4	5	6	7	8	Sum	Rank place	Criterion [pts.]
1	x	1	1	1	1	1	1	1	7	1	0
2	0	x	0	1	0	1	1	1	4	3	50
3	0	1	x	1	0,5	1	1	1	5,5	2	25
4	0	0	1	x	0	0,5	0	1	2,5	4	75
5	0	1	0,5	1	x	1	1	1	5,5	2	25
6	0	0	1	0,5	0	x	0	1	2,5	4	75
7	0	0	1	1	0	1	x	1	4	3	50
8	0	0	0	0	0	0	0	x	0	5	100

Third partial knowledge base KB3 aggregate two attributes: x_7 – number waiting in queue before offset machine {small (S), big (B)}, and x_{456} total evaluation of work in progress in whole system {small (S), middle (M), big (B)}. Output value is total evaluation of work in progress and queues in whole system in points [0,100pts.] x_{4567} – {small (S), middle (M), big (B)}.

Table 5. Distinctive points (rules) for KB3

Rule no.	1	2	3	4	5	6
x_{456}	S	S	M	M	B	B
x_7	S	B	S	B	S	B

Table 6. Pair-wise comparison of distinctive points and their ranking classification for KB3

Rule no.	1	2	3	4	5	6	Sum	Rank place	Criterion [pts.]
1	x	1	1	1	1	1	5	1	0
2	0	x	1	1	1	1	4	2	20
3	0	0	x	1	1	1	3	3	40
4	0	0	0	x	1	1	2	4	60
5	0	0	0	0	x	1	1	5	80
6	0	0	0	0	0	x	0	6	100

Fourth partial knowledge base KB4 aggregate two attributes: x_{123} – total evaluation of waiting time {short (S), middle (M), long (L)}, x_{4567} total evaluation of work in progress and queues in whole system {small (S), middle (M), big (B)}. Output value is total evaluation of system efficiency y in points [0,100pts.] 0 – the best, 100pt – the worst.

Table 7. Distinctive points (rules) for KB4

Rule no.	1	2	3	4	5	6	7	8	9
x_{123}	S	S	S	M	M	M	L	L	L
x_{4567}	S	M	B	S	M	B	S	M	B

Table 8. Pair-wise comparison of distinctive points and their ranking classification for KB4

Rule no.	1	2	3	4	5	6	7	8		Sum	Rank place	Criterion [pts.]
1	x	1	1	1	1	1	1	1	1	8	1	0
2	0	x	1	0	0	1	0.5	1	1	4.5	4	50
3	0	0	x	0	0	1	0	1	1	3	5	66.67
4	0	1	1	x	1	1	1	1	1	7	2	16.67
5	0	1	1	0	x	1	1	1	1	6	3	33.33
6	0	0	0	0	0	x	0	0	1	1	6	83.33
7	0	0.5	1	0	0	1	x	1	1	4.5	4	50
8	0	0	0	0	0	1	0	x	1	3	5	66.67
9	0	0	0	0	0	0	0	0	x	0	7	100

4 Verification and Validation

Developed assessment model will be accurate if it reflects expert tacit knowledge using linguistics rules. The assessment model will be cross-validated using five pairs of test systems. Expert assessment and the model assessment will use data obtained from a discrete-event simulation. The assessment model was implemented in Matlab.

Simulation data are presented in table 9 together with expert opinions on examined systems' efficiency. For the assessment model the lower result the examined system efficiency is higher, i.e. the system is better and should be pointed out by the expert.

In case of pairs 2, 3 and 4 expert opinions' were coherent with the assessment model, moreover the model evaluations were distinctive, the appraisals divers for more than 20%. In the case of first couple, when expert pointed advantage of system S1 over S2, the assessment model also prefers the S1, but a difference between appraisals is very small. In the last, fifth pair both systems were evaluated by an expert as equally efficient although the assessment model decidedly pointed out the 9th system.

Table 9. Verification table

Pair	System	x_1	x_2	x_3	x_4	x_5	x_6	x_7	Expert's opinion	Model evaluation
1	S1	65	192	828	7	31	195	42	+	55.8431
	S2	34	109	515	3	24	156	51	-	55.7151
2	S3	2	13	527	1	18	173	11	+	18.6833
	S4	21	56	1342	2	17	240	72	-	43.1211
3	S5	3	200	999	1	25	207	51	-	55.3782
	S6	2	69	363	0	15	126	48	+	40.9079
4	S7	193	127	606	25	23	182	76	-	65.3648
	S8	43	132	928	4	26	177	8	+	53.5455
5	S9	23	38	169	2	20	102	23	=	19.6898
	S10	7	64	292	1	15	141	0	=	37.6591

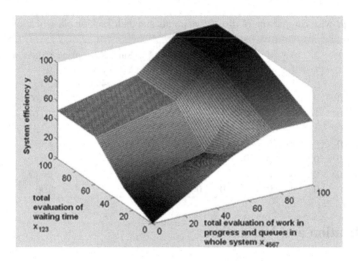

Fig. 6. The efficiency of the offset printing system

In the figure 6 one can observe changes of the efficiency obtained from the developed assessment model for the commercial offices printing system depending on waiting time x_{123} and total evaluation of work in progress and queues in whole system x_{4567}. It demonstrates that expert's tacit knowledge used for systems' evaluation is not linear and the developed assessment model is able to reflect this phenomenon.

5 Conclusions

The developed assessment model fairly correctly imitates expert tacit knowledge what was shown in the paper. As the expert evaluation is based on its own experience and thus it's hard to translate it onto straightly formulated rules, that have often nonlinear characteristics, the assessment model is capable of reflect it using linguistic values, decomposition, distinctive points and pair-wise comparison.

Next step would be to develop an optimization algorithm, which would take results from the assessments model and interactively play with decision parameters (for ex. priority rules) in order to find the system with the best efficiency from the decision-maker point of view.

References

1. Banks, J., Carson, J., Nelson, B.: Discrete-event System Simulation, 3rd edn. Prentice Hall, New York (2001)
2. Buzacott, J.A., Shanthikumar, J.G.: Stochastic Models of Manufacturing Systems. Prentice-Hall, Englewood Cliffs (1993)
3. Dallery, Y., Gerchwin, B.: Manufacturing Flow Line systems: a review of models and analytical results. Queuing systems 12(1-2), 3–94 (1992)
4. Figueira, J., Greco, S., Ehrgott, M. (red.): Multiple criteria decision analysis. Springer, New York (2005)

5. Kelton, W.D.: Simulation with Arena. McGraw-Hill, New York (2006)
6. Korytkowski, P., Wiśniewski, T., Zaikin, O.: Optimal buffer allocation in re-entrant job shop production using simulated annealing. Management and Production Engineering Review 1(3), 30–40 (2010)
7. Li, J., Meerkov, S.M.: Production systems engineering. Springer, New York (2010)
8. Piegat, A.: Lecturers notes for a PhD course on Methods of artificial intelligence, West Pomeranian University of Technology in Szczecin, Poland (2011)
9. Rajendran, C., Holthaus, O.: A comparative study of dispatching rules in dynamic flowshops and jobshops. European Journal of Operational Research 116, 156–170 (1999)

The Motivation Model for the Intellectual Capital Increasing in the Knowledge-Base Organization

Przemysław Różewski[1], Oleg Zaikin[1], Emma Kusztina[1],
and Ryszard Tadeusiewicz[2]

[1] West Pomeranian University of Technology in Szczecin, Faculty of Computer Science
and Information Systems, ul. Żołnierska 49, 71-210 Szczecin, Poland
`{prozewski,ozaikine,ekushtina}@wi.zut.edu.pl`
[2] AGH – University of Science and Technology, Cracow, Poland
`rtad@agh.edu.pl`

Abstract. In the knowledge-base organization the Intellectual Capital is derived from the different form of knowledge owned by the organization. The knowledge is stored mainly in the employee's mind (tacit knowledge) and knowledge repositories (explicit knowledge). The authors assume that the increasing knowledge repository content enlarges the organization's intellectual capital. In the proposed model the moderator and employee work together in order to fill up the repository with high quality content. Proposed motivation model is deigned to take the interest of both sides into consideration and motivate both sides as well. As the reference model the motivation model of cooperation between students and the teacher in educational organization is applied.

1 Introduction

In olden times the value of the company has been mainly determined by possessed material, machine, and capital. Today, due to the increased participation of intangible production in manufacturing processes (which arises from the postulate of the knowledge economy and information society) the potential of a company is located in Intellectual Capital. Moreover, the company (corporate) knowledge is a major component of the company's Intellectual Capital. Based on different scientific studies [20,21] we can say that Intellectual Capital represents the knowledge that can be exploited for some money-making or other useful purpose. Intellectual capital is understood as a set of three factors that are human capital, structural capital, and relational (or customer) capital [10]. The first one represents the value that the employees of a business provide through the application of skills, know how and expertise [7]. The structural capital is a supportive infrastructure, processes and databases of the organization that enable human capital to function [7]. The relational capital is based on the trademarks, licenses, franchises supplemented with customer interactions and relationships [18]. The Intellectual Capital management needs to develop models that included on the one hand specialist's motivating factor for innovation and on the other hand the efficiency of the intellectual content representation in a digital environment. The main task of proposed model is to supervise the growth of the Intellectual Capital by converting human capital into structural capital. From the point of view of information

R. Katarzyniak et al. (Eds.): Semantic Methods, SCI 381, pp. 47–56.
springerlink.com   © Springer-Verlag Berlin Heidelberg 2011

systems, this conversion is represented by the increase of knowledge stored in the organization knowledge repositories. In order to increase the structural part of the Intellectual Capital the knowledge management system should works based on the codification strategy. This strategy is realized based on the externalization process (tacit to explicit) [6], which encodes and stores knowledge in online databases and various repositories where it can be easily used by any knowledge worker [8].

The problem of motivation is one of the more important research subjects of new approach to knowledge economies. The people are willing to maintain knowledge externalization process mainly due to future profits. There are many definitions of motivation [2,4,17] according to which motivation as a phenomenon can be seen as: (a) a system of factors (needs, motives, goals, plans, etc.) determining human actions; (b) a process that supports human activity at a certain level. The motivation model belongs to the class of incentive models and is oriented towards maintaining control of social economic systems. Generally speaking, the incentive model consists of elements induced from a community of actors performing certain actions [11]. The main assumption about the incentive model is that an agent's strategy is defined by a choice of action. The system's strategy is a choice of the incentive function (being the control parameter) – a mapping of agents' actions onto the set of obtainable rewards (remuneration) [12]. So far in the literature the motivation model was developed to supporting activity of students in the process of implementing and using an open and distance learning system [4]. In this case the motivation model is focused on the task of filling the knowledge repository with high quality didactic material. In addition the motivation model was adapted to community-built system context [14]. In community-built system the creator and editor work together operating on a dedicated engine in order to build a massive knowledge repositories (e.g. Wiki). Limitation of these models is the assumption that the benefit in motivation model gets only one page. In this work both the moderator and employee are rewarded.

In this article the authors want to show how the Intellectual Capital grown can be supported by motivation model. We assume that knowledge creation process in organization is similar to the creation of knowledge in educational organizations. Proposed in [4] motivation model will be adapted to the requirements of the knowledge-base organization. The discussed motivation model is aimed to supporting the activity of users in the process of developing and maintaining the structural capital in form of knowledge repository. The motivation model is focuses on the task of filling the knowledge repository with high quality content, in other word the knowledge from employee's mind is transferred to the organization repositories. The advantage of such a form of the Intellectual Capital is the ability to efficient knowledge processing.

2   Problem Description

Preparing and making available the knowledge resources through appropriate computer environment (repository) requires great efforts regarding intelligence and working time of the company employees. The content for repository development involves substantial money capital and consumes a lot of time. Resulting from previous researches on the educational field [4] shows the need to motivate the company employees to supplement his/her duties with developing and monitoring the state of the repository.

Employee refers to the repository in order to search for knowledge, during the process of learning or problem solving [6]. It is also important that after the solution of the task/problem is obtained the relevant feedback should be placed in the repository. These activities are based for the employee's competences, and employee interaction with repository is aimed to increase them. A similar situation occurs in the educational organization, where student works with repository in order to increase their competence in the competence-based learning process [16]. Due to the fact that there is an analogy, when presenting and discussing the motivation model, we combine the idea from the model developed in the past for motivation in learning systems and computer-aided learning [4]. We express the hope that the analogy will help the reader to following the proposed motivation model, which is focused on the intellectual capital issue in knowledge-base organisation.

During the competence-based learning the role of the student is to choose a task from the repository and solve it. The final grade depends on the correctness of the solution and the complexity level of the task. A task solved by a student and highly graded by the teacher is placed in the repository and will serve as an example solution for other students. This way the student participated in the didactic activity added new items to his/her portfolio, meaning that it will be a part of the student's motivation function [16]. At the same time filling the repository with a wide spectrum of high quality solved tasks gives satisfaction to the teacher, for his/her laborious, requiring intelligent efforts primal stage of preparing the repository. Moreover the content of the repository will be increasing.

Similar to student, the employee obtains the task (activit) from the repository and he/she deals with this task in order to acquire specific competencies, which will help to resolve the task. Repository moderator, like teacher, provides the task and sets task complexity. For example, a closed (read-only) task contains the knowledge previously placed in the repository and made up the structural capital part of the Intellectual Capital. Open version of task is focused on the creative process allowing to created new knowledge. An example of building such tasks can be found in [15].

The need for the motivation of the employees to get highly involved in working independently during the repository usage is the biggest challenge. However the high employee's involvement guarantees obtaining a level of competence required for project and knowledge transfer to the repository. In addition the moderator has to be motivated as well. He/She is responsible for knowledge objects/tasks creation.

3   Intelligent Resources: Knowledge Repositories

Knowledge repository is intended for representation of philosophical, scientific, scientific-technical, science-technological state of a given domain in organisation. The elements of the repository, similarly to the data warehouse, are dynamically provided with data, changing their semantic depth with respect to the requests. In order to maintain externalization process in the most efficient way the knowledge repository should guarantee and facilitate the following [5,16]: correspondence with natural language and languages of representation of knowledge originating from varying resources; maintain a sufficient level of unambiguity involved in the process of interpretation; maintain sufficient semantic expressiveness enabling creation of complex structures; provide an ability for structure contents correction, depending on a context; provide an ability for creating complex structures using already available simple structures; provide a mechanism of joining of distinct structures and forming a semantic network of specified and limited semantical linkage; provide means for connecting a single objective domain knowledge model with corresponding method of instruction; enabling domain knowledge reuse. It is worth to note the analogy that exists between the discussed approach and promoted the reuse methodology for the design of information systems and software engineering. In these areas the aim is to collect good practices, as well as elements of knowledge in special libraries and repositories from which the company can use copiously in the implementation of subsequent tasks.

The knowledge repository in organization is the result of mentioned above cooperation. From the knowledge process point of view it is open for everyone storage of knowledge materials, including ontologies, tasks, example solutions etc.; from the scientific point of view it is a copyrighted knowledge resource of a company; from the software-technical point of view it is an information system based on an appropriate network platform.

4   Motivation Model Concept

According to [4] a motif (the reason of action) is a consciously understood need for a certain object, position, situation, etc., therefore we can state that the motif comes from a requirement, becomes its current state and leads to certain actions [1]. During the realization of the mentioned chain "need – motif – action", at each step we are dealing with a decision-situation, meaning: many motifs can lead to a certain action, many needs can make up one motif, many motifs come out of one need. Making a choice is a cognitive process that cannot be observed directly [1]. This means that it is only possible to define the quantitative relationships between the choice parameters through exterior registration of the choice results.

The motivation model can be developed in the form of a certain game scenario, where the activity of an employees and moderators will be supported by their own interests. In order to create environment that support motivation factor following assumptions have to be fulfil [4]: the choice is made in a specific work-related situation, the result of a multiple choice made according to the chain is the obtained competence, the result can be registered, there is a system of assessing the choice results, employee and moderators have access to observations and evaluations of the choices made, the result of a choice has to be evaluated by the employee as a needed and wanted one (usability of the result), the employee has to be certain that the wanted result can be achieved, in a given work-related situation, with probability higher than zero (subjective probability of achieving the result).

Research's discussion about motivation model can be addressed to different area of knowledge-base organization. The conducted analysis of information-processing in judgmental tasks allows to prepare cognitive-motivational model of decision satisfaction [19]. In proposed model confidence serves a role of a major contributing factor of learning motivation. However, more details investigation proves that the motivation is a set of several components. The ARCS Motivation Model [3] is based on four-factor theory. The employee's motivation is hooked up with employee's attention, relevance, confidence, and satisfaction. The ARCS model also contains strategies that can help an instructor stimulate or maintain each motivational element. Other researchers shown that personally valued future goals are core for motivation [9]. Moreover the cultural discontinuities and limited opportunities in employee's learning environment may weaken the motivational force of the future [13].

As a motivation model we consider scenario (see fig. 1) where the moderators and employees are performing actions in a specific work-related situation. The situation is aimed to rising the level of involvement of a employee and moderators in the intellectual capital creation and extending the repository with new tasks and corresponding solutions. External environment and organization activities generate new facts and information, which are prerequisites for the moderator to create new tasks. The tasks are proposed to employee, who selects and solves the tasks in accordance with its motivation function. Correctly solved tasks increase the competence of an employee. In addition the tasks are also included in the repository and become a new part of the Intellectual Capital.

In the discussed approach tasks are created on the basis of the ontology and differ in their complexity level [5]. Employee (knowledge worker) can focus on the open or close task. Moderator (e.g. project manager) analyzes tasks and determines their degree of difficulty. The proposed scenario assumes that the role of the employee is to choose a task and solve it. The final reward depends on the correctness of the solution and the complexity level of the task. A task solved by a employee and highly graded by the moderator is placed in the repository and will serve as an example solution for other employees and become new part of intellectual capital. The moderator is assessed as well. It can be assumed that when he/she actively kepts pace with the changing business environment he/she creates tasks that are more valuable for the company. This way the employee participated in the task solution and we assume that it will raise his/her self-esteem and company position and it will be a part of the employee's motivation function. At the same time filling the repository with a wide spectrum of high quality solved tasks gives profits to the moderator, for his/her effort. Moreover, this will make up the moderator's motivation function.

The moderator's (company) interest is in maximizing the level of repository filling with knowledge. The criterion regarding placing a task in the repository is decided by the moderator on the basis of: complexity level, topicality. The possibility to realize moderator's interests is limited by his/her resources considering time in the quantitative and calendar aspects and other informal preferences.

The employee's motivation function considers obtaining the maximum level of fulfilling one's interests during choosing and solving the task, with given constraints regarding time and the way of grading the resulting solutions.

The statements made above show that both motivation functions depend on the complexity level of a task and have common constraints on time resources. The measure of successfulness of employee and moderator cooperation is the accrual of knowledge in the repository, which can be evaluated through the intensity of its filling with properly solved tasks. When developing a motivation model, one has to consider a very important factor: the stochastic character of employees availability, which is mainly a result of individual working mode and stochastic character of employees' motivation parameters.

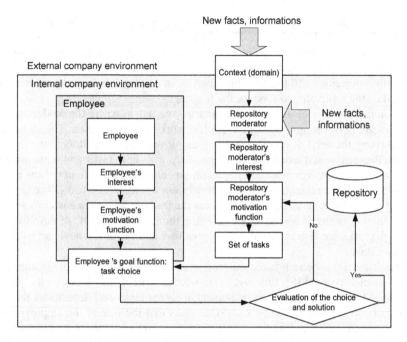

Fig. 1. Scenario of filling and using the repository

5 Formulating the Repository Filling Motivation Model

Let us consider the basic components of the motivation model based on [4].

1. Knowledge-base working process participants

M - Moderator (leads of the subject, disposer of the subject repository),

$E = (e_1, e_2, \ldots, e_j, \ldots)$ - employees coming to choose and solve tasks in real time,

$\pi(e) = \{\chi, \lambda\}$ - parameters of stochastic process of employees availability,

where χ - distribution law, λ - intensity of arrival. We accept the process $\pi(e)$ to be a Markovian one, meaning that it has a stationary, memoryless and sequential character.

2. Domain ontology (based on [22])

$G^D = \{W^D, K^D\}$ – ontology graph, where $W^D = \{w\}$ - nodes of graph (basic concepts),

$K^D = \{k\}$ - arcs of graph (relations between concepts).

3. Tasks set

$R = \{r_i\}, i = 1, 2, \ldots i*$ – tasks set in the frames of a domain D,

$\Pi = \{Q(r_i), A(r_i)\}$ - parameters of task r_i, where $Q(r_i)$ - task r_i complexity level, $A(r_i)$ - task's topicality for the company.

4. Moderator's motivation function

σ^M - moderator's motivation function is a function which depend on tasks parameters,

$$\sigma^M = (Q(r_i), A(r_i)),$$

σ^M defines resources appointed to every task r_i. The resources can be described by vector $X(r_i)$ and mainly covered following items: time of access to telecommunication channels, equipment and software, etc. The moderator's motivation function σ^M is a monotonously rising function of a discrete argument $Q(r_i)$, $i = 1,2,...i*$.

5. Employee's motivation function

σ_j^E - motivation function of each employee.

In general case function σ_j^E depends on parameters of tasks $\sigma_j^E = (Q(r_i), A(r_i))$.

From the point of view of learning objectives the whole group of employees can be divided into groups based on the motivation (ambition, desire for success). For the first group of employees (interesting in achieving the minimal acceptable success level) the motivation function σ^E is a jest monotonously falling function of a discrete argument $Q(r_i)$, $i = 1,2,...i*$, meaning the task complexity. For the second group of employees (interesting in filling the repository with the maximal possible success level) the motivation function σ^E is monotonously rising function of the same argument $Q(r_i)$.

6. Goal function of employee's task choice

Under effectiveness of the decision made we understand maximal satisfaction of employee's and moderator's interests with maximal summary motivation function. The form of satisfying the company interests is placing in the repository a properly solved task of a significantly high level of complexity. The form of satisfying the employee's interests is minimal summary time costs while obtaining a high grade, which also depends on the complexity level of the task. The period of filling the repository is limited by a calendar interval depending on the working situation.

The fact of making the decision can be described by a binary argument, y_i^j

$$y_i^j = \begin{cases} 1, \text{if employee } e_j \text{ chooses task } r_i, \\ 0, \text{otherwise} \end{cases}$$

Then, the goal function has the following structure:

$$\Phi(y_i^j) = \alpha \sigma^M + \beta \sigma_j^E = \max_Y,$$

where $Y = \{y_i^j\}$ $i = 1,2,...i*$, $j = 1,2,...j*$, and $\alpha + \beta = 1$ are waging coefficients and represents importance factor for organization. For example if moderator plays important role in company environment analyzing (outside and inside the company) and tasks creation then $\alpha > \beta$.

7. Moderator goal function of repository filling

The moderator goal function reflects the influence of each task on the accumulation of knowledge in the repository (ΔW). The current state of knowledge in the repository is characterized by two parameters: domain ontology graph G^D and the level of its coverage with properly solved tasks, topical for the company – graph G^P. Each solved task r_i ensures a proper accrual of knowledge in the repository $G(r_i) \subset G^P \subset G^D$, meaning $\Delta W(r_i) = G^D \bigcap G(r_i)$. We assume that tasks with a higher complexity level provide a greater increase of knowledge than tasks with a low complexity level.

Let us consider:

$G^D = \{W^D, K^D\}$ - domain ontology graph,

$E = (e_1, e_2, \ldots, e_j, \ldots)$ - employee coming to choose and solve tasks,

$T(e) = (\tau_1(e_1), \tau_2(e_2), \ldots, \tau_j(e_j), \ldots)$ - stochastic process of employee arrival,

$\tilde{R} = \{r_i(e_j)\}, r_i \in R, e_j \in E$ - set of tasks chosen by employees according to their goal function.

$\Phi(y_j^i) = \alpha \sigma^M + \beta \sigma_j^E = \max_{Y}, \; i = 1,2,\ldots,i^*, \; j = 1,2,\ldots j^*.$

$G(r_i(e_j)) = \{W_i^j, K_i^j\}$ - solved task $r_i(e_j)$ ontology sub-graph,

G^P - summary graph of ontologies of tasks placed in the repository in the interval $\tau \subset [0, T_0]$

$$G^P = G(r_1(e_1)) \bigcup G(r_2(e_2)) \bigcup \ldots \bigcup G(r_i(e_i)) \bigcup \ldots, \text{ where } \tau_1, \tau_2, \ldots, \tau_j, \ldots \in [0, T_0]$$

U^T -accrual of knowledge in the repository in the interval $\tau \subset [0, T_0]$

$$U^T = G^D \bigcap G^P$$

In a certain calendar interval $[0, T_0]$ knowledge accrual in the repository has to be maximal:

$$U^T = G^D \bigcap G^P = \underset{N}{Max}$$

For the given:

Domain D and its ontology graph G^D,
Set of tasks $R = \{r_i\}$ and task parameters: $\Pi = \{Q(r_i), A(r_i)\}$,
Stochastic employees' arrival (availability) pattern $\pi(e)$ and arrival (availability) process parameters $\{\chi, \lambda\}$,
One has to:

a) form the moderator's motivation function σ^M regarding repository filling,

b) optimize each arriving employee's e_j goal function of the task choice $r_i(e_j)$ according to his/her motivation function σ_j^E

$$\Phi(y_j^i) = \alpha \sigma^M + \beta \sigma_j^E = \max_{Y}$$

c) choose properly solved tasks among the ones made by employees according to the grade

$$G^D \bigcap G(r_i(e_j)) \neq \varnothing$$

and supplement the repository with chosen tasks

$$G^P = G(r_1(e_1)) \bigcup G(r_2(e_2)) \bigcup \ldots \bigcup G(r_i(e_i)) \bigcup \ldots$$

Repository filling criterion
U^T - knowledge accrual in the repository on a calendar interval $[0, T_0]$

$$U^T = G^D \bigcap G^P = \underset{T_0}{\max}$$

Constraints:

a) summary resources (time-related, technical, staff) offered to employees for solving tasks:

$$\sum_{e_j \in E} \bar{x}(r_i^j) y_i^j \le \bar{X} \text{ , where}$$

$\bar{x}(r_i^j)$ - resources appointed to task $r_i(e_j)$,

$\bar{X}$ - summary resources for the subject lead by the moderator,

b) calendar interval $\tau \in [0, T_0]$, appointed to employee for choosing and solving tasks

$$\min_j \underline{\tau}(r_i^j) \ge 0, \ \max_j \bar{\tau}(r_i^j) \le T_0,$$

where $\underline{\tau}(r_i^j)$, $\bar{\tau}(r_i^j)$ - appropriate moments to start and end solving task r_i^j by employee e_j.

6 Conclusion

The results given in [4] show the approach to game theory interpretation of proposed motivation model. The proposed model refers to the class of non-cooperative games with a defined number of steps and full information about participants activities in real-time. The equilibrium is obtained as a result of a dominant strategy, what compared to other strategies gives the game participants the possibility to obtain their maximal win regardless of actions of the other participants. Using game theory terminology the motivation model can be seen as a stimulation task, where motivation management signifies direct rewarding an agent (employee) for his actions.

References

1. Anderson, J.R.: Cognitive Psychology and Its Implications, 5th edn. Worth Publishing, New York (2000)
2. Barker, S.: Psychology, 2nd edn. Pearson Education, Boston (2004)
3. Keller, J.M.: Using the ARCS Motivational Process in Computer-Based Instruction and Distance Education. New Directions for Teaching and Learning 78, 37–47 (1999)
4. Kusztina, E., Zaikin, O., Tadeusiewicz, R.: The research behavior/attitude support model in open learning systems. Bulletin of the Polish Academy of Sciences: Technical Sciences 58(4), 705–711 (2010)
5. Kusztina, E.: Concept of open and distance information system. Publisher house of Szczecin University Technology (2006) (in polish)
6. Mack, R., Ravin, Y., Byrd, R.J.: Knowledge portals and the emerging digital knowledge workplace. IBM Systems Journal 40(4), 925–955 (2001)
7. Maddocks, J., Beaney, M.: See the invisible and intangible. Knowledge Management (March 2002)
8. Marwick, A.D.: Knowledge management technology. IBM Systems Journal 40(4), 814–830 (2001)
9. Miller, R.B., Brickman, S.J.: A Model of Future-Oriented Motivation and Self-Regulation. Educational Psychology Review 16(1), 9–33 (2004)

10. Nemetz, M.: A Meta-Model for Intellectual Capital Reporting. In: Reimer, U., Karagiannis, D. (eds.) PAKM 2006. LNCS (LNAI), vol. 4333, pp. 213–223. Springer, Heidelberg (2006)

11. Novikov, D.A., Shokhina, T.E.: Incentive Mechanisms in Dynamic Active Systems. Automation and Remote Control 64(12), 1912–1921 (2003)

12. Novikov, D.A.: Incentives in organizations: theory and practice. In: Proceedings of 14th International Conference on Systems Science, Wroclaw, vol. 2, pp. 19–29 (2001)

13. Phalet, K., Andriessen, I., Lens, W.: How Future Goals Enhance Motivation and Learning in Multicultural Classrooms. Educational Psychology Review 16(1), 59–89 (2004)

14. Różewski, P., Kushtina, E.: Motivation Model in Community-Built System. In: Pardede, E. (ed.) Community-Built Database: Research and Development. Springer ISKM Series, pp. 183–205 (2011)

15. Różewski, P., Małachowski, B.: System For Creative Distance Learning Environment Development Based On Competence Management. In: Setchi, R., Jordanov, I., Howlett, R.J., Jain, L.C. (eds.) KES 2010. LNCS (LNAI), vol. 6279, pp. 180–189. Springer, Heidelberg (2010)

16. Przemysław, R.z., Magdalena, C.: Model of a collaboration environment for knowledge management in competence-based learning. In: Nguyen, N.T., Kowalczyk, R., Chen, S.-M. (eds.) ICCCI 2009. LNCS, vol. 5796, pp. 333–344. Springer, Heidelberg (2009)

17. Skinner, B.F.: Science and Human Behavior. Free Press, New York (1953)

18. Skyrme, D.J.: Valuing Knowledge: Is it Worth it? Managing Information 8(3) (1998)

19. Small, R.V., Venkatesh, M.: A cognitive-motivational model of decision satisfaction. Instructional Science 28(1), 1–22 (2000)

20. Stewart, T.A.: Intellectual Capital. Nicholas Brealey Publishing, London (1997)

21. Sullivan, P.H.: Value-driven Intellectual Capital. John Wiley and Sons, Chichester (2000)

22. Zaikine, O., Kushtina, E., Różewski, P.: Model and algorithm of the conceptual scheme formation for knowledge domain in distance learning. European Journal of Operational Research 175(3), 1379–1399 (2006)

Visual Design of Drools Rule Bases Using the XTT2 Method*

Krzysztof Kaczor, Grzegorz Jacek Nalepa, Łukasz Łysik, and Krzysztof Kluza

AGH University of Science and Technology,
Al. Mickiewicza 30, 30-059 Krakow, Poland
{kk,gjn,llysik,kluza}@agh.edu.pl

Abstract. Drools is one of the most popular expert system frameworks. It uses rules in several formats as knowledge representation. However, Drools has several limitations. It does not support visual modeling of rules. Moreover, it does not assure quality of the knowledge base. In this paper an alternative way of knowledge base designing for Drools is proposed. The method extends the Drools design process by using the XTT2 rule representation and the HQEd visual rule editor. To deal with the differences between the XTT2 and Drools representations, the Drools Export Plugin for HQEd has been implemented. The proposed approach overcomes limitations of the Drools design methodology. Furthermore, it provides a visual rule design editor for Drools and supports formal verification of the model using an optional module.

1 Introduction

An expert system shell is a framework that facilitates a creation of complete expert systems [1]. Such systems are widely used as the Business Rules engines. They often constitute a pluggable software components, separate from the rest of the application. This allows for processing and modifying a knowledge base independently from the other parts of application. These systems can be adapted to the domain-specific problem by development of a proper knowledge base.

However, the knowledge base development is not a trivial task. To build a high quality knowledge base some mechanisms like efficient design or verification methods have to be used. There are techniques that allow for checking knowledge base on a user demand. However, fixing some errors in the knowledge base may introduce other errors. Thus, in order to provide better verification efficiency, the knowledge base should be continuously monitored during the design.

Drools [2], as an advanced framework for the Rule-Based System development, provides a unified and integrated platform for rules and workflow processing. On one hand, Drools provides a mechanism for knowledge base development and management. On the other, these mechanisms have some limitations, which make the creation of a high quality knowledge base difficult.

* The paper is supported by the BIMLOQ Project funded from 2010–2012 resources for science as a research project.

This paper proposes an alternative way of knowledge base development for Drools. The method uses the XTT2 visual knowledge representation and the HQEd rule editor. This approach overcomes several limitations of the existing Drools design methodology. Because of the differences between the Drools and XTT2 representations, the translation algorithm for transforming these two formats has been developed, and a new tool, Drools Export Plugin for HQEd (DEPfH), has been implemented. In the proposed approach, a rule-based system can be designed with a user-friendly tool supporting visual rule modeling. Moreover, the HQEd tool allows for an on-line formal verification of the model, what assures high quality of the Drools knowledge base. The main goal of this paper is to provide a description of the HQEd plugin (DEPfH) as a bridge for this two technologies.

The paper is organized as follows: In Section 2 a short description of the Drools business rules framework is given. Section 3 presents the motivation for this research. Section 4 provides a proposal of a new rule design method for Drools. In this section the XTT2 knowledge representation is described. To demonstrate the XTT2 integration with Drools, as well as the proposed design process a short example is presented in Section 5. The evaluation of the approach with the related research is provided in Section 6. A brief summary is given in Section 7.

2   Drools Rules Framework

Drools[1] is one of the Business Rules engines, which solves many software design problems by separating Business Logic and processes from the software source code. Currently, Drools consists of five modules: *Guvnor* (Business Rules/Processes Management System), *Expert* (Rule Engine), *Fusion* (Complex Event Processing), *Flow* (Process/Workflow) and *Planner*. Our research focuses on using two of them: *Expert*, as the main rules processing module, and *Flow*, which introduces visual modeling of a workflow.

Drools offers Eclipse IDE tools with a textual rule editor, which provides context assistance, syntax highlighting, and syntax error checking. Alternatively, rules can be created in a guided editor, which does not require the knowledge about rule syntax and is more suitable for unexperienced users. Rules, as one of the most basic form of a knowledge representation, follow the *if-then* schema. In Drools they can be complemented by meta information called attributes, and control the inference process. They can specify a rule priority, an agenda, a rule flow, etc. Moreover, a user can create Java functions, and include them into a rule base if needed. Especially, helper functions can be applied to rules or decision tables. They can be used in rule conditions inside `eval()` conditional element, e.g.: `Person(eval( validAge(age) ), sex = 'M')`.

Decision tables are one of the Drools mechanisms to manage sets of rules having the same schema. However, Drools does not provide any design tool for creating decision tables. They can be created in MS Excel, OpenOffice or

---

[1] See: `http://www.jboss.org/drools` for Drools project from JBoss.org.

as comma-separated values (CSV) file. It is important to mention that during inference process, decision tables are transformed into separate rules.

A single decision table contains rules of the same schema. Each rule, stored as a table row, consists of columns marked as "CONDITION" (left-hand side part of a rule) and "ACTION" (right-hand side of a rule). Each attribute and logical operator takes one column header while table cells contain values corresponding to appropriate table headers.

Expert, the core of Drools, has mechanisms that make declarative programming possible. It moves the responsibility for the logic from Java objects to Drools. Consequently, the knowledge maintenance process is much easier.

Another Drools module – Drools Flow – contains a graphical editor for designing the inference. The inference flow constitutes a graph consisting of special blocks, as well as transitions that can be joined and splitted. There is also a ruleflow-group block that contains sets of rules. When process reaches such a block, the rules in this block are evaluated.

Despite the variety of features, Drools does not provide knowledge verification. Although rule syntax is checked, many errors can still be undetected during the design process, and they can appear during the execution. This can be partially solved by using tools, such as Drools Verifier or VALENS (described in Section 6).

3   Motivation

There are several problems in the existing approaches to the rule design in Drools. They become most evident in the case of designing complex systems which contain large rule bases. It is difficult to provide the high quality design as well as the refinement of such systems. These problems are related to:

1. *Flat rule base* – the knowledge base is flat. Although Drools provides two rule representation methods, none of them is appropriate to create a structured knowledge base. They are not able to represent the complex knowledge base clearly. The large set of rules becomes unclear and the dependencies between rules are unreadable. This makes the gradual system design impossible.
2. *Lack of formalization* – the high quality rule base cannot contain errors especially on the logical level. There are some mechanisms that attempt to verify the existing knowledge. However, the lack of the formal rule definition prevents its full formal verification, and this decreases rule base quality.
3. *Lack of a dedicated visual rule editor* – Drools provides a few methods of rules development (text editor, guided editor, as well as extracting rules from the decision tables). However, it does not provide any dedicated decision table editor. A decision table have to be created with the help of a spreadsheet, and this makes Drools not user-friendly and inconvenient for this purpose.

Next section provides a proposal of a new approach to rule modeling in Drools. This method tries to solve the main problems following from the above analysis:

1. *Structured rule base* – introducing internal structure of the rule base. The similar rules that work together in a specific context are grouped into

single decision table. This allows for compact and transparent knowledge base representation in case of large rule bases.

2. *Formal rule definition* – This issue is crucial for the formal verification task.
3. *Dedicated visual rule editor* – allows for gradual and visual modeling of the decision tables and the inference flow between them.

Section 4 introduces the method and tools, which are proposed as an alternative way for rule design method for Drools.

4   Solution Proposal

The problems considered in the previous section can be solved with the help of the Extended Tabular Trees (XTT2) method. This method is a part of the Hybrid Knowledge Engineering (HeKatE) [3] methodology for designing, implementing and verifying production knowledge-based systems.

Because the rule base in Drools is flat and rules are created as separated and, modeling of the systems with a large knowledge base is very complicated. The XTT2 method supports designing of a hierarchical structure of a knowledge base. Thanks to that, the design and maintenance of the complex systems is more efficient.

Although Drools provides many rule design features, rule verification is not supported. On the other hand, the XTT2 method offers on-line formal rule verification during the design process.

The integration of Drools and XTT2 is a solution to the limitations of Drools mentioned above. In this paper we propose to extend the Drools modeling process by using HQEd, a visual editor for XTT2 rule designing. A new Drools Export Plugin for HQEd has been implemented for this purpose. The XTT2 method, HQEd as well as Drools Export Plugin are described in the next subsections.

4.1   XTT2 Rule Representation

The main goals of the XTT2 representation are to provide an expressive formal logical calculus for rules as well as a compact, structural and visual knowledge representation and design. The representation allows for advanced inference control and formal analysis of the production systems.

The XTT2 method provides formal rule representation language based on the Attributive Logic with Set Values over Finite Domains (ALSV(FD)) logic [4]. The ALSV(FD)-based knowledge representation makes the XTT2 language more expressive than the propositional logic. Formal definition of the XTT2 language makes the formal verification and analysis possible. During the design process, the model of the system is continuously checked against syntax and logical errors. This allows for discovering the errors at the time they appear. On the logical level the model is checked for several logical anomalies, such as consistency, redundancy, determinism and completeness [5].

The advantage of the method is the internal structure of the XTT2 knowledge representation. The rule base is not flat, and the rules that work together are

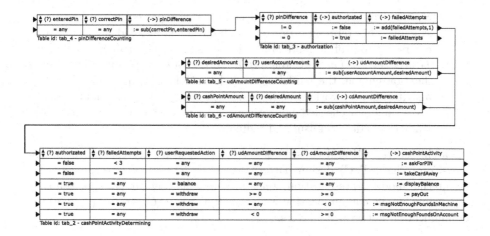

Fig. 1. An example of the XTT2 diagram for CashPoint example

grouped into contexts. Hence, the inference mechanism can work more efficient because only the rules from the focused context are evaluated. Moreover, the internal representation and structure makes the knowledge base suitable for the visual editors. The knowledge base can be depicted as the graph (see: Fig. 1). The tables constitute the graph nodes and correspond to the contexts in the knowledge base. The rules that belong to a specific context are placed in a single table in a network of tables.

Within the HeKatE project a number of tools supporting the visual design and implementation of the XTT2-based systems have been developed (see [6]). The two most important, HQEd and HeaRT, are described in next section.

4.2   HQEd Visual Editor

HeKatE Qt Editor (HQEd) [7] is a editor for XTT2-based expert systems. In the proposed approach, HQEd is a key tool, because it provides all the necessary mechanisms for the rules visual modeling as well as for integration with the external systems. This tool allows a user to design a rule base in a visual way using network of the decision tables. It also provides an interface for plugins which can extend its functionality and allow for integration with other systems.

The HQEd tool provides two mechanisms that assure the high quality of the knowledge base:

– *Visual design* – the visual knowledge representation makes the knowledge base more transparent what allows for more effective finding and fixing errors and anomalies.
– *Verification* – the design is supported by two types of the verification, that can discover the anomalies and errors that were omitted by designer. The HQEd tool supports two types of verification, which are provided by the XTT2 methodology:

- *Syntax verification* – HQEd stores an internal representation of a created system. This representation is continuously monitored against syntax errors, e.g. out of domain values, inaccessible rules, orphan tables, etc.
- *Logical verification* – Although HQEd does not have a built-in logical verification engine, the model is verified by HeKatE Run Time (HeaRT), an external tool working as the HQEd plugin. With HeaRT [8] the model is continuously checked against logical errors, such as completeness, contradictions, determinism and redundancy.

The architecture of HQEd consists of three layers, which correspond to Model-View-Controller (MVC) design pattern (see: Fig. 2). *Model* is responsible for the internal model representation. *View* provides the user interface as well as is responsible for the communication with the application. *Controller* provides the communication between the layers.

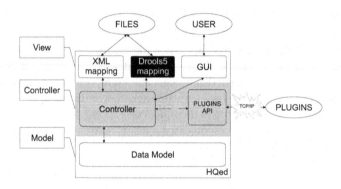

Fig. 2. HQEd architecture

In our approach, the *Data model* stores the internal representation of the XTT2 knowledge base. The remaining layers are included in the *View* that maps the *Model* to the other formats:

- GUI – the visual model representation, appropriate for a user. This layer generates the graphical XTT2 diagram.
- XML mapping – translates the *Model* data into an XML-based language and allows for saving and reading model to/from files.
- Drools mapping – translates the *Model* data into the Drools files.
- Plugins API – provides the TCP/IP-based communication between HQEd and other services.

HQEd has been implemented with the use of the Qt library[2] that makes it a cross-platform tool. Moreover, the tool allows for integration with other technologies by using plugins. Next subsection provides a description of a Drools Export Plugin, which has been implemented to integrate Drools and XTT2.

---

[2] See: www.trolltech.com.

4.3   Drools Export Plugin for HQEd

In Section 3 several problems of existing approaches to the rule design in Drools have been described. These limitations can be overcome by using the XTT2 method. Although both Drools and XTT2 use rules as a knowledge representation, they store them in different formats. Therefore, a translation from XTT2 to Drools model is necessary. Drools Export Plugin for HQEd (DEPfH) implements an algorithm, which performs this non-trivial task.

The basic element of the XTT2 structure is a decision table. Such a table consist of rules of the same schema. A header of the table contains attributes, which are common for all the rules in the table. The remaining parts of the rules are stored in the table cells. On the other hand, Drools does not support decision tables directly. Sets of rules can be stored in spreadsheets in a table-like form. However, the structure differs significantly from the XTT2 tables. In this case, the logic operators are stored with attributes in table header, and are common for all the column cells.

Another difference between the XTT2 and Drools representations is the method of decision table linking. In XTT2 each rule can have a link to another rule in the network of tables. These connections are used during the inference process and indicate a rule execution order. In Drools, in turn, rules in tables can not be connected directly. The connections are allowed only between ruleflow-groups, which can contain sets of rules or spreadsheet decision tables. It is worth mentioning that the Drools decision tables are used only during the design process and they are not used during the inference process.

These differences have been taken into consideration during the DEPfH implementation. In the exported Drools model additional control elements have been introduced. Each rule has an additional action and condition, which influence the inference process. The action sets a global variable `rule_to_execute` to the ordinal number of the rule in the next table, which is to be executed. The condition provides the mechanism for executing the proper rule.

The runtime environment for Drools is Java. The DEPfH plugin uses the dedicated Java helper function to map ALSV(FD) operators to Java expressions. The function takes three parameters: a value of the attribute, a literal symbol of the logical operator, and the value to be compared to. Using this function the operator can be stored in the table cell, despite the fact that in original Drools model the logic operators are stored with attributes in table header.

The XTT2 and Drools models differ in the file structures as well. The XTT2 model stores all the model elements (rules, tables, attributes and attribute types) in a single XML-based file. Drools, in turn, keeps the model elements in separate parts: Drools Flow, decision tables, and additional Java classes (see: Fig. 3).

To deal with this problem, the DEPfH plugin splits an XTT2 model into three files: one with the XTT2 model attributes, one with the decision tables, and additional file containing the inference flow.

Due to space limitation of this paper we omit the translation algorithm description and implementation details of the DEPfH plugin.

64   K. Kaczor et al.

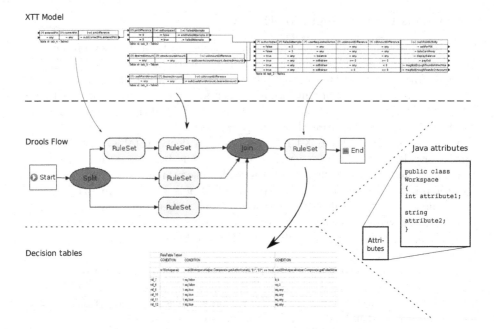

Fig. 3. An example of the XTT2 model and its Drools equivalent

5 CashPoint Example

To demonstrate the design and integration scenario of XTT2 and Drools, a Cash-Point example has been selected (see Fig. 1). The example presents a simple, yet real-life system, studied as one of the benchmark cases in the HeKatE project. CashPoint is a device that allows a person to access to his/her savings. This can be achieved by inserting a card and typing in a Personal Identification Number (PIN). Then, PIN is compared with a code stored in the card. After successfully identification, a customer may withdraw a cash or display the balance.

The design starts with specification of the attribute types. Next, the model attributes are identified. When all the attributes are specified, the user can define the XTT2 tables. First, the table schema has to be indicated. Then, the table can be filled in with rules (each row of the table corresponds to one rule). After filling the XTT2 tables with rules, connections between tables can be defined. The complete XTT2 model (shown in Fig. 1) can be exported using the DEPfH plugin to the Drools format (as depicted in Fig. 3).

There are two main advantages of using the presented XTT2 approach for modeling a knowledge base for Drools. The model has been designed in a convenient visual way, using a user-friendly tool. Moreover, the formal verification of the CashPoint model has been provided by the HeaRT plugin. Therefore, the exported model has high quality assured.

6   Related Research and Evaluation

Although Drools does not provide any built-in rule verification mechanisms, there are some tools, which can be helpful in checking Drools rule base, such as Drools Verifier[3] and VALENS [9] from LibRT[4].

Drools Verifier is a dedicated tool for verifying the Drools knowledge base. It can discover several kinds of errors, such as rules redundancy, subsumption, equivalence, incoherence, or always true expressions. It can work continuously during the design process, as well as on a user demand. The detected errors are simultaneously reported with information about the quality of a knowledge base.

Unlike Drools Verifier, VALENS is not dedicated only for Drools. It allows for maintaining and verifying rule bases in different formats. In addition, a user can also import knowledge in the various representation formats such as production rules and decision trees. The tool is run as a standalone application, and allows for checking rules against consistency, completeness and correctness.

The tools mentioned above verify an original Drools rule base. However, they are not based on formal methods, so any formal verification of the rule base cannot be performed. In our approach, a Drools rule base is designed with the help of the XTT2 method. Thanks to formal rule representation language based on ALSV(FD) logic the formal verification and analysis during the design process is possible.

When it comes to rule visual designing, Drools does not provide any dedicated tool for creating the decision tables. Although Drools Flow supports visual modeling of the inference process, and decision tables can be created in spreadsheets or text files, the whole design process is not really visualized.

There are also attempts to use some rule visual modeling language, such as URML, however, these techniques allow for modeling only single rules, not the decision tables and the inference structure [10]. HQEd supports the gradual visual design method for decision tables, which is more convenient to use than spreadsheets or text files used for Drools. Then, the DEPfH plugin, which implements a translation algorithm from XTT2 to Drools representation, can be used. This allows for executing the created knowledge base in Expert, the Drools rule engine.

The advantage of the proposed approach of knowledge base development for Drools is that the system is created with a user-friendly dedicated tool, which uses the formally defined method, and supports the visual design process. In this case the quality of the knowledge is assured by formal verification.

7   Summary

In this paper an alternative way of knowledge base development for Drools using the XTT2 method is proposed. The new modeling approach uses the HQEd rule editor, which overcomes limitations of the Drools design methodology. To deal

---

[3] See: http://community.jboss.org/wiki/DroolsVerifier.

[4] See: http://www.librt.com.

with the differences between the XTT2 and Drools representations, the Drools Export Plugin for HQEd has been implemented.

There are two main advantages of the proposed approach. The system can be designed with a user-friendly tool that supports visual modeling of the knowledge base. Furthermore, the HQEd tool allows for on-line formal verification of the model. This assures high quality of the Drools knowledge base.

References

1. Giarratano, J.C., Riley, G.D.: Expert Systems. Thomson (2005)
2. Browne, P.: JBoss Drools Business Rules. Packt Publishing (2009)
3. Nalepa, G.J., Ligęza, A.: HeKatE methodology, hybrid engineering of intelligent systems. International Journal of Applied Mathematics and Computer Science 20(1), 35–53 (2010)
4. Ligęza, A.: Logical Foundations for Rule-Based Systems. Springer, Heidelberg (2006)
5. Ligęza, A., Nalepa, G.J.: Proposal of a formal verification framework for the XTT2 rule bases. In: Tadeusiewicz, R., Ligęza, A., Mitkowski, W., Szymkat, M. (eds.) CMS 2009: Computer Methods and Systems: 7th Conference, Kraków, Poland, Kraków, November 26-27, pp. 105–110. AGH University of Science and Technology, Cracow, Oprogramowanie Naukowo-Techniczne (2009)
6. Kaczor, K., Nalepa, G.J.: HaDEs – presentation of the HeKatE design environment. In: Baumeister, J., Nalepa, G.J. (eds.) 5th Workshop on Knowledge Engineering and Software Engineering (KESE2009) at the 32nd German Conference on Artificial Intelligence, Paderborn, Germany, September 15, pp. 57–62 (2009)
7. Kaczor, K., Nalepa, G.J.: Design and implementation of HQEd, the visual editor for the XTT+ rule design method. Technical Report CSLTR 02/2008, AGH University of Science and Technology (2008)
8. Nalepa, G.J.: Architecture of the heaRT hybrid rule engine. In: Rutkowski, L., Scherer, R., Tadeusiewicz, R., Zadeh, L.A., Zurada, J.M. (eds.) ICAISC 2010. LNCS, vol. 6114, pp. 598–605. Springer, Heidelberg (2010)
9. Gerrits, R., Spreeuwenberg, S.: Valens: A knowledge based tool to validate and verify an aion knowledge base. In: ECAI 2000, Proceedings of the 14th European Conference on Artificial Intelligence, Berlin, Germany, August 20-25, pp. 731–738 (2000)
10. Pascalau, E., Giurca, A.: Can URML model successfully drools rules? In: Giurca, A., Analyti, A., Wagner, G. (eds.) ECAI 2008: 18th European Conference on Artificial Intelligence: 2nd East European Workshop on Rule-based applicat ions, RuleApps2008, July 22, pp. 19–23. University of Patras, Patras (2008)

New Possibilities in Using of Neural Networks Library for Material Defect Detection Diagnosis

Ondrej Krejcar

University of Hradec Kralove, FIM, Department of Information Technologies,
Rokitanskeho 62, Hradec Kralove, 500 03, Czech Republic
Ondrej.Krejcar@ASJournal.eu

Abstract. Use of neural network as a tool for a data selection or data mining is well known and discussed many times. Use of neural networks and new software libraries for diagnostics is however not widely used. Our described project deal with a phenomenon of new possibilities in using neural networks implemented in software library for material defect diagnostics which can save a huge amount of money in several special cases. This paper describe also a design and realization of a test application developed in object orientated programming language C#. Developed application classifies data obtained from industrial processes of metallurgical plant. Data used in this project was received during measurement of material diagnostics and its structural defects. Core of developed application is based on neural networks design, which is capable to classify whether the material has defect or not. Basically this project's aim is to substitute commonly used *Statistica* software by a special application software solution developed in C#.

Keywords: AForge.Neuro library, material diagnostics, Statistica software, supervised learning.

1 Introduction

Neural networks phenomenon in sense of their use in various industrial application is a very interesting possibility to learn a strong tool to detect any predefined state of surveyed device or environment.

Neural networks as a part of artificial intelligence, are quickly evolving branch of computer engineering, which allows solving tasks that would have (when used conventional computing algorithms) high performance requirements on computing power because of the complex mathematical description. Generally many real tasks can be solved by linear models described by complicated mathematics, usually systems of inhomogeneous differential equations. Realization of this kind of solutions is often time consuming, sometimes even impossible because of too complex mathematical description [3]. To design solution even for this complicated tasks, neural networks were developed during 20th century. Neural networks are a non-linear models that allows to solve specific tasks from almost all areas of engineering. Many artificial neural networks architectures were developed since the first moments

R. Katarzyniak et al. (Eds.): Semantic Methods, SCI 381, pp. 67–80.

of their introduction. Practical usage of these various architectures is very broad and many tasks may be solved by them.

In principle each of these neural networks architectures has basically one element in common. This element is called neuron. Neuron in computer environment is very similar to its biological pattern. It can be imagined as object with many inputs and only one output. Every input has its own weight value which is important for defining of the priority during neural processing. Another basic element of the neuron is its threshold which determines input values needed to activate neuron processing so that it can work as designated [3].

Last basic element of the neuron is its transfer function which determines the way of neuron processing. Output value of the neuron is computed by this transfer function which transfers input values combined by threshold. There are many kinds of transfer functions used in application based on neural networks. The most common transfer function is threshold function or sigmoid. Details of this transfer function will be provided in other chapters of this project.

Architecture of the artificial neural network which uses only one neuron is called perceptron. This architecture can classify only to two classes, that's why it is used rarely in practical applications [1]. For the real purposes the architecture composed of more neurons is used. These neurons are set in layers. In principle there are three layers: input layer, hidden layer and output layer. Input layer's task is to adjust raw input data so that it can be computed by the rest of the neural network afterwards. Hidden and output layers are computing power of the application [4]. Details of this kind of neural network architecture will be provided hereinafter.

Support for the solution of this project in .NET framework is C# neural network library which was introduced by the British programmer Andrew Kirillov in 2006 [5].

This library contains a huge amount of classes and objects that can be used for design of the various architectures of the artificial neural networks. Programmer also attached some demonstrative application developed in library which shows the principles of neural networks design.

2 Problem Definition

Many methods and algorithms are applied in material diagnostics issue nowadays. Some of them are very precise and are able to diagnose materials correctly; some are less efficient when determining required material properties. An important subset of diagnostics is material defects diagnostics. We previously proposed a solution for a defect diagnostics which uses the emission of pulses which are fetching to the diagnosed machinery [2]. Accelerometers are here used to measure a response to these pulses. Obtained data is transformed by Fast Fourier Transformation and modified by filters in MATLAB afterwards [2].

Adjusted data is needed to be computed by an instrument, which allows researchers to classify the material as convenient or inconvenient. Statistica software developed by Statsoft Company has been this classification instrument so far. This commercial software contains neural network's design and can be used for classification purposes.

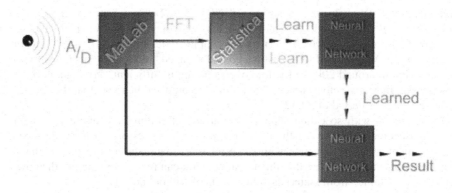

Fig. 1. Measurement network scheme [2].

Aim of this project is to substitute this commercial software by application that works in .NET framework as less expensive possibility of classification of the measured, adjusted and modified data.

Essence of the whole problem is classification of the material diagnostics. It is required that from the application outputs is possible to assess whether the material has defects and if so, what kind of defect that would be.

Schema of measuring chain [Fig. 1] contain measuring part by accelerometers which produce an A/D data. Values from accelerometer goes to Matlab environment, where they are processed using Fast Fourier Transformation (FFT). Statistica SW is used to creat an neural network with coefficients to learn and detect cracks or errors in inspected material.

Vibrations are scanned by use of accelerometer (Bruel&Kjaer company type marking 4332). After the signal forsing (sensor offers only tens of mV) the signal is

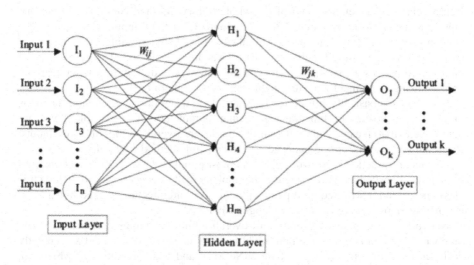

Fig. 2. Three-layer neural network

digitalized with the help of multi I/O card NI PCI 6221 from National Instrument firm. This card disposes of 16 analogue inputs with joint multiplex which runs on 250kHz. Only one canal which works with sample frequency 100kHz was used for given purpose. With regard to the estimated and scanned frequencies of order of tens kHz (maximum around 20kHz) fivefold oversampling is sufficient. Driving algorithm was built up in Matlab environment – Simulink and adjusted in a way that the trapped data are saved for period of 20. [2]

Architectures with so called supervised learning algorithms are implemented for the classification purposes. In these architectures it is necessary to learn designed neural network for the first time with some reference data (samples of measurement materials with no defects), so that the neural network can recognize whether the other samples are obtained from materials with defects of without defects.

3 New Solution for Use of Neural Networks

This chapter is devoted to detailed description of the C# neural network library, which, as aforesaid, is possible to use for design of application that classifies outputs of spectral analysis of material diagnostics.

Library AForge.Neuro.dll is based on Microsoft Visual Studio Solution called AForge, which contains elementary classes, methods, components and other elements that can be used not only for design of neural networks, but even for design genetic algorithms, fuzzy systems, image processing, robotics and many other practical applications. AForge.Neuro.dll namespace contains interface and classes for computing purposes of neural networks. Following description of all classes that are contained in the library helps for better insight of neural network architectures design.

3.1 Classes in AForge.Neuro Library

Neuron class is elementary class of the whole library. Method Neuron has only one input parameter in its argument which is count of neuron's inputs. This class also contains vector of all neuron inputs weights and one output variable. After neuron is created, all its input weight values are set by Randomize method. The range of these randomized values is from 0 to 1. Weight values are during following learning process changed, so that the results and its errors are mineralized.

Layer class is basically similar to Neuron Class and summarizes common functions of all neural layers. In this class particular parameters of the whole layer are determined, that means count of neurons in layer, overall count of layer inputs. Resulting vector of layer outputs is generated by using method Compute.

In standard practice is Network class the most popular class. If some specific kind of neural network architecture ought to be implemented, it is necessary to include this class and extend it with required parameters of given architecture, e.g. Kohonen self-organizing maps that are capable to recognize colors.

Network class contains declaration of overall number of inputs to neural network, number of layer in network and number of neurons in each layer. From this generally designed class inherits classes ActivationNetwork and DistanceNetwork, which can be used for particular design of applications.

4 Design of Proposed Solution

Design of application which substitutes Statistica programme is possible by using ActivationNetwork class.

This class differs from its parent in input parameter of method ActivationNetwork. It is necessary to set type of activation function which, after calculation of the weighted sum of inputs and threshold values of neuron, transfers potential of each neuron and eventually generate output signal.

Description of most common non-linear transfer functions that are used for classification purposes follows.

ActivationNetwork class is generally used for design of multi-layer neural network architectures, which is also this project's case.

DistanceNetwork class differs from ActivationNetwork class in way of calculation of output values.

Output of each neuron is defined as the difference (distance) between values of weights and values of inputs. DistanceNetwork is composed of one layer and is usually used for solving elastic networks or self-organizing maps tasks.

As Network class have Neuron and Layer classes its Activation and Distance variations which also differs in input parameters of its methods. However these classes are rarely used in practical applications. Nevertheless for design of specific kinds of neural network architectures library AForge.Neuro.dll contains these classes.

IActivationFunction interface forms the basis of transfer functions of all neurons. Among method defined in this interface is mathematical definition, then first and second differentiation of defined function.

From this interface come out three activation functions that can be used for programming of neural networks:

- ThresholdFunction
- SigmoidFunction
- BipolarSigmoidFunction

ThresholdFunction class implements this simple mathematical function into neural network architecture. This function acquires zero ouput values in negative domain and value 1 in positive domain. However this function can only be used for the simplest applications.

4.1 SigmoidFunction Class

SigmoidFunction class which implements non-linear sigmoid with 0 to 1 range is much more common. Precision of neural network learning depends on its steepness that is defined as $\lambda = tg\ \alpha$. In this class default setting of λ value is 2. It is possible to change this value if the neural network is not trained in a good quality.

Bipolar sigmoid differs from common sigmoid only in range of output value which is from -0.5 to 0.5. During neural network training it is recommended to try both non-linear transfer functions and depending on errors in the outputs choose the one that achieves better results.

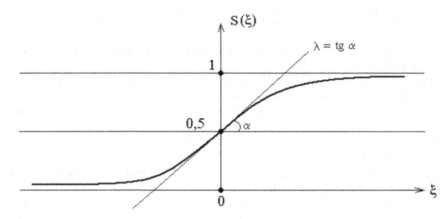

Fig. 3. Sigmoid progression [7].

4.2 Neural Network Learning Algorithms

AForge.Neuro.Learning namespace contains classes and interfaces, in which user defines types of learning algorithms that is used in required application.

There are two basic interfaces in the library:

• ISupervisedLearning (supervised learning algorithms), from which BackPropagationLearning class, DeltaRuleLearning class and PerceptronLearning class inherits.

• IUnsupervisedLearning (unsupervised learning algorithms), from which ElasticNetworkLearning class and SOMLearning class inherits.

During unsupervised learning no required output values are submitted to neural network model. Training process is based on finding similarities among samples that are submitted to network. This kind of learning is often used in applications implementing self-organizing map when colorful samples are sorted by RGB values. However unsupervised learning is not issue of this project.

For classification purposes it is necessary to use supervised learning algorithms. Samples of adjusted data obtained from spectral analysis of material diagnostics (material without defects) is submitted to neural network.

It is possible to use PerceptronLearning class for design and development of application that classifies spectral analysis data.

Principles of supervised learning are described in second chapter of this project. The only input argument of PerceptronLearning method is name of neural network.

Library also provides supervised learning algorithm called Delta Rule, which in one-layer neural networks runs method of determining addition of gradient of error function and following minimalization of this error. Library then provides Back Propagation algorithm, which is basically an extension of Delta Rule learning. Back Propagation is nowadays the most popular algorithm used for training of neural networks.

Other information relevant to AForge.Neuro.dll library is provided in well-arranged help in HTML format, which is attached to the library.

5 Implementation of Proposed Solution

Application which is able to classify data obtained from spectral analysis of diagnostics of internal material defects can be programmed in object orientated language C#. Support for this design is AForge.Neuro.dll library which is described in previous chapters. Project is designed in Microsoft Visual Studio 2008. This chapter is devoted to development of source code of entire application.

First of all it is necessary to include beside common libraries also AForge, AForge.Neuro and AForge.Neuro.Learning libraries and also Threading which enables usage of thread in the project.

At the beginning all needed variables are declared, memory array allocated and delegation for thread is created. Memory buffer which may contain previous data is cleaned by Dispose method. All buttons and check boxes are disabled before the input data is loaded.

User loads reference data by using button LOAD. Data is then loaded by File.OpenText(openFileDialog.FileName) method to allocated array, which server as training set for the neural network model.

If data is in wrong format, it has to be parsed (it depends on particular input data – outputs of MATLAB spectral analysis). It is necessary to adjust parse algorithm to specific task to achieve compatibility with submitted input data. All imported data is shown in the table so that used can check whether the import process ran through without any problems. This can be performed by using ShowTrainingData method. In this method date arrays are allocated and filled with imported data. If there is too much data, used can save them to file, so that i will be easier to check import correction. After importing data all buttons and control components are enabled.

User sets required parameters of neural network in GUI before training process starts:

• **learning rate** – this parameter determines the sensitivity of finding the global minimum of the error function, in other words how precisely will be model of neural network trained. If value of learning rate is too high, it is possible that neural network will not be trained in a sufficient quality because of not finding the global minimum, but only local minimum of error function. If value of learning rate is too low, it can overload computer, so that the model of neural network will not be trained at all. Default setting of learning rate is usually 0.1.

• **lambda value** – this value influences steepness of transfer function, which determines how vector of neuron's output signal is generated, default setting of lambda value in AForge.Neuro library is 2.

• **hidden layer neurons count**
• **iterations count**
• **transfer function**

User changes these parameters to achieve best results when training the neural network.

SearchSolution method is the heart of the application. Input and output data arrays are declared and initialized. Output array has as many elements as the number of classified classes.

After that it is necessary to create the neural network architecture itself. For classification purposes it is convenient to use ActivationNetwork method, which is described in previous chapter. In the method's argument is chosen transfer function and numbers of neurons in each layer. User can only set number of neurons in hidden layer. ActivationLayer method creates layers of the network (command ActivationLayer layer[i]=network[i], where i is layer of the network).

For supervised learning purpose PerceptronLearning method is used. In its argument there is name of trained neural network. Command teacher.LearningRate = learningRate sets sensitivity of global minimum of error function finding to the teacher.

Statistical data is saved by using StreamWriter method (creates two files: Error function file and Weight set file). Command error = teacher.RunEpoch (input, output) starts learning process of the neural network. Variable Error is saved into array. As soon as it reaches value within range of tolerance, command break terminated training process. Current iteration is shown in textbox, so that user can change network parameter or number of iterations if training process lasts for too long time.

After training process is finished, progression of error function is shown in graph.

After training phase can validation and testing phases be performed. By pressing button SOLVE all submitted input sample runs through the trained network and output values are calculated. Difference between training and testing is in substitution of command teacher.RunEpoch by Compute method. All buttons of output graphs, from which user interprets achieved results, are now enabled. Graphs are shown by using of ZedGraph library afterwards.

6 Testing of Developed Application

Testing of the outputs of the neural networks is generally quite complicated procedure, because it's important to interpret the results correctly. Before the beginning of the learning process it's convenient to remove irrelevant data from the input set. Irrelevant data can be e.g. non-numerical or nominal variables. Nominal variables (two-position – e.g. man and woman or multiple-position) have to be transformed to numerical variables by vectors, for example man=(0,1), woman=(1,0) [4]. When input data is adjusted to designated shape, it's divided into three sets – learning set, validating set and testing set. Default setting of the ratio in Statistica programme is following: 70% of the input data is learning set, 15% validating set and 15% testing set. To achieve better results when modeling the neural network it is possible to reset the percentage ratio.

User sets all required settings of the neural network before the learning process (chapter IV.) When the neural network is learned for the first time, user checks values of the error function (the difference between required values submitted by the teacher and outputs of the network), which determines the quality of the learning process. If the network is learned insufficiently, there are many possibilities of achieving better results:

- • change counts of the neurons in hidden layer – this count has a great influence on the learning
- • change of the transfer function of the neurons – in case of achieving insufficient results try to change sigmoid to bipolar sigmoid of the simple threshold function (threshold fcn. is used only when simple input data is provided)
- • change steepness of the transfer function (lambda value) – steepness influences computed transferred potential (output signal) of each neuron
- • change of the iterations count
- • change learning rate value – this value determines the size of the step when searching the global minimum of the error function
- • another adjustment of the input data – scaling of input data (usually linear scaling). To achieve better results it is recommended to transfer submitted data to a certain range of values that fits better to the not learned network.

If these structural changes of the neural network don't help to achieve better outputs, it is recommended to think about using another neural network architecture or changing the learning method. It is impossible to get no errors of the output in the practical applications.

Statistica programme uses for the classification tasks three layer perceptron architecture, nevertheless it is necessary to take into account that Statsoft company who develops this software for many years has its own Know-How, that can definitely help to improve the quality of the learning process as far as empirical changes of the network structural parameters are concerned, as so as the learning algorithms and adjustment of the output data.

As soon as the error function of learned neural network achieves range of tolerance, validation phases starts. Validation phase validates the quality of learning. Validating set of data is submitted to the neural network and processed by the network without the learning again. After output set of data is computed, it is compared whether the error function achieves tolerated values.

When the neural network is ready, it is possible to submit data obtained during industrial process and interpret outputs of the classification afterwards.

In case of diagnostics of the material defects the network classifies whether the material has a defect or not. If there is a defect, it classifies which kind of defect it is.

When designing the neural network, it's necessary to keep in mind, that neuron count of the output layer determines the number of classes in which the input data will be classified.

6.1 Testing Environment

Proposed solution was tested during last 6 month at smelting-house in plant of leader metallurgical company in the world. Firstly we realized a basic tests set to recognize a system parameters and possibilities as well as boundary values of the system. We realized almost hundreds of testing measurements in showed real testing environment [Fig. 4].

Fig. 4. Resl testing environment wth desktop PC and accelerometers sensors.

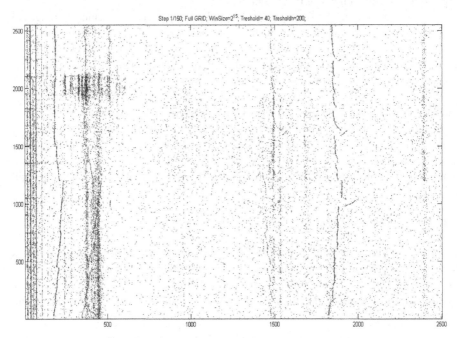

Fig. 5. Graph from real testing – threshold set to 40. Axis X represents a frequency; Axis Y represents sequence number of Spectral Density.

Graph [Fig. 5] represents a sequence number of Spectral Density versus frequency. Every point in the graph represents a presence of frequency in a specific Spectral Density. Difference between both graphs [Fig. 5 and Fig. 6] is in different tresholding level.

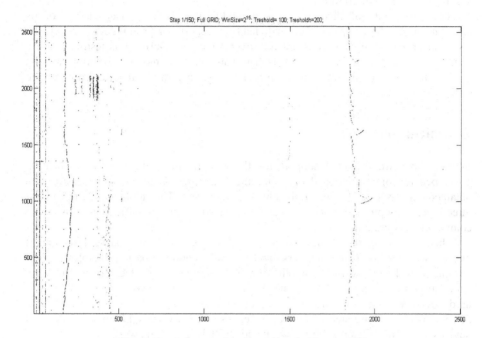

Step 1/150; Full GRID; WinSize=2^{15}; Treshold= 100; Tresholdh=200;

Fig. 6. Graph from real testing – threshold set to 100.

Both graphs [Fig. 5 and Fig 6] show tresholding option for measurement processing in offline stage. From real environment process we received only bmp figures with spectrum including all frequencies. After fine grained adjustment a requested figure was received. From such adjusted figure all points representing curves need to be detected and stored in developed software as an array of values. These "mined" values are base boundary values for neural networks learning process.

Table 1. Input data used for teaching the neural network [2].

Number of measurement	Characteristics frequencies of several measurements (measured at same conditions)			
1	459	488	550	0
2	459	485	488	550
3	459	485	488	489
4	459	485	488	489
5	459	485	487	488
6	459	485	487	488
7	459	485	487	488

Based of previous knowledge about neural network library Aforge we did one series of measurements, where from 7 measurements 6 were detected correctly (85.71%). For verification of authenticity of state determination another check measurement were did on sample of 20 measurements. Successfulness of detection was in this case 16 of 20 (80%).

The future state of this project is a modularization and encapsulation of its hardware and software parts. This will lead to creation of prototype which will be tested in proposed test environment for several months. During this testing phase a neural network parameters as well as hardware parts of modular solution will be verified to reach a full autonomous product with only minimal need from service staff.

7 Conclusions

In this paper we provided support for the design and realization of application developed in object orientated programming language C#, which will classify data obtained on basis of the material defects diagnostics. The application implements three layer perceptron architecture that was so far used for classification purposes in commercial programme Statistica.

When compiling the theoretical part of this work, we used particularly expert materials focused on neural networks and we understood the basic principles of their design. Developed software application was coded in C Developed software application was coded in C# language within .NET Framework. Such solution is hardware independent because of used managed code of .NET platform. Due to this fact we await a future step of project in recoding a software solution for embedded devices with .NET Compact Framework and .NET Micro Framework.

As far as our opinion is concerned, we think that the future of neural network is very perspective, because very complicated and complex tasks can be solving by these algorithms and libraries. This idea is partially confirmed by several mentioned tests which were made on our testing hardware in lab as well as in real environment.

Acknowledgement. This work was supported by „SMEW – Smart Environments at Workplaces", Grant Agency of the Czech Republic, GACR P403/10/1310, We also acknowledge support from student Jakub Hlavica in development of software application and from Robert Frischer in hardware realization of testing environment [2].

References

1. Brad, R.: Satellite Image Enhancement by Controlled Statistical Differentiation. In: Innovations and Advances Techniques in Systems, Computing Sciences and Software Engineering, International Conference on Systems, Computing Science and Software Engineering, ELECTR NETWORK, December 3-12, pp. 32–36 (2007)
2. Krejcar, O., Frischer, R.: Non Destructive Defects Detection by Performance Spectral Density Analysis. Journal Sensors, MDPI Basel 11(3), 2334–2346 (2011)

3. Krejcar, O.: Problem Solving of Low Data Throughput on Mobile Devices by Artefacts Prebuffering. EURASIP Journal on Wireless Communications and Networking, Article ID 802523, 8 pages (2009), doi:10.1155/2009/802523
4. Neural Networks in Statistica program (2010), http://www.statsoft.com/textbook/ neural-networks/ [Quoted 3/11/2010]
5. Neural Networks on C# - The Code Project (19/11/2006), http://www.codeproject.com/KB/recipes/aforge_neuro.aspx [Quoted 25/10/2010]
6. Introduction to neural networks. Study support FELK CVUT (2010), http://cyber.felk.cvut.cz/gerstner/biolab/bio_web/teach/ FunBio/neuron/neursite.html [Quoted 7/11/2010]
7. Mikulecky, P.: Remarks on Ubiquitous Intelligent Supportive Spaces. In: 15th American Conference on Applied Mathematics/International Conference on Computational and Information Science, Univ Houston, Houston, TX, pp. 523–528 (2009)
8. Predicting using neural networks (1999), http://www.obitko.com/tutorials/ predpovidani-neuronovou-siti/zaver.html [Quoted 23/11/2010]
9. Brida, P., Machaj, J., Duha, J.: A Novel Optimizing Algorithm for DV based Positioning Methods in ad hoc Networks. Elektronika IR Elektrotechnika (Electronics and Electrical Engineering) 1(97), 33–38 (2010) ISSN 1392-1215
10. Tucnik, P.: Optimization of Automated Trading System's Interaction with Market Environment. In: Forbrig, P., Günther, H. (eds.) BIR 2010. Lecture Notes in Business Information Processing, vol. 64, pp. 55–61. Springer, Heidelberg (2010)
11. Labza, Z., Penhaker, M., Augustynek, M., Korpas, D.: Verification of Set Up Dual-Chamber Pacemaker Electrical Parameters. In: 2010 The 2nd International Conference on Telecom Technology and Applications, ICTTA 2010, Bali Island, Indonesia, March 19-21, vol. 2, pp. 168–172. IEEE Conference Publishing Services, NJ (2010) ISBN 978-0-7695-3982-9, doi:10.1109/ICCEA.2010.187
12. Brida, P., Machaj, J., Benikovsky, J., Duha, J.: An Experimental Evaluation of AGA Algorithm for RSS Positioning in GSM Networks. Elektronika ir Elektrotechnika 104(8), 113–118 (2010) ISSN 1392-1215
13. Chilamkurti, N., Zeadally, S., Jamalipour, S., Das, S.K.: Enabling Wireless Technologies for Green Pervasive Computing. EURASIP Journal on Wireless Communications and Networking 2009, Article ID 230912, 2 pages (2009)
14. Chilamkurti, N., Zeadally, S., Mentiplay, F.: Green Networking for Major Components of Information Communication Technology Systems. EURASIP Journal on Wireless Communications and Networking 2009, Article ID 656785, 7 pages (2009)
15. Dondio, P., Longo, L., Barrett, S.: A translation mechanism for recommendations. In: IFIPTM 2008/Joint iTrust and PST Conference on Privacy, Trust Management and Security Trondheim, Nórway, June 18-20. IFIP Proc, vol. 268, pp. 87–102 (2008)
16. Korpas, D., Halek, J.: Pulse wave variability within two short-term measurements. Biomedical papers of the Medical Faculty of the University Palacký, Olomouc, Czechoslovakia 150(2), 339–344 (2006) ISSN: 12138118
17. Kasik, V.: Acceleration of Backtracking Algorithm with FPGA. In: 2010 International Conference on Applied Electronics, Pilsen, Czech Republic, pp. 149–152 (2010) ISBN 978-80-7043-865-7, ISSN 1803-7232

18. Stankus, M., Penhaker, M., Srovnal, V., Cerny, M., Kasik, V.: Security and reliability of data transmissions in biotelemetric system. In: XII Mediterranean Conference on Medical and Biological Engineering and Computing 2010. IFMBE Proceedings, vol. 29, Part 2, pp. 216–219., doi:10.1007/978-3-642-13039-7_54
19. Penhaker, M., Kasik, V., Korpas, D.: Electrical parameters measurement and testing of implantable cardioverters – defibrillators. Przegląd Elektrotechniczny 87(6), 186–189 (2097) ISSN 0033-2097
20. Pindor, J., Penhaker, M., Cernohorsky, J., Korpas, D., Vancura, V.: Evaluation of Effect Cardiac Resynchronization Therapy. In: Proceedings of the First Middle East Conference on Biomedical Engineering, MECBME 2011, Sharjah, United Arab Emirates, February 21-24 (2011) ISBN 978-1-4244-7000-6
21. Penhaker, M., Kasik, V., Stankus, M., Kijonka, J.: User adaptive system for data management in home care maintenance systems. In: Nguyen, N.T., Kim, C.-G., Janiak, A. (eds.) ACIIDS 2011, Part II. LNCS, vol. 6592, pp. 492–501. Springer, Heidelberg (2011), doi:10.1007/978-3-642-20042-7_50

Intransitivity in Inconsistent Judgments

Amir Homayoun Sarfaraz[1] and Hamed Maleki[2]

[1] Industrial engineering faculty, South Tehran Branch, Islamic Azad University,
Tehran, Iran
dr.ahsarfaraz@yahoo.com
[2] Young Researchers Club, South Tehran Branch, Islamic Azad University, Tehran, Iran
St_h_maleki@azad.ac.ir

Abstract. In this paper we address one type of criticisms of the AHP. This is about inconsistent judgments and their effect on aggregating such judgments or on deriving priorities from them. Intransitivity is possible when the original AHP or REMBRANDT are used. A numerical example is examined using the original AHP and REMBRANDT to demonstrate intransitivity. This example shows that intransitivity is possible when inconsistent pairwise comparisons are used.

1 Introduction

The analytic hierarchy process (AHP) which was developed by Saaty [9, 10, 14]. Within its framework, a decision maker (usually a complex one) is decomposed into a hierarchy of the goal, criteria, sub-criteria, and finally the alternatives lying at the bottom of the hierarchy. Although Analytic Hierarchy Process gained almost immediate acceptance by many academics and practitioners, it was a new methodology that threatened established methodologies in particular, Multi Attribute Utility Theory (MAUT). This has led to numerous academic debates about AHP. Since most of the debates originated from academics schooled in MAUT, it was natural for them to base their arguments on their own paradigms and axioms a phenomenon described in Kuhn's [3] The Structure of Scientific Revolution. Of course, there is no reason why AHP should satisfy the axioms of MAUT, especially since the reasonableness of those axioms have been called into question from MAUT academics themselves (as discussed below).

Another reason for the academic controversy, in our opinion, is because AHP is not just a methodology for choice problems, but is applicable to *any* situation where structuring complexity, measurement, and synthesis are required. The application of AHP to forecasting problems, for example, requires a different perspective than for some choice problems.

We need to highlight to the reader the importance of transitivity in this context. There are two kinds of transitivity: one is ordinal and the other is cardinal. The first is that if A is preferred to B and B to C, then A must be preferred to C. The second is that if A is preferred to B three times and B to C twice, then A must be preferred to C six times, a much stronger requirement which implies consistency. A consistent matrix is

R. Katarzyniak et al. (Eds.): Semantic Methods, SCI 381, pp. 81–89.
springerlink.com   © Springer-Verlag Berlin Heidelberg 2011

cardinally transitive and, therefore, ordinally transitive. An inconsistent matrix need not be either [15].

A group in the Netherlands, led by F.A. Lootsma, developed a system which uses Ratio Estimation in Magnitudes or deci-Bells to Rate Alternatives which are Non-DominaTed [4, 5]. The decision alternatives may be judged via a method of pairwise comparisons or via direct rating. We only consider the parirwise comparisons method. In the basic pairwise comparison step the Decision Maker (DM) states whether he/she has a weak, strict, strong or very strong preference for one of the two alternatives presented to him/her, or whether he/she is different between the two. Thereafter, his/her verbal judgment is converted into a numerical value on a geometric scale. Next, the impact scores of the alternatives are estimated under the respective criteria via logarithmic regression, and the final scores of the alternatives are computed via the geometric mean aggregation rule. The advantage of the REMBRANDT system is that it would better enable one to enforce transitivity $(A \succ B, B \succ C$ imply $A \succ C)$ if judgment matrices are perfectly consistent.

2 Transitivity vis-à-vis MAUT

The AHP has been criticized as it does not adhere to the multi-attribute utility theory axioms of transitivity. We challenge those who raise this criticism with the following questions:

- Must we use a particular axiomatic (normative) model that tells us "what ought to be?"
- Should we strive for a descriptive method ("what is") even though we do not really know how humans make decisions?
- Must the axiomatic foundations of the so-called rational model of expected utility apply to all decision methodologies?
- Must we use an axiomatic-based model that forces transitivity to hold even though we know of situations where it need not hold?

Expected utility theory (and rational decision making theory, in general) is grounded on the axiom of transitivity, that is, if A is preferred to B, and B is preferred to C, then A is preferred to C (or if A is three times as preferable than B and B is twice as preferable as C, then A is six times as preferable as C).

3 Transitivity

The AHP comparison mode allows for inconsistent transitivity relationships. In the example, a DM checking preferences may conclude that A is 8 times more preferable than C (or perhaps even that C is preferred to A). Under a single-criterion rubric, we might not ordinarily expect to have intransitive relations (Fishburn [2] discusses how intransitive situations can arise for this case). But, for multi criteria problems, it is often impossible not to have intransitivities, as the DM cannot simplify the complexities of the problem to achieve true transitivity. As pointed out by Tversky [17]: "...the fact remains that, under the appropriate experimental conditions, some people are intransitive and these intransitivities cannot be attributed to momentary fluctuations or random variability;'' and "When faced with complex multidimensional alternatives, such as job offers, gambles, or candidates, it is extremely difficult to utilize properly all the available information. Instead, it is contended that people employ various approximation methods that enable them to process relevant information in making a decision.'' Fishburn [2], states: ''Transitivity is obviously a great practical convenience and a nice thing to have for mathematical purposes, but long ago this author ceased to understand why it should be a cornerstone of normative decision theory.'' Fishburn [2] offers the following reasonable real life decision situation. ''Professor P is about to change jobs. She knows that if two offers are far apart on salary, the salary will be the determining factor in her choice. Otherwise, factors such as prestige of the university will come into play. She eventually receives three offers, described in part as follows:

Table 1. Example

Alternative	Salary	Prestige
x	$65,000	Low
y	$50,000	High
z	$58,000	Medium

On reflection, P concludes that $x > y$, $y > z$, and $z > x$, (where > means preferred to).'' Fishburn [2] concludes his discussion on nontransitive preferences with the following:

"Transitivity has been the cornerstone of traditional notions about order and rationality in decision theory. Three lines of research during the past few decades have tended to challenge its status. First, a variety of experiments and examples that are most often based on binary comparisons between multiple factor alternatives suggest that reasonable people sometimes violate transitivity, and may have good reasons for doing this. Second, theoretical results show that transitivity is not essential to the existence of maximally preferred alternatives in many situations. Third, fairly elegant new models that do not presume transitivity have been developed, and sometimes axiomated, as alternatives to the less flexible traditional methods (emphasis added)."

Interestingly enough, Luce and Raiffa [7] state:

"No matter how intransitivities exist, we must recognize that they exist, and we can take only little comfort in the thought that they are an anathema to most of what constitutes theory in the behavioral sciences today".

The Analytic Hierarchy Process is not inhibited by the need for transitive relationships and instead of ignoring such relationships, provides a measure of inconsistency so that the decision maker can proceed accordingly.

Sugihara *et al.* [16] proposed an interval preference relation. Since a pairwise comparison matrix in the AHP is based on human intuition, the given matrix will often include inconsistent elements that violate transitivity. Interval judgment data are used to express uncertainty in human pairwise comparison judgments from which interval weights are obtained that represent inconsistency in the judgment data especially so for cardinal data.

4 Sources of Uncertainty in Modeling a Decision Problem

In the decision-making context, uncertainty may be categorized into one of three distinct classes [18].

4.1 Imprecision or Vagueness

The linguistic or semantic qualifications describing each category of preference intensity are rather vague or imprecise; so that the DM has difficulty deciding which category best describes his/her feelings about a comparison or rating. In his/her mind, several of the linguistic qualifications or categories might be more or less appropriate, some more so than others. For example, the DM might know that he/she prefers alternative A over alternative B, but the categories 'definite preference' and 'strong preference' both seem more or less acceptable descriptions to him/her. An example of this type of uncertainty might arise when a decision is to be taken by a project co-ordinator, based on written inputs from experts in a number of distinct fields (e.g. an environmental impact assessment). The co-ordinator has to interpret the experts' written responses and convert these into subjective judgments.

4.2 Inconsistency

The categories of preference intensity offered to the DM are distinct (i.e. with crisp, non- overlapping endpoints) and are well understood by the DM. However, if the DM is asked to provide a number of replications of a specific pairwise comparison under different environmental conditions, his/her responses are not unique; the range of conditions and his/her recent experiences create an inability for him/her to classify his/her preference intensity into the same single category each time. Psychological evidence exists to support this: experimentation has shown that preferences may vary even when the underlying decision context appears to be the same on all identifiable counts [18].

4.3 Stochastic Judgment

The level of preference intensity depends on some event whose outcome is not known with certainty at the time of the decision and is not under the control of the DM. The stochastic nature thus reflects either subjective probabilities that a particular alternative better achieves a given goal or objective probabilities that reflect uncertain consequences of selecting a particular alternative. For example, a DM is asked to choose one of two investment alternatives that require an identical one-time investment at the beginning of the planning period. The choice is likely to depend on the interest rate ruling over the entire investment period, which may be unknown at the time of the investment decision.

Although each of the above classes signify a different form of indecision, we will loosely label the above classes under the general heading of "uncertainty". In each case, however, the DM is likely to be reluctant to supply a single value or category to represent the intensity of his/her preference or rating.

5 Rank Reversal

The first category of rank reversal has to do with Triantaphyllou's observation that the ranking of the alternatives based on the sub-problems is different from their original ranking based on the complete problem. The second category of rank reversal is the violation of the transitivity principle. That is, if the first sub-problem suggests that $A_2 \succ A_1$ and the second sub-problem suggests that $A_1 \succ A_3$, then the third sub-problem can, despite perfect consistency, suggest that $A_3 \succ A_2$ [20]. We deal with second category of rank reversal in this paper.

6 REMBRANDT

The REMBRANDT system is intended to adjust for three contented flaws in AHP. First, direct rating is on a logarithmic scale, which replaces the fundamental 1-9 scale presented by Saaty. Second, the Perron-Frobenius eigenvector method of calculating weights is replaced by the geometric mean, which avoids potential rank reversal. And third, aggregation of scores by arithmetic mean is replaced by the product of alternative relative scores weighted by the power of weights obtained from analysis of hierarchical elements above the alternative [8]. The REMBRANDT scale is longer than Saaty's fundamental scale with the respective values $...,1/5,1/3,1,3,5,...$. It is easy to see now why the last-named scale is controversial. Saaty [11, 12] invokes Fechner's law to explain the choice of the echelons $3,5,7,...$, although Stevens' power law is now generally accepted in psychophysics, and he chooses the echelons $1/3,1/5,1/7,...$ in order to maintain reciprocity. Such a scale, neither geometric nor arithmetic, may introduce inconsistencies which are not necessarily present in the mind of decision maker. Consider three stimuli S1, S2 and S3, for instance. Weak preference for S1 over S2, estimated by r12=3 on the Saaty scale, and weak preference for S2 over S3, estimated by r23=3, would consistently lead to r13=9, a value which represents absolute preference for S1 over S3 on the Saaty scale. These preference intensities (weak, weak and absolute) are not likely to be compatible, however. A geometric scale yields a more plausible relationship between preference intensities (weak, weak and strict) [6].

Synthesizing priorities derived in REMBRANDT by raising them to the power of the priority of the corresponding criterion and then multiplying the outcome [1, 6] has the shortcoming that $0 < x < y < 1$ and $0 < p < q \Rightarrow x^p > y^q$ for some p and q. This means that an alternative that has a smaller value under a less important criterion is considered to be more important than an alternative that has a larger value under a more important criterion, which is absurd.

One can also show the absurdity of this process of synthesis because it yields wrong known results. By considering alternatives with known measurements under two or more criteria which then inherit their importance from the measurements under them, raising them to the power of the priority of their corresponding criterion and multiplying, one obtains a different outcome than simply adding the measurements and then normalizing them [19].

7 Example 1: An Application of the Original AHP

We consider an illustrative problem with two criteria and three alternatives. The a_{ij} data are numbers from the continuous interval 1 to 9.

Table 2. Pairwise comparison matrix relative to C1 criteria

Alternative	A_1	A_2	A_3
A_1	1	2	0.125
A_2	0.5	1	8
A_3	8	0.125	1

Table 3. Pairwise comparison matrix relative to C2 criteria

Alternative	A_1	A_2	A_3
A_1	1	3	0.11
A_2	0.33	1	9
A_3	9	0.11	1

The three alternatives are compared in a pairwise fashion. At first, it is assumed that the criteria weights change according to Wijnmalen and Wedley method. Then, a set of smaller problems are formed by considering two alternatives at a time. In general, there are $m(m-1)/2$ such smaller problems (where m is the number of alternatives). The pairwise comparisons described here, refer to small problems that are comprised of two alternatives and the new criteria. Wijnmalen and Wedley showed that cause of the rank reversal in Triantaphyllou's observation is AHP's independence axiom, which prohibits adjusting the criteria weights when the set of alternatives or type of normalization change. They also believed the adjustment factor should be equal to that fraction of each original unity sum that is defined by the new totality of remaining alternatives but they consider no inconsistent circumstance.

Suppose that one uses the original AHP. When alternatives A_1 and A_2 are compared as above, Then it is derived that: $A_1 \succ A_2$. Similarly, when alternatives A_1 and A_3 are compared, then it is derived that: $A_3 \succ A_1$. That is, these relations suggest that the ranking of the three alternatives must be as follows: $A_3 \succ A_1 \succ A_2$. However, when the two alternatives A_3 and A_2 are compared, then the derived ranking is: $A_2 \succ A_3$. That is, a logical contradiction (i.e. failure to follow the transitivity property in the derived rankings) is reached. The below data have been derived from above inconsistent judgment matrices with pairwise comparisons.

Table 4. Problem 1

Alternative	Criteria		Overall weights
	C_1 (0.50)	C_2 (0.50)	
A_1	0.67	0.75	0.71
A_2	0.33	0.25	0.29

Table 5. Problem 2

Alternative	Criteria		Overall weights
	C_1 (0.49)	C_2 (0.51)	
A_1	0.11	0.10	0.11
A_3	0.89	0.90	0.89

Table 6. Problem 3

Alternative	Criteria		Overall weights
	C_1 (0.51)	C_2 (0.49)	
A_2	0.89	0.90	0.89
A_3	0.11	0.10	0.11

8   Example 2: An Application of the REMBRANDT

Consider the data used in Example 1. When the original AHP was applied, the ranking derived when two alternatives were considered at a time, had some internal contradictions.

In the following paragraphs the three alternatives are compared in a pairwise fashion. At first, it is assumed that the criteria weights remain the same as before. Then, a set of smaller problems are formed by considering two alternatives at a time. When formula $P_i = \prod_{j=1}^{n}(w_j)^{o_j}$ with longer scale is applied on the previous decision matrix, we have:

The first sub-problem (which considers the pair of alternatives A_1 and A_2) is described by following decision matrix:

Table 7. Problem 1

Alternative	Criteria		Final scores
	C_1 (0.5)	C_2 (0.5)	
A_1	1.41	2	1.68
A_2	0.71	0.50	0.59

Similarly, the second and third such sub-problem (which consider the pair of alternatives (A_1 and A_3) and (A_2 and A_3), respectively) are as follows:

Table 8. Problem 2

Alternative	Criteria		Final scores
	C_1 (0.5)	C_2 (0.5)	
A_1	0.09	0.06	0.07
A_3	11.31	15.99	13.45

Table 9. Problem 3

Alternative	Criteria		Final scores
	C_1 (0.5)	C_2 (0.5)	
A_2	11.31	15.99	13.45
A_3	0.09	0.06	0.07

The two previous pairwise results, when are taken together, indicate that the ranking of the three alternatives must be: $A_3 \succ A_1 \succ A_2$. That is, this ranking is different of the last sub-problem derived.

9 Conclusion

In general people can only compare stimuli in a limited range where their perception is sensitive enough to make distinctions. The range cannot be too wide or they will err. When the range is too wide elements that are close together tend to be summarily lumped together. The AHP uses the fundamental scale of absolute values 1-9 to represent paired comparison judgments to keep measurement within the same order of magnitude. It is well known from cognitive psychology that people are unable to make accurate comparisons past a certain upper bound. Any scale that extends beyond that upper bound (and power scales do so very rapidly) is likely to be inaccurate. When applied to physical phenomena beyond our ability to perceive or respond to, the scale would lead to failure [13].

It can very well be the case that the decision maker may never know the exact ranking of the alternatives in a given multi-criteria decision-making problem when the original AHP or any other Multi-Criteria Decision Making (MCDM) method, are used.

It should be emphasized at this point that the fact that both original AHP and REMBRANDT possess intransitivity in this paper. However, it is obvious that the reverse argument should always be true: That is, a perfect MCDM method should never possess intransitivity. Clearly, this is a very important and fascinating issue in decision analysis, and more research is required.

References

1. Barzilai, J., Lootsma, F.: Power Relations and Group Aggregation in the Multiplicative AHP and SMART. Journal of Multi-Criteria Decision Analysis 6, 155–165 (1997)
2. Fishburn, P.: Nontransitive Preferences in Decision Theory. Journal of Risk and Uncertainty 4, 113–134 (1991)
3. Kuhn, S.T.: The Structure of Scientific Revolution. University of Chicago Press (1962)
4. Lootsma, F., Mensch, T., Vos, F.: Multi-criteria analysis and budget reallocation in long-term research planning. European Journal of Operational Research 47, 293–305 (1990)
5. Lootsma, F.: The REMBRANDT system for multi-criteria decision analysis via pairwise comparisons or direct rating. Report 92-05, Faculty of Technical Mathematics and Informatics (1992)
6. Lootsma, F.: Scale Sensitivity in he Multiplicative AHP and SMART. Journal of Multi-Criteria Decision Analysis 2, 87–110 (1993)
7. Luce, R., Raifa, H.: Games and Decisions. John Wiley and Sons, Inc., New York (1957)
8. Olson, D., Fliedner, G., Currie, K.: Comparison of the REMBRANDT system with analytic hierarchy process. European Journal of Operational Research 82, 522–539 (1995)
9. Saaty, T.: A scaling method for priorities in hierarchical structures. Journal of Mathematical Psychology 15, 234–281 (1977)
10. Saaty, T.: The Analytic hierarchy process. McGraw-Hill, New York (1980)
11. Saaty, T.: What is the analytic hierarchy process? In: Mitra, G. (ed.) Mathematical Models for Decision Support, pp. 109–122. Springer, Berlin (1988)
12. Saaty, T.: Physics as a decision theory. European Journal of Operational Research 48, 98–104 (1990)
13. Saaty, T.: Highlights and critical points in the theory and application of the analytic hierarchy process. European Journal of Operational Research 74, 426–447 (1994)
14. Saaty, T.: Decision making for leaders. RWS Publication, Pittsburgh (1995)
15. Saaty, T., Hu, G.: Ranking by Eigenvector Versus Other Methods in the Aanalytic Hierarchy Process. Appl. Math. Lett., 121–125 (1998)
16. Sugihara, K., Ranaka, H.: Interval evaluations in the analytic hierarchy process by possibility analysis. Computational Intelligence 17(3), 567–579 (2001)
17. Tversky, A.: Intransitivity of Preferences. Psychological Review 76, 31–48 (1969)
18. Van Den Honert, R.: Stochastic pairwise comparative judgments and direct ratings of alternatives in the REMBRANDT system. Journal of Multi-Criteria Decision Analysis 7, 87–97 (1998)
19. Vargas, L.: Why the multiplicative AHP is invalid: A practical counterexample. Journal of Multi-Criteria Decision Analysis 6, 169–170 (1997)
20. Wijnmalen, D.J., Wedley, W.: Correcting Illegitimate Rank Reversals: Proper Adjustment. Journal of Multi-Criteria Decision Analysis, 135–141 (2009)

Part II
Computational Collective Intelligence
in Knowledge Management

A Double Particle Swarm Optimization
for Mixed-Variable Optimization Problems

Chaoli Sun[1], Jianchao Zeng[1], Jengshyang Pan[2],
Shuchuan Chu[3], and Yunqiang Zhang[1]

[1] Complex System and Computational Intelligence Laboratory,
Taiyuan University of Science and Technology, Taiyuan, Shanxi, 030024, China
clsun1225@163.com, zengjianchao@263.net, zyq19840120@163.com
[2] Department of Electronic Engineering,
National Kaohsiung University of Applicated Sciences, Kaohsiung, 807, Taiwan
jspan@cc.kuas.edu.tw
[3] School of Computer Science, Engineering and Mathematics, Flinders University, Australia
scchu@bit.kuas.edu.tw

Abstract. A double particle swarm optimization (DPSO), in which MPSO proposed by Sun et al. [1] is used as a global search algorithm and PSO with feasibility-based rules is used to do local searching, is proposed in this paper to solve mixed-variable optimization problems. MPSO can solve the non-continuous variables very well. However, the imprecise values of continuous variables brought the inconsistent results of each run. A particle swarm optimization with feasibility-based rules is proposed to find optimal values of continuous variables after the MPSO algorithm finishes each independent run, in order to obtain the consistent optimal results for mixed-variable optimization problems. The performance of DPSO is evaluated against two real-world mixed-variable optimization problems, and it is found to be highly competitive compared with other existing algorithms.

Keywords: Particle swarm optimization, mixed-variable optimization problems, feasibility-based rules.

1 Introduction

Mixed-variable optimization problems, which often involve discrete, integer and zero-one variables as well as continuous ones, are widely encountered in engineering design. For example, the size of standard diametric pitch of a gear must be defined as discrete variables, the number of the teeth of a gear much be chosen as integers, and the switch selection may defined as a zero-one variable. Generally, a mixed-variable optimization problem can be expressed in mathematical form as follows:

R. Katarzyniak et al. (Eds.): Semantic Methods, SCI 381, pp. 93–102.
springerlink.com   © Springer-Verlag Berlin Heidelberg 2011

$$\min \quad f(\vec{x})$$
$$\begin{aligned}
\text{s.t.} \quad & g_i(\vec{x}) \leq 0 \quad i = 1, 2, \ldots, m \\
& h_j(\vec{x}) = 0 \quad j = 1, 2, \ldots, l \\
& \vec{x} = (\vec{x}^c, \vec{x}^z, \vec{x}^{i'}, \vec{x}^d)^T \\
& \quad = (x_1^c, \ldots, x_{n_c}^c, x_1^z, \ldots, x_{n_z}^z, x_1^{i'}, \ldots, x_{n_i}^{i'}, x_1^d, \ldots, x_{n_d}^d)^T \quad (1) \\
& x_{k1}^{cl} \leq x_{k1}^c \leq x_{k1}^{cu}, k1 = 1, 2, \ldots, n_c \\
& x_{k2}^z \in \Re^z, k2 = 1, 2, \ldots, n_z \\
& x_{k3}^{i'} \in \Re^{i'}, k3 = 1, 2, \ldots, n_i \\
& x_{k4}^d \in \Re^d, k4 = 1, 2, \ldots, n_d
\end{aligned}$$

where $f(\vec{x})$ is the objective function, $\vec{x} = (\vec{x}^c, \vec{x}^z, \vec{x}^{i'}, \vec{x}^d)^T$ is the vector of solutions, $\vec{x}^c = (x_1^c, \ldots, x_{n_c}^c), \vec{x}^z = (x_1^z, \ldots, x_{n_z}^z), \vec{x}^{i'} = (x_1^{i'}, \ldots, x_{n_i}^{i'})$ and $\vec{x}^d = (x_1^d, \ldots, x_{n_d}^d)$ are the vector of n_c continuous, n_z zero-one, n_i integer and n_d discrete variables, respectively. The total number of variables $n = n_c + n_z + n_i + n_d$. The problem also accounts for m inequalities $g(\vec{x})$ and l equalities $h(\vec{x})$. $g(\vec{x}) \leq 0$ and $h(\vec{x}) = 0$ are called performance constraints, and $x_{k1}^{cl} \leq x_{k1}^c \leq x_{k1}^{cu}, k1 = 1, 2, \ldots, n_c$ is called boundary constraints (or upper and lower bounds constraints) of each continuous variable. $\Re^z, \Re^{i'}$ and $\Re^d$ are respectively represented the set of zero-one, integer and discrete variables. The set composed by all solutions which satisfy all boundary and performance constraints is called the feasible region. A mixed-variable problem is a mathematical programming problem with multiple variables and nonlinearities in the objective function and/or the constraints. Generally, equality constraints are hard to satisfy in the mixed-variable optimization problems. Therefore, an equality constraint is often transformed into inequality constraint using tolerance allowed $|h_j(\vec{x})| \leq \delta$ in order to find the right solution of the problem as much as possible.

Some research work have been reported to solve the mixed-variable optimization problems. In general, there are two classes of optimization methods for solving mixed-variable optimization problems: deterministic search methods and stochastic search methods. The branch and bound method (BBM) has been proposed by Land and Doig [2] to solve mixed-variable optimization problems. In BBM, sub-problems are created by partitioning the feasible domain to force the integer variables to take integer values. The partitioning is done perpendicular to the axis of the integer chosen for the branching. Bremicher et al. [3] combined the well established branch and bound method with a sequential linearization procedure to find global solutions for the mixed-variable problems. The branch and bound algorithm is applied within a sub-problem that is based on a linearization of the original problem. Praharah et al. [4] proposed a two-level decomposition strategy in which the entire domain of variables is partitioned into two levels, one involving the continuous variables and the other involving the discrete variables. Variables in one level are optimized for fixed values of the variable from the other level. The drawbacks of the deterministic search methods include that they make strong assumptions on the continuity and differentiability of objective functions [1]. This has provided an opportunity for stochastic evolutionary algorithms, such as genetic algorithm, evolutionary programming, particle swarm optimization and differential evolution, to solve mixed-variable optimization problems. Rao et al. [5] proposed a new hybrid genetic algorithm for the solution of mixed-variable problems, in which the genetic algorithm was used mainly to determine the optimal feasible region that contains the global optimum point and the hybrid negative sub-gradient method was used to find the final optimum solution. Jouni Lampinen et al. [6, 7] proposed a novel DE-based algorithm for mixed-variable problems. The required handling techniques in this algorithm included the boundary constraint handling technique as well as those of non-linear and non-trivial constraint functions conflicts. Particle swarm optimization was proposed by Kitayama et al. [8] to solve mixed discrete nonlinear problems, in which the penalty function approach was employed to handle the discrete design variables. N. Salam et al. [9] proposed a hybrid particle swarm branch-and-bound (HPB) optimizer for mixed-discrete non-linear programming. The fly-back mechanism was used as constraint violation handling technique in the HPB and the truncation operator was selected to handle discrete variables.

Sun et al. [1] proposed a modified particle swarm optimization (MPSO) to solve mixed-variable problems, in which the value of each kind of variable was decided by the direction of the velocity, and a turbulence operator was proposed so as to overcome the premature convergence brought by the constraint-handling mechanism. The experimental results showed that the non-continuous variables handling mechanism in MPSO was excellent, and the MPSO algorithm had a fast convergence speed and good stability in finding the global optimal results for non-continuous optimization problems. However, for those mixed-variable problems which include continuous variables, the consistency of the results obtained by MPSO is not well. The difference between each independent run is mainly resulted from the values of continuous variables. Therefore, in this paper, particle swarm optimization is used as a local search algorithm in order to obtain the best consistent optimal results for mixed-variable optimization problems.

The rest of this paper is organized as follows. Section 2 reviews the standard particle swarm optimization. In Section 3, a double particle swarm optimization for mixed-variable optimization problems is presented. Section 4 provides experimental results and gives some discussions. Finally, Section 5 outlines conclusions and some suggestions for future work.

2   Standard Particle Swarm Optimization

Particle Swarm Optimization (PSO) was proposed as a global evolutionary algorithm by Kennedy and Eberhart in 1995 [10-12]. The idea was based on the simulation of simplified social models such as bird flocking and fish schooling. In PSO, it assumes that in a D-dimensional search space $S \subseteq R^D$, the swarm consists of n particles each of which has no volume and no weight, holds its own velocity and denotes a solution. The trajectory of each particle in the search space is dynamically adjusted by updating the velocity of each particle according to its own flying experience as well as experiences of neighbor particles. Particle i is in effect a D-dimensional vector

$\vec{x}_i = (x_{i1}, x_{i2}, \ldots, x_{iD}) \in S$. Its velocity is also a D-dimensional vector $\vec{v}_i = (v_{i1}, v_{i2}, \ldots, v_{iD}) \in S$. The best historical position visited by particle i itself is a point in S, denoted as $\vec{p}_i = (p_{i1}, p_{i2}, \ldots, p_{iD})$ or p_{best}, and the best historical position that the entire swarm has passed is denoted as $\vec{p}_g = (p_{g1}, p_{g2}, \ldots, p_{gD})$ or g_{best}. To particle i, the new velocity and position on dimension $d(1 \leq d \leq D)$ are updated as follows:

$$v_{id}(t+1) = \omega v_{id}(t) + c_1 r_1(p_{id}(t) - x_{id}(t)) + c_2 r_2(p_{gd}(t) - x_{id}(t)) \qquad (2)$$

$$x_{id}(t+1) = x_{id}(t) + v_{id}(t+1) \qquad (3)$$

where ω is a parameter called the inertia weight, c_1 and c_2 are positive constants respectively referred as cognitive and social parameters, r_1 and r_2 are random numbers uniformly distributed between 0 and 1.

3 Double Particle Swarm Optimization (DPSO)

A modified particle swarm optimization (MPSO) has been proposed by Sun et al. [1] for solving mixed-variable optimization problems. The experimental results showed that the MPSO algorithm can find best results for non-continuous variables. However, the capability of MPSO to find the best consistent optimal solutions for those mixed-variable problems with at least one continuous variable is not good. The analysis on the experimental results showed that the difference is mainly resulted from the value of continuous variables. Therefore, in this paper, a particle swarm optimization with feasibility-based rules is proposed to find the optimal values of continuous variables after each independent run of MPSO, so as to improve the precision and stability of MPSO to find the best global solutions.

3.1 Treatment of Mixed-Variables

Particle swarm optimization is a global stochastic algorithm proposed originally to solve continuous optimization problems. Therefore, it cannot be directly used to solve mixed-variable problems which include non-continuous variables as well as continues ones. In the MPSO algorithm, the values of non-continuous variables were obtained according to the velocity. The experimental results showed that the proposed method for value of non-continuous variables are good, so in our DPSO algorithm, the handle mechanism of non-continuous of MPSO is also continue used. For detailed information please refer to Ref. [1].

3.2 Constraint-Handling Mechanism

The main approaches for stochastic evolutionary algorithms to handle constraints are the penalty methods, the constraint-preserving methods and the feasibility-based rules methods. In the DPSO algorithm, the feasibility-based rules methods was proposed to handle constraints both in MPSO and the local search algorithm, because it does not need to tune additional penalty parameters, nor does it need to guarantee all particles be in the feasible region at any time. According to the feasibility-based rules proposed in Ref. [13], the best historical position of particle i on iteration $t+1$ ($\vec{p}_i(t+1)$) would be set to $\vec{x}_i(t+1)$ at any of the following scenarios:

(1) $viol(\vec{p}_i(t)) > 0$, while $viol(\vec{x}_i(t+1)) = 0$;
(2) $viol(\vec{p}_i(t)) = 0$, $viol(\vec{x}_i(t+1)) = 0$, but $f(\vec{x}_i(t+1)) < f(\vec{p}_i(t))$;
(3) $viol(\vec{p}_i(t)) > 0$, $viol(\vec{x}_i(t+1)) > 0$, but $viol(\vec{x}_i(t+1)) < viol(\vec{p}_i(t))$;

where the constraint violation $viol(\vec{x})$ is calculated as follows:

$$viol(\vec{x}) = \sum_{i=1}^{m} \max(0, g_i(\vec{x})) + \sum_{j=1}^{l} \max(0, abs(h_j(\vec{x}))) \qquad (4)$$

3.3 The Evolutionary Mode of DPSO

According to the feasibility-based rules, feasible solutions are always considered to be better than infeasible ones, which may lead to pressure on selection of feasible solutions and result in premature convergence. Considered that in the standard particle swarm optimization, each particle adjusts its trajectory only according to its own flying experience and the best position of the swarm, without considering the velocity of other particles. This seems not in accordance with the reality. Therefore, in this paper, the velocity of other particles are supposed to be considered in the updating of each particle's velocity in order not to fall into local position prematurely. Generally, the average velocity of the swarm can reflect the fly direction and speed of the swarm, especially with the increasing iterations. So in this paper, the reversed average velocity of the swarm is considered in the updating of each particle's velocity, which is expected to expand the search ranch so as to improve the diversity of the swarm and to avoid the premature converge. The new updating equations for velocity and position are shown as follows:

$$v_{id}(t+1) = \omega(v_{id}(t) - \omega'\bar{v}_d(t)) + c_1 r_1(p_{id}(t) - x_{id}(t)) + \\ c_2 r_2(p_{gd}(t) - x_{id}(t)) \qquad (5)$$

$$x_{id}(t+1) = x_{id}(t) + v_{id}(t+1) \qquad (6)$$

where $\bar{v}_d(t)$ represents the average velocity of the swarm on dimension d which is calculated as follows:

$$\bar{v}_d(t) = \sum_{i=1}^{n} v_{id}(t) \qquad (7)$$

ω' is called turbulence parameter.

3.4 Description of DPSO

The proposed double particle swarm optimization is mainly consisted by two parts. One is the MPSO which is used as a global search algorithm, and the other is the particle swarm optimization with feasibility-based rules which is used as a local search algorithm to improve the precision of the results MPSO obtained. The description of both algorithms are shown as follows.

(1) The procedure of the DPSO algorithm
    Begin
        Parameter setting, including the swarm size n, inertia weight ω, cognitive parameter c_1, social parameter c_2 and turbulence parameter ω', $t = 0$;
        Initialize positions: For each dimension of particle i, random generate a value in the upper and lower bounds. Set $int(x_{id}(t))$ as the value of integer variable and the closest discrete value as the value of discrete one;
        Initialize velocity for each particle;
        Calculate the violation and fitness of each particle;
        Generate the best history position of each particle: $\vec{p}_i(t) = \vec{x}_i(t), i = 1, 2, \ldots, n$;
        Generate the best history position of the swarm: find min $f(\vec{p}_i(t))$ where $viol(\vec{p}_i(t)) = 0$, set $\vec{p}_g(t) = \vec{p}_i(t)$ if $f(\vec{p}_i(t))$ is minimal;
        While the stopping criterion is not met
            For each particle i in the swarm
                Updating the velocity and position using (5) and (6);
                Call the handling procedure for different kinds of variables;
                Calculate the violation value using (4);
                Calculate the fitness value;
                Update the best history position of particle i according to feasibility-based rules;
            End for
            Update the best history position of the swarm: find min $f(\vec{p}_i(t))$ where $viol(\vec{p}_i(t)) = 0$, set $\vec{p}_g(t) = \vec{p}_i(t)$ if $f(\vec{p}_i(t))$ is minimal;
        End While
        call local search procedure;
    End

(2) The procedure of the local search algorithm
    Begin
        Parameter setting, including the swarm size n_1, inertia weight ω_1, cognitive parameter c_1' and social parameter c_2', $t = 0$;
        Initialize positions: random generate a position in the upper and lower bounds for each particle.
        Random generate a velocity for each particle;
        Calculate the violation and fitness of each particle;
        Generate the best history position of each particle: $\vec{p}_i(t) = \vec{x}_i(t), i = 1, 2, \ldots, n$;
        Generate the best history position of the swarm: find min $f(\vec{p}_i(t))$ where $viol(\vec{p}_i(t)) = 0$, set $\vec{p}_g(t) = \vec{p}_i(t)$ if $f(\vec{p}_i(t))$ is minimal;
        While the stopping criterion is not met
        For each particle i in the swarm

Updating the velocity and position according to equation (5) and (6);
Calculate the violation value using (4);
Calculate the fitness value;
Update the best history position of particle i according to feasibility-based
rules;
End for
Update the best history position of the swarm: find min $f(\vec{p}_i(t))$ where
$viol(\vec{p}_i(t)) = 0$, set $\vec{p}_g(t) = \vec{p}_i(t)$ if $f(\vec{p}_i(t))$ is minimal;
End While
End

Pay attention to that in the fitness calculation of the local search algorithm, all values of non-continuous variable is treated as constants, which are just same to the optimal results obtained by the MSPO algorithm.

4   Experiments and Some Discussions

Two real-world mixed-variable optimization problems which included continuous and non-continuous variables and frequently employed in the literature are used to evaluate the performance of the proposed DPSO algorithm. The parameters in DPSO are set as follows. The inertia weight ω is decreased linearly from 0.9 down to 0.4; the perturbation coefficient is changed sinusoidally, with its amplitude of 0.5. The cognitive parameter is set to decrease linearly from 3.6 down to 1.8 and the social parameter increase linearly from 0.2 up to 1.8 so as to fight against prematurity with the reversed average velocity of the swarm. Total number of individuals is 40, the maximum number of iterations is 10,000 per run and 30 independent runs are executed for each example. In the local search algorithm, the parameters are set same as DPSO except the total number of individuals is 20, the maximum number of iterations is 1000 and only one run is executed.

Example 1: Pressure Vessel Design
Pressure vessel design is a common benchmarking problem for mixed-variables optimization problems presented by Sandgren [14]. The objective of this problem is to minimize the manufacturing cost of the pressure vessel, which is a combination of material cost, welding cost and forming cost. The design variables are specified as

$$\vec{x} = (x_1, x_2, x_3, x_4)^T = (T_s, T_h, R, L)^T$$

where x_1 is the spherical head thickness, x_2 is the shell thickness, x_3 and x_4 are the radius and length of the shell, respectively. In the design variables, x_3 and x_4 are continuous, while x_1 and x_2 are discrete values and integer multiplies of 0.0625 inch respectively.

The best optimal results of this problem obtained by different PSO-based algorithms are shown in Table 1. It can be seen all PSO-based algorithms obtained better results than other stochastic algorithms. Table 2 shows the results comparison on all PSO-based algorithms, though the best results of MPSO and DPSO are not good as that of HPB, the mean results and the standard deviations are all better than

HPB, especially our proposed DPSO obtained consistent best optimal solutions for each run.

Example 2: Spring Design

The second example is to minimize the volume of a compression spring under constant loading. It is a common real-world optimization problem which involves discrete, integer and continuous design variables. The design variables are specified as

$$\vec{x} = (x_1, x_2, x_3) = (d, D, N)$$

The spring wire diameter, x_1, may have only discrete values according to available stand spring steel wire diameters shown as follow:

$$x_1 \in \{0.009, 0.0095, 0.0104, 0.0118, 0.0128, 0.0132,$$
$$0.014, 0.015, 0.0162, 0.0173, 0.018, 0.020, 0.023,$$
$$0.025, 0.028, 0.032, 0.035, 0.041, 0.047, 0.054,$$
$$0.063, 0.072, 0.080, 0.092, 0.105, 0.120, 0.135,$$
$$0.148, 0.162, 0.177, 0.192, 0.207, 0.225, 0.244, 0.263,$$
$$0.283, 0.307, 0.331, 0.362, 0.394, 0.4375, 0.500\}$$

The outside diameter, x_2, is a continuous variable and the number of spring coils, x_3, is an integer variable. For its mathematical model please referred to Ref. [9].

Table 1. Optimal solutions of the pressure vessel design

	EP[15]	EA[16]	GA[17]	PSO[18]	HPB[9]	MPSO[1]	DPSO
x_1	1.000	0.9345	0.8125	0.8125	0.8125	0.8125	0.8125
x_2	0.625	0.5000	0.4375	0.4375	0.4375	0.4375	0.4315
x_3	51.1958	48.3290	40.097398	42.0984456	42.09893	42.098446	42.098446
x_4	90.7821	112.6790	176.654047	176.636595	176.6305	176.636792	176.636792
g_1	-0.0119	-0.00475	-0.00002	0.000000	0.00000	0	0
g_2	-0.1366	-0.038941	-0.035891	-0.0358808	-0.03587	-0.03588083	-0.0388083
g_3	-13584.5631	-3652.87683	-27.886075	0.000000	0.00000	-1.166400×10^6	-1.166400×10^6
g_4	-149.2179	-127.321	-63.345953	-63.363404	-63.69484	-63.36321	-63.36321
$f(\vec{x})$	7108.6160	6410.3811	6059.94634	6059.7143	6059.65457	6059.718932	6059.718932

The optimal results of this problem obtained by different algorithms are shown in Table 3, and Table 4 shows the comparison results between different PSO-based algorithm. Seen from these two Tables, it can obviously be seen that all stochastic algorithms can find the best optimal solutions except the genetic algorithm [16]. The comparison results between different PSO-based algorithm showed that our proposed DPSO can obtain consistent optimal results for each run, the mean result and the standard deviation for this problem is much better than other PSO-based algorithms.

The computational cost, measured in the number of evaluations of the fitness function (FFE), performed by DPSO, is $40 \times 10000 + 20 \times 1000 = 420,000$ FFE which is higher than MPSO. It is because another PSO is proposed to do local searching, which is aimed to obtain coincident results for each mixed-variable problems.

Table 2. Comparison of DPSO with other PSO-based approach

	PSO[18]	HPB[9]	MPSO[1]	DPSO
Best	6059.7143	6059.65457	6059.718932	6059.718932
Mean	6289.92881	6060.08	6059.727988	6059.718932
Worst	-	-	6059.764215	6059.718932
Standard deviation	305.87	0.02194	0.003559	2.761309×10^{-10}

Table 3. Optimal solutions of the coil compression spring design

	GeneAS[16]	DE[6]	PSO[18]	HPB[9]	MPSO	DPSO
x_1	0.283	0.283	0.283	0.283	0.283	0.283
x_2	1.226	1.22304101	1.22304101	1.22301421	1.223041	1.223041
x_3	9	9	9	9	9	9
g_1	-713.510	-1008.8114	-1008.8114	-1011.6168	-1008.629	-1008.653
g_2	-8.933	-8.9456	-8.9456	-8.9457	-10.55277	-10.55278
g_3	-0.083	-0.083	-0.083	-0.0830	-0.083	-0.083
g_4	-1.491	-1.777	-1.777	-1.7769	-1.776938	-1.776959
g_5	-1.337	-1.3217	-1.3217	-1.3216	-1.321774	-1.3217
g_6	-5.461	-5.4643	-5.4643	-5.4643	-5.946424	-5.946429
g_7	0.0000	0.0000	0.0000	0.0000	0.000000	0.000000
g_8	-0.009	0.0000	0.0000	0.0000	-0.000115	0.000000
$f(\vec{x})$	2.665	2.65856	2.65856	2.65850	2.658555	2.658555

Table 4. Comparison of DPSO with other PSO-based approach

	PSO	HPB	MPSO	DPSO
Best	2.65856	2.65850	2.658555	2.658555
Mean	2.738024	2.6985	2.658601	2.658555
Worst	-	-	2.659459	2.658555
Standard deviation	0.107061	0.0239	0.000030	1.621585×10^{-16}

5 Conclusion

A double particle swarm optimization, in which two modified particle swarm optimization are cooperated to find the best insistent optimal solutions for mixed-variable optimization problems. Feasibility-based rules are used as the handling mechanism for constraint conflict. The proposed DPSO make good use of the method of MPSO for non-continuous variables' valuing, and the fast convergence of PSO for finding the continuous optimization problems.

Experimental results showed that our proposed DPSO can obtain the best coincident optimal results for mixed-variable optimization problems. In the future, a deterministic algorithm will be considered to substitute PSO with feasibility-based rules so as to reduce the number of fitness calculation and save the time on the same optimal results for solving mixed-variable optimization problems.

Acknowledgments. This work is supported in part by Youth Foundation of Taiyuan University of Science and Technology under Grant No. 20103012.

References

1. Sun, C., Zeng, J., Pan, J.: A modified particle swarm optimization with feasibility-based rules for mixed-variable optimization problems. International Journal of Innovative Computing, Information and Control 7(6), 3081–3096 (2011)
2. Land, A.M., Doig, A.G.: An automatic method of solving discrete programming problems. Econometrica 26, 497–520 (1960)
3. Bremicher, M., Papalambros, P.Y., Loh, H.T.: Solution of mixed-discrete structural optimization problems with a new sequential linearization algorithm. Computers & Structures 37(4), 451–461 (1990)
4. Praharah, S., Azarm, S.: Two-Level Nonlinear Mixed Discrete-Continuous Optimization-Based Design: An Application to Printed Circuit Board Assemblies. Journal of Electronic Packaging 114, 425–435 (1992)
5. Rao, S.S., Xiong, Y.: A Hybrid Genetic Algorithm for Mixed-Discrete Design Optimization. Transactions of the ASME 127, 1100–1112 (2005)
6. Lampinen, J., Zelinka, I.: Mixed Integer-Discrete-Continuous Optimization By Differential Evolution - Part 2: a practical example. In: Proceedings of MENDEL 1999, 5th International Mendel Conference on Soft Computing, pp. 77–81 (1999)
7. Lampinen, J., Zelinka, I.: Mixed integer-discrete-continuous optimization by differential evolution. Part 1: the optimization method. In: Proceedings of MENDEL 1999, 5th International Mendel Conference on Soft Computing, pp. 71–76 (1999)
8. Kitayama, S., Arakawa, M., Yamazaki, K.: Penalty function approach for the mixed discrete nonlinear problems by particle swarm optimization. Struct. Multidisc. Optim. 32, 191–202 (2006)
9. Nema, S., et al.: A Hybrid Particle Swarm Branch-and-Bound (HPB) Optimizer for Mixed Discrete Nonliear Programming. IEEE Transactions on Systems, Man and Cybernetics–Part A: Systems and Humans 38(6), 1411–1424 (2008)
10. Kennedy, J., Eberhart, R.: Particle swarm optimization. In: Proc. IEEE Int'l. Conf. on Neural Networks, pp. 1942–1948 (1995)
11. Eberhart, R., Kennedy, J.: A new optimizer using particle swarm theory. In: Proceedings of the Sixth International Symposium on Micro Machine and Human Science, pp. 39–43 (1995)
12. Hsu, C.-H., Shyr, W.-J., Kuo, K.-H.: Optimizing Multiple Interference Cancell-ations of Linear Phase Array Based on Particle swarm Optimization. Journal of Information Hiding and Multimedia Signal Processing 1(4), 292–300 (2010)
13. Pulido, G.T., Coello, C.A.C.: A constraint-handling mechanism for particle swarm optimization. In: Congress on Evolutionary Computation (2004)
14. Sandgren, E.: Nonlinear integer and discrete programming in mechanical design optimization. Journal of Mechanical Design 112(2), 223–229 (1990)
15. Cao, Y., Wu, Q.: A mixed variable evolutionary programming for optimisation of mechanical design. Int. J. Eng. Intell. Syst. Electic. Eng. Commun. 7(2), 77–82 (1999)
16. Deb, K.: Gene AS: A robust optimal design technique for mechanical component design. Springer, Heidelberg (1997)
17. Coello, C.A.C., Montes, E.M.: Use of Dominance-Based Tournament Selection to Handle Constraints in Genetic Algorithms. In: Intelligent Engineering Systems through Artificial Neural Networks (ANNIE 2001), pp. 177–182 (2001)
18. He, S., Prempain, E., Wu, Q.H.: An improved particle swarm optimizer for mechanical design optimization problems. Engineering Optimization 36(5), 585–605 (2004)

Particle Swarm Optimization with Disagreements on Stagnation

Andrei Lihu and Ştefan Holban

Department of Computer Science, Politehnica University of Timişoara,
Bd. Vasile Pârvan, 300223 Timişoara, Romania
andrei.lihu@gmail.com, stefan@cs.upt.ro

Abstract. This paper introduces a new family of particle swarm optimizers that model social disagreements among particles when stagnation occurs inside the swarm. After a short introduction into particle swarm optimization, we describe theoretically the new concept and define a special type of particle swarm optimization with disagreements that generates riots on stagnation. After testing it on several popular optimization benchmarks, we conclude that it can help any PSO escape stagnation. This work illustrates one of the many benefits of using the particle swarm optimization with disagreements.

Keywords: particle swarm optimization, disagreements, riot, stagnation, swarm intelligence.

1 Introduction

The *particle swarm optimization* (PSO) is a top competitor among optimization algorithms. PSO is a population based optimization technique that implements "the social mind" metaphor by simulating the social behavior exhibited by various animal communities like birds or fish.

PSO's main advantage is the ability to operate without the need of gradient information. Initially described in [1], it models a swarm of particles flying iteratively through the hyperspace of solutions until a termination condition is met. Particles improve their positions during iterations based on their own personal and their group's best experience. The ability to be influenced by neighbors represents the social component of the algorithm and the ability to be influenced by its own experience represents the cognitive component.

Depending on the problem, when searching through the space of solutions, optimizers can stagnate — they cannot find a better solution within a specified amount of time. Some of the reasons stagnation occurs are because the algorithm has no means to escape a local minimum it is currently trapped in – thus leading to premature convergence, or because it moves along a large plateau, or maybe it jumps on an equally sized densely spiked region. For a population-based algorithm like PSO, in order to avoid stagnation, the diversity inside the swarm should be increased. Simultaneously, the algorithm should still converge and it should converge in a reasonable amount of time.

R. Katarzyniak et al. (Eds.): Semantic Methods, SCI 381, pp. 103–113.
springerlink.com   © Springer-Verlag Berlin Heidelberg 2011

We propose a simple mechanism to increase the diversity inside the swarm only when stagnation occurs. Since PSO is a social algorithm, we modeled and added a new social behavior to it: the disagreements that naturally occur in a social group, which we applied only to the social component of the algorithm. Whenever stagnation is detected, the particles from the swarm can oppose their group's way by exhibiting different opinions with a given probability. This way the particles' positions change their original path and have more chances to disrupt the current stagnation period. For some generations full-blown riots can appear and change the evolution of the algorithm. In order to preserve the final capacity of convergence, we apply the disagreements following a linearly decreasing probability rule.

After conducting relevant empirical tests, we concluded that this new enhancement might help any PSO escape from local minima when needed, making it suitable for searching multi-modal functions. The extra-added computational cost is minimal.

Following our previous experience with *the disagreements metaphor* in [2], this paper presents another application where the disagreements concept is helpful: stagnation mitigation.

As an overview, in Section 2 we present the standard particle swarm optimization and how to measure stagnation. In Section 3 we explain the theoretical foundation of disagreements as previously introduced in [2]. In Section 4 we introduce a practical approach to handle stagnation — *the riot-on-stagnation operator* (RS-PSOD) and in Section 5 we conduct some tests using the new operator and analyze their results. In Section 6 we conclude that the new approach can have real benefits in stagnation-prone environments.

2   Stagnation in Particle Swarm Optimization

2.1   Particle Swarm Optimization

As Bergh in [3], we define PSO as follows:

Let n be the dimension of the solution hyperspace H^n, let s be the number of particles from the swarm, and let i be the index of a particle, such that $i \in \overline{1 \ldots s}$. Each particle i has the following variables: x_i – the current position, v_i – the current velocity, y_i – the current best position, $\hat{y}$ – neighborhood's best. The function f is the function to be minimized. PSO has three phases: initialization, iterations, termination.

In the initialization phase, the particles' positions and velocities are randomly spread in the search space. Best positions are initially updated using (3) and (4).

In the iterations' phase, the updating principle for velocities and positions is given in (1) and (2):

$$v_{ij}(t+1) = wv_{ij}(t) + c_1 r_{1j}(t)[y_{ij}(t) - x_{ij}(t)] + c_2 r_{2j}(t)[\hat{y}_j(t) - x_{ij}(t)] \ , \quad (1)$$

$$x_i(t+1) = x_i(t) + v_i(t+1) \ , \quad (2)$$

where c_1 is the personal coefficient, c_2 is the social coefficient, $c_1, c_2 \in (0, 2]$. r_1 and r_2 are random vectors, such that: $r_1, r_2 \sim \mathcal{U}(0, 1)$. The first term of (1) is the previous velocity influenced by an inertial weight w. The second term is the personal component that makes the particle move toward its best personal position found so far and the third term orients the particle toward neighborhood's best position found so far.

At each iteration, y_i and $\hat{y}$ are updated using the formulae:

$$y_i(t+1) = \begin{cases} y_i(t) & \text{if } f(x_i(t+1)) \geq f(y_i(t)) \\ x_i(t+1) & \text{if } f(x_i(t+1)) < f(y_i(t)) \end{cases} . \tag{3}$$

$$\hat{y}(t) \in \{y_0(t), y_1(t), \ldots, y_s(t) | f(\hat{y}(t))\} = $$
$$min \{f(y_0(t)), f(y_1(t)), \ldots, f(y_s(t))\} . \tag{4}$$

When an *a priori* established criterion is met, the algorithm terminates (a number of fitness calls or generations have elapsed, stagnation, etc.).

2.2 Stagnation

Stagnation is defined as a situation in which there is no improvement of current solution for an amount of time. It is defined in [4] as follows:

"The particle swarm system is thought to be in stagnation, if arbitrary particle i's history best position P_i and the total swarm's history best position P_g keep constant over some time steps."

To measure the stagnation in PSO, [5] takes into consideration also the velocities and defines the improvement ratio as:

$$R = \left| \frac{1 - f_c/f_p}{1 - v_c/v_p} \right| , \tag{5}$$

where f_c is the current fitness value of the best particle, f_p is the previous and v_c is the current average velocity of all particles, while v_p is the previously recorded one. Stagnation is detected when R drops under a preset value ϵ.

The approach from [5] assumes that when stagnation occurs the velocities tend to 0. However, this is valid only in situations when particles get trapped into local minima, while for very densely spiked or plateau functions this could not true. To decide for all cases if no improvement took place in an amount of time Δt_h between two iterations, t and $t + \Delta t_h$, we decided to measure the Euclidean distance between the current fitness of best particle at iterations $t + \Delta t_h$ and t. To achieve a relative measurement, this is compared with the t's best fitness norm multiplied with a preset threshold value, ϵ. Therefore, the following relation must hold true:

$$\|\hat{y}_{t+\Delta t_h} - \hat{y}_t\| < \epsilon \cdot \|\hat{y}_t\| . \tag{6}$$

3 Disagreements

For a better understanding and for completeness, we will introduce the disagreements concept as we did in [2]. Disagreements are an ubiquitous social phenomenon that leads to greater diversity and heightened awareness of current problems. We can apply this concept in a social algorithm like PSO to increase the diversity of the swarm when stagnation is detected.

We replaced the first term from (1) with a generic $C_v(t)$ — the velocity component, the second term with $C_c(t, x_i, y_i)$ — the cognitive component, the third term with $C_s(t, x_i, \hat{y})$ — the social component. Making the substitution in (2), where we also change the position component with a generic one, C_x, we obtain the generalized updating equation:

$$x_i(t+1) = C_x(t, x_i) + C_v(t) + C_c(t, x_i, y_i) + C_s(t, x_i, \hat{y}) + \zeta \ ,$$
$$C_c(t, x_i, y_i) \rightarrow y_i, C_s(t, x_i, \hat{y}) \rightarrow \hat{y} \ . \tag{7}$$

$C_c(t, x_i, y_i) \rightarrow y_i$ should be read as "the result of C_c tends to y_i" and $C_s(t, x_i, \hat{y}) \rightarrow \hat{y}$ should be read as "the result of C_s tends to $\hat{y}$". ζ is usually 0 and can accommodate any other more elaborate variant of PSO that may consist of other components too.

A disagreement is defined as a function that takes values in H^n. One can define the disagreements as the family of D functions for which the following property holds true:

$$F_D = \{D : H^n \rightarrow H^n | \forall z \in H^n. \ D(z) \neq z\} \ . \tag{8}$$

Let P_{all} be all PSO algorithms that contain the social component C_s in the updating principle. We define the powerset Δ_{all} as:

$$\Delta_{\text{all}} = \mathcal{P}(\{id\} \cup F_D) \ , \tag{9}$$

where id is the identity function, also written as $\emptyset_D$, which is a no-op (no disagreement).

Let ρ be a *decision function* called "disagreement selector" that takes a set of disagreements $\Delta_v \subseteq \Delta_{\text{all}}$ as argument at iteration t (from all iterations t_{all}) and decides which disagreement is invoked for a particle i (from the total of s particles):

$$\rho : \{\Delta_{\text{all}} \times t_{\text{all}} \times s\} \rightarrow \Delta_{\text{all}}, \ \rho(\Delta_v, t, i) = D_i, \ D_i \in \Delta_v \ . \tag{10}$$

The "disagreement injector" on any social PSO is defined as:

$$\Psi_{\text{PSO}} : \{P_{\text{all}} \times \Delta_{\text{all}} \times \rho_{\text{all}}\} \rightarrow P_{\text{all}}, \ \Psi_{PSO}(P_i, \Delta_v, \rho) = P_{i\text{D}} \ , \tag{11}$$

where P_i is one of the many PSO variants with social component, and $P_{i\text{D}}$ is the resulting particle swarm optimization with disagreements.

After we apply the injection operator Ψ_{PSO}, the updating principle of the newly obtained PSO, now called *particle swarm optimization with disagreements* (PSOD) is transformed from (5) to:

$$x_i(t+1) = C_x(t, x_i) + C_v(t) + C_c(t, x_i, y_i) + D_i(C_s(t, x_i, \hat{y})) + \zeta \ ,$$
$$C_c(t, x_i, y_i) \rightarrow y_i, C_s(t, x_i, \hat{y}) \rightarrow D_i(\hat{y}) \ . \tag{12}$$

The concept of disagreements is a special operator that can be applied to any social PSO without modifying the internals of the algorithm. The social component can vary in implementation from algorithm to algorithm, but the injection operator can be applied in any case. A disagreement operator can affect only the social component of PSO.

4   Riot-on-Stagnation Operator

To demonstrate how PSOD can resolve stagnation, we defined in terms of (11) the *riot-on-stagnation operator* (RS-PSOD) as follows:

$$\Psi_{RS-PSOD}(P_i) = \Psi_{PSO}(P_i, \Delta_{RS}, \rho_{RS}) \ . \tag{13}$$

The subset of disagreements (the Δ_v that contains the disagreements, D_{RSi}) is defined by:

$$\Delta_{RS} = \{\emptyset_D, D_{RS}\} \ . \tag{14}$$

We modeled an "extreme disagreement", D_{RS}, which multiplies member-wise (a Hadamard product, $\otimes$) the social component C_s (e.g. $c_2 r_{2j}(t)[\hat{y}_j(t) - x_{ij}(t)]$) by a vector r containing random uniformly distributed values in the intervals $[-\lambda_u, -\lambda_l]$ and $[+\lambda_l, +\lambda_u]$, with $\lambda_l, \lambda_u \in \mathbb{R}_+^*, \lambda_u > \lambda_l$ and $\lambda_l \geq 1$:

$$D_{RS}(z) = r_i \otimes z, \ r_i = r_{i_1} + sgn(r_{i_1}) \cdot \lambda_l, \ r_{i_1} \sim \mathcal{U}(-(\lambda_u - \lambda_l), +(\lambda_u - \lambda_l)) \ , \tag{15}$$

where r_i is the i–th component of r and r_{i_1} is a random number for each r_i.

Fig. 1 is a very simple visualization of the concept in two dimensions:

Let $\theta_{RS}(t, i) \sim \mathcal{U}(0, 1)$ be an uniformly distributed random variable that is generated at each iteration t for each particle i. Let $\delta = \frac{t}{t_{max}} \in [0, 1]$ be the current execution progress indicator, where t_{max} is the total number of iterations. The selector function is defined as follows:

$$\rho_{RS}(\Delta_{RS}, t, i) = \begin{cases} \emptyset_D & \text{if } \theta_{RS}(t, i) < \delta \\ D_{RS} & \text{if } \theta_{RS}(t, i) \geq \delta \text{ and (6) holds true} \end{cases} \ . \tag{16}$$

The updating principle from (11) becomes:

$$x_i(t+1) = C_x(t, x_i) + C_v(t) + C_c(t, x_i, y_i) + D_{RSi}(C_s(t, x_i, \hat{y})) + \zeta \ ,$$
$$C_c(t, x_i, y_i) \rightarrow y_i, C_s(t, x_i, \hat{y}) \rightarrow D_{RSi}(\hat{y}) \ . \tag{17}$$

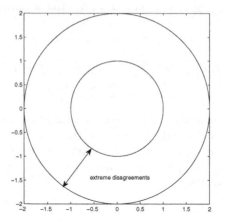

Fig. 1. Area between circles shows where r_i — that produces "extreme disagreements" — can be generated for the case when $\lambda_l = 1$ and $\lambda_u = 2$.

5   Experimental Results

5.1   Setup

The experimental setup is similar to the one we used to test the 6σ-PSOD operator in [2], but we have chosen a single high dimension: 30. We have selected two algorithms for testing: the standard PSO of Clerc ([6]), but with constriction, called SPSO, having the following configuration: $\chi = 0.729$ with $c_1 = c_2 = 2.05$ as in [7]; the second chosen algorithm is recent, the PSO-VG ([8]), a social-only inertial PSO, with the following configuration: $w = 0.729$ with $c_2 = 1.494$.

As in [2], we transformed these two algorithms into their disagreement-enabled counterparts applying the RS-PSOD operator: $\Psi_{\text{RS-PSOD}}(SPSO) = SPSOD_{\text{RS}}$ and $\Psi_{\text{RS-PSOD}}(PSO - VG) = PSO - VGD_{\text{RS}}$, with $\lambda_l = 1$ and $\lambda_u = 2$. Both algorithms use the grid topology. For each algorithm, we will measure the mean best fitness value with its standard deviation across 50 runs and the average number of riots (how many generations disagreed) for a swarm made of 50 particles. We run the tests on 12 optimization benchmarks in 30 dimensions, but we pick only four popular benchmark problems to report results on. Algorithms terminate after 30000 fitness evaluations. Stagnation is detected when $\epsilon = 0.005$ after $\Delta t_h = 10$ generations passed.

Inspired by Chen and Li's work in [9], we selected the following test problems: Generalized Rosenbrock (F_2), Shifted Rastrigin ($F_1 4$), Shifted Schwefel (F_{21}) and Griewank (F_5). The reasons for picking each one, the same as in [9], are explained along this enumeration:

1. We started with Generalized Rosenbrock to observe how our new PSODs behave on plateau functions:

$$F_2(X) = \sum_{i=1}^{s-1} (100(x_{i+1} - x_i^2)^2 + (1 - x_i)^2), \;\; X \in \mathbb{R}^n \; . \tag{18}$$

2. Shifted Rastringin was used to check for behavior of convergence:

$$F_{14}(X) = \sum_{i=1}^{s} (z_i^2 - 10\cos(2\pi z_i) + 10), \;\; X \in [-5, +5]^s, \;\; Z = X - o \; . \tag{19}$$

3. To see if the new PSODs retain the original PSO robustness, we tested with Shifted Schwefel:

$$F_{21}(X) = \sum_{i=1}^{s} \left(\sum_{j=1}^{i} z_j \right)^2 , \;\; X \in [-100, +100]^s, \;\; Z = X - o \; . \tag{20}$$

4. Finally, with Griewank we looked if the new behavior helps escaping local minima:

$$F_5(X) = \sum_{i=1}^{s} \frac{x_i^2}{4000} - \prod_{i=1}^{s} \cos\left(\frac{x_i}{\sqrt{i}} \right) + 1, \;\; X \in [-600, 600]^s \; . \tag{21}$$

We used the evolutionary framework Java EvA2 ([10]) to test the algorithms and implement the above described RS-PSOD. A modified version of the framework that we develop in parallel contains the "RiotOnStagnation" variant of PSOD. Sources are currently available online at [11].

5.2 Results

Table 1 and Table 2 contain the results of the performed tests:

Table 1. Benchmark results for F_2 and F_{14}.

Algorithm	F_2			F_{14}		
	Mean	Std. dev.	Riots	Mean	Std. dev.	Riots
SPSO	57.1260	40.4190	-	64.0869	18.2855	-
SPSOD$_{RS}$	43.2560	29.5548	16.25	59.3327	15.9307	24.20
PSO-VG	67.3753	48.8016	-	112.9689	34.4212	-
PSO $-$ VGD$_{RS}$	62.5470	37.6536	12.80	96.0294	31.0198	31.45

As we can notice in Table 1, applying correction with disagreements in F_2's case works both on SPSO and PSO-VG. Fig. 2 is a graphical representation of the convergence graphs we obtained when we tested F_2 and concludes that the

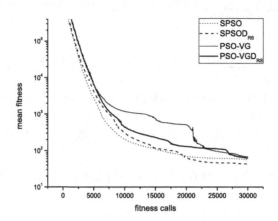

Fig. 2. Convergence graph for F_2 (30 dimensions; $s = 50$). PSODs find better solutions on plateau functions than their original PSO counterparts.

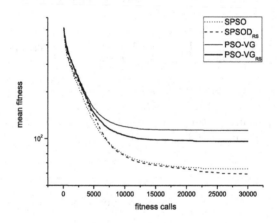

Fig. 3. Convergence graph for F_{14} (30 dimensions; $s = 50$). Comparing in pairs, PSOD variants have slight better convergence than original PSOs.

riot-on-stagnation operator is beneficial because it successfully mitigates stagnations on plateau functions.

In F_{14}'s case, for the selected values of ϵ and Δt_h, the convergence of both PSOs is improved after applying the operator. Fig. 3 clearly demonstrates that PSODs converge towards a better solution in the presence of a quite high number of riots (an average of 24.20 and 31.45). PSO-VGD$_{RS}$ also performs slightly better in this stagnation prone environment.

Table 2. Benchmark results for F_{21} and F_5.

Algorithm	F_{21}			F_5		
	Mean	Std. dev.	Riots	Mean	Std. dev.	Riots
SPSO	1327.1377	1403.8522	-	0.0090	0.0116	-
SPSOD$_{RS}$	1122.9062	1012.4270	0.33	0.0075	0.0103	5.16
PSO-VG	5583.0198	4478.2125	-	0.0147	0.0149	-
PSO $-$ VGD$_{RS}$	4431.4144	4195.0482	7.45	0.0101	0.0078	11.38

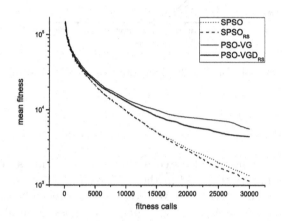

Fig. 4. Convergence graph for F_{21} (30 dimensions; $s = 50$). PSODs retain the original PSO robustness.

When we tested for Shifted Schwefel (F_{21}), we obtained better results for our transformed PSOs than for the original variants; with the chosen parameters, the conclusion is that new PSODs are slightly more robust, as can be seen in Fig. 4. Correlating the number of average riots from this test with those obtained for F_{14} or F_2, we understand that a smaller but more pinpointed number of riots was generated only when needed, not to introduce too much randomness in the swarm.

For F_5, as it can be seen in Table 2 and from the convergence graph in Fig. 5, we got marginally improved results.

A small amount of disagreements gives better results than no amount. The range of minimum and maximum acceptable riots needs to be investigated more thoroughly. This simple stagnation recovery method proves itself useful in multi-modal environments and the empirical analysis confirms that for the considered parameters a slight overall improvement is obtained.

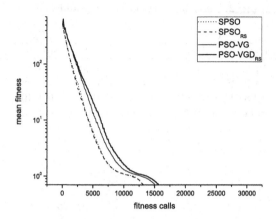

Fig. 5. Convergence graph for F_5 (30 dimensions; $s = 50$). Marginally improved convergence.

6   Conclusion

In this paper, we introduced and tested with improved results a new family of particle swarm optimizers that add in their behavior a common characteristic of social groups: the disagreements. We described theoretically the metaphor of disagreements and exemplified it with one of its possible applications: the mitigation of swarm stagnation by using the riot-on-stagnation operator (RS-PSOD). This operator transforms a given PSO to a PSOD that is rioting when stagnation is detected.

We wanted to find out how the newly resulted algorithms behave on plateau test functions, if they keep their convergence rates, if they are still as robust as before and how they can help to escape local minima. Therefore, we employed several test functions from the related scientific literature to capture information about all these aspects.

Even though we obtained better results for our fixed set of stagnation parameters, this is only a preliminary study. Future work should consist in studying the optimum amounts of disagreements (number of riots) that can be safely employed and how they should be applied in order to improve furthermore the PSO.

Overall, we have concluded that adding disagreements when a PSO stagnates can help it to tackle local minima and find a better way towards the solution in multi-modal and plateau environments.

Although not staggering, the results are promising and open new research directions for the disagreements metaphor. The added randomness does not impede the convergence rate and the robustness of the PSO. The extra computationally involved cost is minimal.

Acknowledgments. This work was developed in the frame of PNII-IDEI-PCE-ID923-2009 CNCSIS–UEFISCSU grant and was partially supported by the strategic grant POSDRU 6/1.5/S/13-2008 of the Ministry of Labor, Family and Social Protection, Romania, co-financed by the European Social Fund – Investing in People.

References

1. Kennedy, J., Eberhart, R.C.: Particle Swarm Optimization. In: Proceedings of IEEE International Conference on Neural Networks, pp. 1942–1948 (1995)
2. Lihu, A., Holban, Ş.: Particle Swarm Optimization with Disagreements. In: Tan, Y., Shi, Y., Chai, Y., Wang, G. (eds.) ICSI 2011, Part I. LNCS, vol. 6728, pp. 46–55. Springer, Heidelberg (2011)
3. Bergh, F.: An Analysis of Particle Swarm Optimizers (PhD thesis). University of Pretoria, Pretoria (2001)
4. Jiang, M., Luo, Y., Yang, S.: Stagnation Analysis in Particle Swarm Optimization. In: IEEE Swarm Intelligence Symposium (SIS 2007), pp. 92–99 (2007)
5. Worasucheep, C.: A Particle Swarm Optimization with stagnation detection and dispersion. In: IEEE Congress on Evolutionary Computation (CEC 2008), pp. 424–429 (2008)
6. Clerc, M., Kennedy, J.: The particle swarm - explosion, stability, and convergence in a multidimensional complex space. IEEE Transactions on Evolutionary Computation 6, 58–73 (2002)
7. Parsopoulos, K., Vrahatis, M.: Particle Swarm Optimization and Intelligence: Advances and Applications, pp. 38–38. IGI Global, New York (2010)
8. Pedersen, M.E.H., Chipperfield, A.J.: Simplifying particle swarm optimization. Applied Soft Computing 10, 618–628 (2010)
9. Chen, X., Li, Y.: A Modified PSO Structure Resulting in High Exploration Ability With Convergence Guaranteed. IEEE Transactions on Systems, Man, and Cybernetics, Part B: Cybernetics 37(5), 1271–1289 (2007)
10. EvA2 Project Homepage, http://www.ra.cs.uni-tuebingen.de/software/EvA2 (accessed in March 2011)
11. EvA2-AL Project Homepage (a modified version of EvA2), https://github.com/andrei-lihu/Eva2-AL (accessed in March 2011)

Classifier Committee Based on Feature Selection Method for Obstructive Nephropathy Diagnosis

Bartosz Krawczyk

Wroclaw University of Technology, Department of Systems and Computer Networks,
Wybrzeze Wyspianskiego 27, 50-370 Wroclaw, Poland
bartosz.krawczyk@pwr.wroc.pl

Abstract. The article presents a multiple classifiers approach to the obstructive nephropathy recognition - a disease posing a significant threat to newborns. Nature of the data reflects a problem known as high dimensionality small sample size. In presented approach a feature space division amongst number of classifiers is used to balance the relation between the number of objects and the number of features. Methods of feature selection are apllied for optimum splitting the feature space for classifier ensemble. The optimal size of subspaces and selection of classifier for ensemble is thoroughly tested. Complex performance test are presented to highlight the most efficent tuning of parameters for the presented approach, which is then compared to the classical solutions in this field.

Keywords: pattern recognition, classifier ensemble, feature selection, genetic algorithm, bioinformatics, high dimensionality small sample size problem.

1 Introduction

Recently health– and bioinformatics are one of the most vital fields for introducing new ideas and technologies, such as machine learning[3] or hybrid methods [5]. Unfortunately many diseases that are known to traditional medicine had not been the object of computer data analysis. One of them is the obstructive nephropathy the most frequently appearing kidney disorder among newborns and children.

Obstructive nephropathy is a pathology characterized by the presence of an obstacle in the urinary tract, e.g. stenosis or abnormal implantation of the urethra in the kidney. The urine cannot flow properly and accumulates within the kidney leading to progressive alterations of the renal parenchyma, development of renal fibrosis and loss of renal function, what is shown in the figure 1.

An analysis of the urinary miRNA, protein, and metabolite profiles of newborns are required to properly detect the case of this disease. The miRNAs control the level of repression of mRNAs thus controlling the level of expression of twotypes of proteins (Antibody Arrays and LC-MS/MS) to which the mRNA is translated. Then the expression level of these proteins determines the levels of expression of different metabolites, as it is shown in figure 2.

R. Katarzyniak et al. (Eds.): Semantic Methods, SCI 381, pp. 115–125.
springerlink.com   © Springer-Verlag Berlin Heidelberg 2011

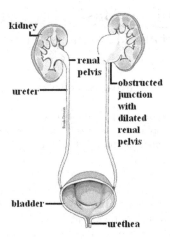

Fig. 1. Comparison between healthy kidney and an obstructed one with dilated pelvis.

The miRNA attributes were associated to the proteins whose expression levels are measured by the Antibody Arrays and LC-MS/MS by first passing to the mRNAs that these miRNAs target, with the help of miRBase and Targetscan databases. Subsequently these mRNAs are translated to their corresponding proteins and associated to the attributes of the antibody arrays and LC-MS/MS datasets. Proteins that appear within the LCMS/MS and Antibody Arrays are associated with some of the measured metabolites with the help of the Kyoto Encyclopedia of Genes and Genomes (KEGG) and The Human Metabolome Database (HMDB).

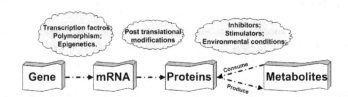

Fig. 2. Dependencies between different biological levels.

Obstructive nephropathy is the first cause of end-stage renal disease in children and it is treated by dialysis or transplantation. Thus, it is important to understand the complex pathological mechanisms involved in the progression of this nephropathy and create computational methods for fast and precise prediction and/or detection of it.

The paper focuses on the problem of the high dimensionality of the feature space. For the problem under consideration it resolves around one thousand. This is even further aggravated by the limited availability of samples, because of example acquisition cost - too expensive to be repeated many times. Also time needed for such studies plays an important role. This problem is usually known as the High Dimensionality Small Sample Size problem. To deal with this inconvenience a multiple classifier approach is proposed. Although this method was initially created as a task-specific there We hope that it can find a more broad usage for dealing with all kinds of high dimensionality, small sample size problem.

The raw data was provided by Inserm U858, Toulouse, France; the preprocessing of the data was performed by the University of Geneva, Switzerland and the University of Manchester, United Kingdom. The problem and data were provided by e-LICO Project[1].

The content of this work is as follows. In the next section related works in this field are discussed. Then the pattern recognition background is presented shortly in section 3. Following section describes the presented approach. In section 5 results of computer experiments are discussed. The last section concludes the paper.

2   Related Works

The problem of high dimensionality, small sample size is a challenge that often occurs in bioinformatics. Some solutions were presented in last years. Ramaswamy et al. presented a Linear classifers with quadratic regularization solution to microarray analysis [21] . Guo et al. showed the advanced usage of RDA with shrinkage of centroids[9] . Both aftermentioned works returned good results, unfortunately authors did not report about stability of proposed methods. Tibshirani and Hastie proposed the margin tree classifier, in which support vector classifiers are used in a binary tree, much as in CART[28], very efficent method but without any mentioning of the stability of the results. Zou used Least absolute shrinkage and selection operator (Lasso) for such problems[32], which is often used as a refference algorithm.

Presented approach is derived from the random subspace method, proposed by Ho[12]. He showed it as a generalization of random forest method. One of the main differences is that random forest uses only unpruned, fully-grown trees, where random subspace can be applied to any type of classifier. Skurichina[25] showed the application of this method for Basic linear classifiers. Tao[27] was one of the first to apply random subspace to support vector machine in his case for image retrieval problem. Bryll[4] presented a different approach to the same idea, calling it attribute bagging. Basic idea behind it is however the same as in Ho's work.

Previous methods based on random selection of features for subspaces, while approach presented in this paper uses methods of feature evaluation. Guerra-Salcedo et al.[10] presented an genetic algorithm-based approach to feature selection for creating ensembles. Li et al.[16] focused on the sensitivity of feature selection of k-NN. Xing et al.[30] proposed an interesting and stable modification of typical algorithms for dealing with high dimensional data. Inza et al.[13]

presented a detailed comparison between filter and wrapper methods applied for genetic data. It is worth noticing that recently a new approach to feature selection for bioinformatics data, based on pathways activation, was introduced. It goes away from statistical approach to this problem, using gene ontology databases instead to find the meaningful combination of cooperating genes. Sun et al.[26] used this ontology for breast cancer prediction. Rapaport et al.[22] focused of incorporating whole gene networks into the classification process. Enthusiasts of this method put a strong emphasis on the stability of selected features, yet up to the Author's knowledge no exhaustive comparison between pathway and cutting-edge statistical algorithms was made so far.

3   Pattern Recognition Task

The aim of the pattern recognition algorithm Ψ is to classify a given object to the one of the predefined categories from set of class labels M on the basis of observation of the features describing it. All data concerning the object and its attributes are presented as a feature vector $x \in X$.

$$\Psi : X \to M. \tag{1}$$

The mapping (1) is established on the basis of examples included in a learning set or rules given by experts. The learning set consists of learning examples, i.e. observation of features described object and its correct classification. This paper focus on the binary classification problem with majority voting case. Let's assume that we have n classifiers $\Psi^{(1)}$, $\Psi^{(2)}$, ..., $\Psi^{(n)}$. For a given object $x \in X$ each of them decides if it belongs either to a positive class by giving response equal to 1 or to a negative class by giving response equal to -1. The combined classifier $\bar{\Psi}$ makes decision on the basis of the following formulae:

$$\bar{\Psi}\left(\Psi^{(1)}(x),\ \Psi^{(2)}(x),\ ...,\ \Psi^{(n)}(x)\right) = sgn\left(\sum_{l=1}^{n} \Psi^{(l)}(x)\right). \tag{2}$$

4   Application of Multiple Classifier System

The main problem addressed in this paper is the disproportion between the number of objects we have at our disposal for the training process and the dimensionality of feature space. In the problem under consideration we have 24 objects, each consisting of about one thousand features. In this area classical approaches deliver poor performance and are prone to overfitting, because of the low generalization for small, but highly complex data sample.

Therefore a new approaches must be proposed. This paper concentrates on the usage of multiple classifier systems[14][24] . A feature space is partitioned into the much smaller, disjoint subspaces. Each of them is created by usage of feature selection algorithm[15] and then it is used to train the classifier. One proposes to use the random subspaces method, which deliver good results. By the usage of feature selection algorithms for this task it is possible to ensure that created subspaces consist of relevant features. Therefore we propose to divide N–dimensional feature space into L subspaces, where $N = N_1 \cup N_2 \cup ... \cup N_L$. And the example of such division, for nine features and three target subspaces, is presented in figure 3. Then most relevant classifiers are selected and their outputs are fused to deliver the final decision. They are ranked according to their individual accuracy $p^{(l)}$.

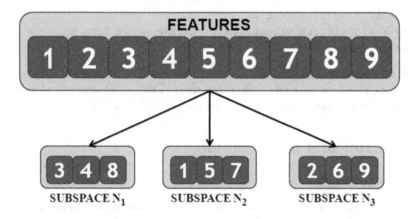

Fig. 3. Idea of splitting feature space between 3 diffrent subspaces/classifiers.

In statistical approach to pattern recognition some of the attributes are irrelevant and bring only biological noise to data set, thus resulting in less precise and stable classifier model. Before any training it is a common and valuable practice to preprocess acquired data in order to eliminate features that do not have any or very small discriminate power. To achieve this a paired difference t-test[19] is conducted, measuring the difference between the means. By the usage of it features that are useless in classification terms - with similar means and variances, are easily discarded.

A pseudocode of proposed procedure is presented below:

INPUT:

```
A = feature set
L = number of subspaces
U = size of the subspace
V = maximum size of the ensemble
W = set of trained classifiers
```

ALGORITHM:

```
Use t-test to discard irrelevant features from A
Partition the feature space:
FOR i=1 TO i=L
use feature selection algorithm to choose U features from A
remove chosen features from A

Train L classifiers, each on diffrent subspace
Rank each classifier from W
Choose V classifiers with highest ranking score from W
Create an ensemble
```

Another issue that should be addressed is the fusion of classifiers outputs. Normally a weighted or majority voting are used. This paper concentrates on the majority voting, tuning it to achieve clinically significant results.

5 Experimental Investigation

5.1 Goal of Experiments

The goal of experiments presented below was to examine the behavior of proposed method for creating multiple classifier system for obstructive nephropathy dataset. For the feature selection we used filter methods (Fast Correlation-Based Filter[31] and ReliefF[23]) and wrapper methods (with tabu search[7] and genetic algorithm[8][20]). A most efficient method from them will be chosen. Additionally the optimal size of the subspaces and size of the ensemble are examined. Also voting method modification was tested. Received results are compared to classical solutions in this field. Tests were concluded separately on each of the biological levels datasets (mRNA, metabolites and two sets of proteins).

5.2 Set-Up

As a classifie **SVM**[29] was chosen, because this classifier behaves very well for small training sets[6]. Kernel function was set to Gaussian Radial Basis, sigma parameter set to 0.1 and cost parameter C was set to 10 (those are the recommended values).

Random subspace method also was used with SVM algorithm, with the same tuning parameters. For genetic feature selection algorithm a tournament selection and two-point crossover operator were used. A mutation was concluded using simple bit string mutation operator. An elitism selection was implemented for creating new population. As a termination condition we set the 100 iterations without improvement of the final result. Additionally after 65 iterations without improvement a macromutation was conducted.

All experiments were carried out in R environment, with SVM and feature selections algorithms taken from dedicated packages[17] [18] , thus ensuring that results achieved the best possible efficiency and that performance was not decreased by a bad implementation. Kernlab contains the ksvm() function, a kernlab implementation of SVM. It includes the C-SVM classification algorithm. Experiments were repeated ten times and average values are presented. Classifiers were evaluated on the basis of the 5 x 2 cv F Test[2].

5.3   Results

Results below are shown for mRNA biological level dataset, as it is the most representative one. Figure 4 presents the efficency of tested feature selection algorithms, tested on diffrent subspace sizes with ensemble size fixed to 30 and with standard majority voting . Figure 5 shows the optimal size of ensemble for most efficent algorithm. Figure 6 proves, that 50%+1 majority voting does not return the best results in this specific case. Table 1 compares the best results from proposed methods with classical soutions in this field, where MCSFS stand for presented method. Table 2 shows the results of the statistical significance tests.

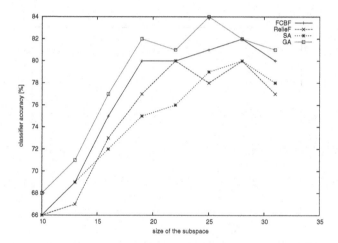

Fig. 4. Comparison of different feature selection methods and optimal subspace size for them.

5.4   Analysis of Results

As previous section shown, the efficiency of feature selection algorithms were quite close to each other. But genetic algorithm has shown best properties, especially when combined with size of each subspace set to 25. Using these setting next experiments showed that optimal number of classifiers in ensemble is 22. Additionally typical majority voting setting did not returned best results. By experimental investigation the optimal number was fixed to 65%+1. It is very important to note, that this applies to pathological class. This means, that conditions for classifying new object as suffering from obstructive nephropathy are

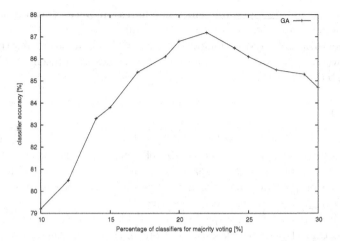

Fig. 5. Comparison of different ensemble sizes for best feature selection algorithm (GA).

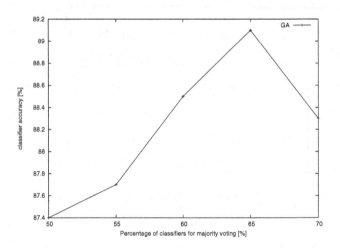

Fig. 6. Comparison of different percentage of classifiers for majority voting.

more strict. This lead to only small increase in classification accuracy, but will have great impact in further research, when to biomedical real-value parameters must additionally be estimated, but this is beyond the scope of this paper. Presented approach was compared to three classical approaches in this field : Random Subspace[12] , LASSO[11] and simple SVM. In most cases it delivered better results than any one of them. In case of the metabolites dataset it returned slightly worse results, but statistical test showed there is no statistically significant difference between them.

Table 1. Comparison of classification prediction performances for different approaches (MCSFS stand for the presented solution).

Classifier type	MCSFS	Random Subspace	Simple SVM	LASSO
mRNA dataset [%]	**89.1±1.1**	87.6±1.9	76.9±2.5	87.8±1.5
Antibody array dataset [%]	**88.3±1.3**	88.1±1.6	78.2±2.7	87.2±1.4
LC-MS/MS dataset [%]	**83.1±1.9**	81.9±2.2	69.5±2.9	80.8±1.7
Metabolites dataset [%]	83.1±1.6	**83.4±1.9**	72.4±2.5	82.7±1.5

Table 2. Probabilities of Rejecting the Null Hypothesis (MCSFS stand for the presented solution and RS stands for Random Subspace).

Type II error	MCSFSvs.RS	MCSFSvs.SVM	MCSFSvs.LASSO
mRNA dataset	0.125	0.005	0.092
Antibody array dataset	0.594	0.008	0.130
LC-MS/MS dataset	0.113	0.007	0.102
Metabolites dataset	0.790	0.008	0.240

6 Conclusions and Future Work

In the article a new approach for predicting obstructive nephropathy, based on algorithm-driven feature space division for multiple classifier system was shown. Based on a random subspace method, the modification using feature selection algorithms for disjoint subspaces was presented. Disjoint subspaces with fixed size and elimination of the randomness from the feature space division allowed to create an improved version of mentioned earlier algorithm. By dividing the feature space a certain degree of stable classification of nephropathy was achieved. The efficiency of this method was proven on the basis of computer experiments and statistical analysis of received results.

It is worth noticing that this approach was initially created for a specific task. However promising results encourage us to test presented approach on more diverse database set to get the more complex view on usefulness of this method.

There is still much work to be done in the future. First thing will be to merge the results from four different biological levels. Next step will consist of the prediction of mentioned two biomedical parameters, which are valuable for clinical treatment of the disease. Interesting research direction would be to incorporate algorithms for evaluating classifiers in competency space. It will be very important to carefully test the idea of subspaces - check the differences if they are not always of the same size and if they are not always disjoint ones. Additionallythe computational complexity of the presented method should be examined.

Acknowledgement. This work is supported in part by The Ministry of Science and Higher Education under the grant for the period 2011-2014.

References

1. Website of the data provider, http://www.e-lico.eu/
2. Alpaydin, E.: Combined 5 x 2 cv F Test for Comparing Supervised Classification Learning Algorithms. Neural Computation 11, 1885–1892 (1998)
3. Alpaydin, E.: Introduction to Machine Learning, 2nd edn. The MIT Press, London (2010)
4. Bryll, R.: Attribute bagging: improving accuracy of classifier ensembles by using random feature subsets. Pattern Recognition 20(6), 1291–1302 (2003)
5. Burduk, R.: Classification error in Bayes multistage recognition task with fuzzy observations. Pattern Analysis and Applications 13(1), 85–91 (2010)
6. Christianini, N., Shawe-Taylor, J.: An introduction to Support vector machines and other kernel-based learning methods. Cambridge University Press, Cambridge (2000)
7. Duda, R.O., Hart, P.E., Stork, D.G.: Pattern Classification. Wiley Interscience, Hoboken (2001)
8. Goldberg, D.E.: Genetic Algorithms in Search, Optimization and Machine Learning. Addison-Wesley Longman Publishing Co., Inc., Boston (1989)
9. Guo, Y., Hastie, T., Tibshirani, R.: Regularized linear discriminant analysis and its application in microarrays. Biostatistics 8, 86–100 (2006)
10. Guerra-Salcedo, C., Whitley, D.: Feature selection mechanisms for ensemble creation: a genetic search perspective. In: Freitas, A.A. (ed.) Proceedings of the AAAI 1999 and GECCO 1999 Workshop on Data Mining with Evolutionary Algorithms: Research Directions, AAAI, Menlo Park (1999)
11. Hastie, T., Tibshirani, R., Friedman, J.: The Elements of Statistical Learning: Data Mining, Inference, and Prediction, 2nd edn. Springer, Heidelberg (2009)
12. Ho, T.K.: The random subspace method for constructing decision forests. IEEE Transactions on Pattern Analysis and Machine Intelligence 20(8), 832–844 (1998)
13. Inza, I., Larranaga, P., Blanco, R., Cerrolaza, A.: Filter versus wrapper gene selection approaches in DNA microarray domains. Artificial Intelligence in Medicine 31(2), 91–103 (2004)
14. Kuncheva, L.I.: Combining Pattern Classifiers: Methods and Algorithms. Wiley, Chichester (2004)
15. Guyon, I., Gunn, S., Nikravesh, M., Zadeh, L. (eds.): Feature extraction, foundations and applications. Springer, Heidelberg (2006)
16. Li, L., Weinberg, C.R., Darden, T.A., Pedersen, L.G.: Gene selection for sample classification based on gene expression data: study of sensitivity to choice of parameters of the GA/KNN method. Bioinformatics 17(12), 1131–1142 (2001)
17. Karatzoglou, A., Smola, A., Hornik, K., Zeileis, A.: Kernlab An S4 Package for Kernel Methods in R. Journal of Statistical Software 11(9) (2004)
18. Karatzoglou, A., Meyer, D., Hornik, K.: Support Vector Machines in R. Journal of Statistical Software 15(9) (2006)
19. Markowski, C.A., Markowski, E.D.: Conditions for the effectiveness of preliminary test of variance. The American Statistican 44, 322–326 (1990)
20. Michalewicz, Z.: Genetic algorithms + data structures = evolution programs, 3rd edn. Springer, London (1996)
21. Ramaswamy, S., Tamayo, P., Rifkin, R., Mukherjee, S., Yeang, C., Angelo, M., Ladd, C., Reich, M., Latulippe, E., Mesirov, J., Poggio, T., Gerald, W., Loda, M., Lander, E., Golub, T.: Multiclass cancer diagnosis using tumor gene expression signature. PNAS 98, 15149–15154 (2001)

22. Rapaport, F., Zinovyev, A., Dutreix, M., Barillot, E., Vert, J.-P.: Classification of microarray data using gene networks, BMC Bioinformatics 8, article 35 (2007)
23. Robnik-Sikonja, M., Kononenko, I.: Theoretical and empirical analysis of relieff and rrelieff. Machine Learning 53, 23–69 (2003)
24. Rokarch, L.: Pattern Classification using ensemble methods. World Scientific Publishing Co. Pte. Ltd., Singapore (2010)
25. Skurichina, M.: Bagging, boosting and the random subspace method for linear classifiers. Pattern Analysis and Applications 5(2), 121–135 (2002)
26. Sun, Y., Goodison, S., Li, J., Liu, L., Farmerie, W.: Improved breast cancer prognosis through the combination of clinical and genetic markers. Bioinformatics 23(1), 30–37 (2007)
27. Tao, D.: Asymmetric bagging and random subspace for support vector machines-based relevance feedback in image retrieval. IEEE Transactions on Pattern Analysis and Machine Intelligence (2006)
28. Tibshirani, R., Hastie, T.: Margin trees for high-dimensional classification. Journal of Machine Learning Research 8, 637–652 (2007)
29. Vapnik, V.: Statistical Learning Theory. Wiley, Chichester (1998)
30. Xing, E.P., Jordan, M.I., Karp, R.M.: Feature Selection for High-Dimensional Genomic Microarray Data. In: Proceedings of the Eighteenth International Conference on Machine Learning, pp. 601–608 (2001)
31. Yu, L., Liu, H.: Feature selection for high-dimensional data: A fast correlation-based filter solution. In: 12th Int. Conf. on Machine Learning (ICML 2003), Washington, D.C, pp. 856–863. Morgan Kaufmann, San Francisco (2003)
32. Zou, H.: The adaptive lasso and its oracle properties. Journal of the American Statistical Association 101, 1418–1429 (2006)

Construction of New Cubature Formula of Degree Eight in the Triangle Using Genetic Algorithm

Grzegorz Kusztelak and Jacek Stańdo

Centre of Mathematics and Physics, Technical University of Lodz,
Al. Politechniki 11, 90-924 Lodz, Poland
{grzegorz.kusztelak,jacek.stando}@p.lodz.pl

Abstract. In the following paper we present new cubature formulas in the triangle for polynomial of degree eight. They have positive weights, containing no points outside the domain. With the use of genetic algorithm we have created a special algorithm effectively searching for cubatures.

Keywords: Cubature formula, genetic algorithm.

1 Introduction

In 1971, Stroud published his paper about numerical integration [1]. In this work, he presented theoretical and practical aspects of numerical integration. In 1981 Mysovski continued futher research on this topic [2] In 1993, which saw the advent of rapid development of compuers, his research started to.develop even faster,. In 1997 Cools started to upload information on the topic onto the website [6], which has become the basic source of information with reference to the results obtained in the numerical integration.

There are different methods applied to search for cubatures. For example: analytical or Monte Carlo [8, 9]. So far, nobody has even tried to apply genetic algrithms to search for new cubatures. In our work we present the results obtained for cubatures of degree eight over triangle.

1.1 Definition and Property of Cubature

Let us assume that the domain $D \subset \Re^2$ is a region of integration given by unions of triangles T.

Consider $I = \int_D f(x, y)\, dxdy = \sum_{T \in D} \int_T f(x, y)\, dxdy$. The integrals $\int_T f(x,y)dxdy$, are approximated by finite sums of the form $\sum_{i=1}^{N} w_{i_T} f(x_{i_T}, y_{i_T})$, where w_{i_T} are weighs and (x_{i_T}, y_{i_T}) are nodes.

$$\int_T f(x, y)\, dxdy \approx Q[f] := \sum_{i=1}^{N} w_{i_T} f(x_{i_T}, y_{i_T}) \tag{1}$$

R. Katarzyniak et al. (Eds.): Semantic Methods, SCI 381, pp. 127–134.
springerlink.com   © Springer-Verlag Berlin Heidelberg 2011

A cubature formulae is said to be of order d if (1) is exact for all polynomials of degree less than or equal to d and it is not exact for at least one polynomial of degree $d+1$.

Let v_i for $i=1,2,3$, be the vertices of T. The barycentric coordinates (ξ_1,ξ_2,ξ_3) of the point P with respect to T is determined by: $P = \sum_{i=1}^{3}\xi_i v_i$ and $\sum_{i=1}^{3}\xi_i = 1$.

Let (ξ_1,ξ_2,ξ_3) be a cubature point with weight w, then cubature formula is said to be symmetric if for any permutation i_1,i_2,i_3 of the indices 1,2,3 , the point $(\xi_{i_1},\xi_{i_2},\xi_{i_3})$ is also a cubature point with the same weight [4].

A cubature formula of degree d is minimal, if its number of nodes is minimal among all cubature of degree d. As given in [3, 5, 7] , for triangle it is known that the lower bounded for number of nodes is described by:

$$v(d) \geq \frac{(d + 2)(d + 4)}{8} \quad , \text{if d is even,} \tag{2}$$

$$v(d) \geq \frac{(d + 1)(d + 3)}{8} + \left[\frac{d + 1}{4}\right], \quad \text{if d is odd} \tag{3}$$

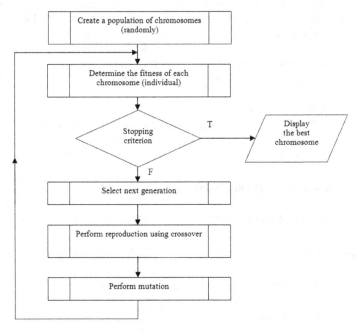

Fig. 1. The graph presenting the scheme of genetic algorithm

1.2 Genetic Algorithms

Evolutionary algorithms, including genetic algorithms, constitute a class of artificial intelligence methods. They are effective and frequently used tools for optimization. They are particularly useful for problems where the exact method, due to the enormous computational complexity, cannot be applied.

Genetic algorithms are a way of solving problems by mimicking the same processes mother nature uses. They use the same combination of selection, recombination and mutation in order to customize the solution to the problem.

2 Motivation

For triangle and degree 8, the minimal number of points is 15. A cubature formula that attains this bound was obtained by Cools and Haegemans [3].

Let $T = \{(x, y) \in \mathfrak{R}^2 : x \geq 0, y \geq 0, x + y \leq 1\}$ be the standard triangle. We give the points and weights in Table 1. The cubature has the form

$$\sum_{i=1}^{5} w_i (f(x_i, y_i) + f(y_i, 1 - x_i - y_i) + f(1 - x_i - y_i, x_i)) \tag{4}$$

Table 1. 15-point formula of degree 8 for triangle

i	w_i	x_i	y_i
1	0.066940767639916174192	0.51334692063945414949	0.28104124731511039057
2	0.043909556791220782402	0.31325121067172530696	0.63062143431895614010
3	0.029285717640165892159	0.65177530364879570754	0.31347788752373300717
4	0.026530624434780379347	0.06510199345893916638	0.87016510156356306078
5	0.16058343856681218798e-9	0.34579201116826902882	3.6231682215692616667

We have 3 points outside the region, which is presented in the Fig.2 below

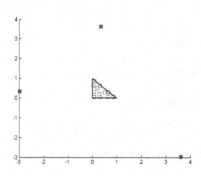

Fig. 2. Cools's 15-point formula of degree 8 for triangle

The location of nodes over the integration area seems to be very unnatural. We have posed 2 questions then:

-   Is it possible to search for a cubature, which will be exact or at least will produce a sufficient precision for polynomials of degree 8 and, at the same time all its nodes will be located in the area of integration?
-   If we accept the error of certain level, what number of nodes would be sufficient enough then?

In order to find answers to these questions we used genetic algorithms.

3   Methodology of Searching for New Cubatures

Denote by P^d the set of polynomials of degree less than or equal to d. The problem of finding an n-point cubature formula of order d consists of finding the cubature points (x_i, y_i) and weight w_i such that

$$\int_T f(x, y)\, dxdy = \sum_{i=1}^{n} w_i f(x_i, y_i) \quad \text{for all } f \in P^d \tag{5}$$

where T is a standard triangle. In order to apply a genetic algorithm to search for new cubatures it was necessary to define the objective function. Defining the objective function limits the issue of finding new cubatures to the concept of minimization of the objective function with the help of genetic algorithm.

3.1   Objective Function

We have defined the objective function in the following way:

$$F(\mathbf{w}, \mathbf{x}, \mathbf{y}) = \sum_{0 \le m+n \le d} |EV(m,n) - AV(\mathbf{w}, \mathbf{x}, \mathbf{y}, m, n)| \tag{6}$$

where

$$EV(m,n) := \int_T x^m y^n\, dxdy = \frac{m! \cdot n!}{(m+n+2)!} \tag{7}$$

the exact value of the integral,

$$AV(\mathbf{w}, \mathbf{x}, \mathbf{y}, m, n) := \sum_{i=1}^{N} w_i \cdot x_i^m \cdot y_i^n \tag{8}$$

the approximate value of the integral,

$$\mathbf{w} = [w_1, w_2, ..., w_N], \ \mathbf{x} = [x_1, x_2, ..., x_N], \ \mathbf{y} = [y_1, y_2, ..., y_N] \tag{9}$$

vector of weights, vector of prime and secondary coordinates of nodes respectively.

Function [4] represents the total error resulting from the approximation of the integral from the polynomial of degree less than or equal to d over the standard triangle with the use of the cubature consisting of weights $\mathbf{w}$ and nodes $(\mathbf{x}, \mathbf{y})$.

Let us realise that, if $F(\mathbf{w}, \mathbf{x}, \mathbf{y}) = 0$, then the cubature will be exact for polynomials of degree d. In order to refer to the results obtained by Cools, we took polynomials of degree $d=8$ into account. According to Cools's results $N=15$.

3.2 The Representation and Parameters of the Genetic Algorithm

A single chromosome has the form

$$Chrom := [\mathbf{w}, \mathbf{x}, \mathbf{y}] = [w_1, w_1, ..., w_N, x_1, x_2, ..., x_N, y_1, y_2, ..., y_N] \tag{10}$$

We assumed commonly used settings of the genetic algorithm:

- real coding;
- standard mutation;
- probability of mutation $p_m = 0.05$;
- uniform crossover;
- probability of crossover $p_c = 0.7$;
- population size $pop\_size = 100$.

We used several selection methods. The best results we obtained using the difference selection operator [10].

3.3 Constraints and the Penalty Function

We want to search for nodes (x_i, y_i) inside the area of integration. Also we need normalized weights. Therefore, the following constraints must be satisfied:

$$0 \leq x_i \leq 1 \ \wedge \ 0 \leq y_i \leq 1 \wedge \ 0 \leq w_i \leq 1 \tag{11}$$

$$x_i + y_i - 1 \leq 0 \tag{12}$$

The condition (12) we can write in the following way

$$h(x_i, y_i) \leq 0 \quad \text{where } h(x_i, y_i) := x_i + y_i - 1 \tag{13}$$

Using the constraint (15) we have defined the penalty function

$$p(\mathbf{x}, \mathbf{y}) = \max\left\{0, \ \sum_{h(x_i, y_i) > 0} h(x_i, y_i)\right\} \tag{14}$$

When we search for new cubatures using genetic algorithms we optimize in fact the following function

$$G(\mathbf{w}, \mathbf{x}, \mathbf{y}) := F(\mathbf{w}, \mathbf{x}, \mathbf{y}) + p(\mathbf{x}, \mathbf{y}) \tag{15}$$

3.4 The Average Error

$$AE(\mathbf{w}, \mathbf{x}, \mathbf{y}) := F(\mathbf{w}, \mathbf{x}, \mathbf{y}) \cdot \frac{1}{\alpha} \qquad (16)$$

where

$$\alpha = \frac{(d+1) \cdot (d+2)}{2} \text{ - the number of polynomial of degree } d \qquad (17)$$

For $d=8$ α equals 45.

4 Results

Applying genetic algorithms to search for new cubatures we have obtained the following results:

Table 2. Our 15-point formula of degree 8 for triangle, average error $AE \approx 1.4 \cdot 10^{-4}$

i	w_i	x_i	y_i
1	0,04531574	0,762424	0,2200894
2	0,001124382	0	0
3	0,06103504	0,3334608	0,3332757
4	0,05841462	0,2657961	0,6963439
5	0,0007762909	0	0

All nodes are located in the area of integration. It is presented in the Fig.3 below

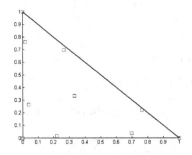

Fig. 3. Our 15-point formula of degree 8 for triangle with the average error $AE \approx 1.4 \cdot 10^{-4}$

We have noticed that some nodes overlap. Therefore, we have limited the number of nodes from 15 to 12. The level of error have remained the same.

Table 3. Our 12-point (in fact 10-point) formula of degree 8 for triangle, average error $AE \approx 1.7 \cdot 10^{-4}$

i	w_i	x_i	y_i
1	0,019092424	0,33333433	0,33333173
2	0,053967767	0,45441757	0,088590848
3	0,089840304	0,72253243	0,13625087
4	0,0037664228	0	0

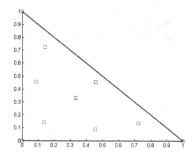

Fig. 4. Our 12-point (in fact 10-point) formula of degree 8 for triangle with the average error $AE \approx 1.7 \cdot 10^{-4}$

All nodes are located in the area of integration. The real number of nodes equals 10. It is presented in the Fig.4.

5 Conclusions

The presented results show, that finding new cubatures with the use of genetic algorithms should be further researched and continued. We already know that many problems in this thematic field are still to be solved [7]. Solutions to these problems can be found with the help of genetic algorithms. Also parameters tuning and operators tuning of GA for those problems seems interesting.

References

1. Stroud, A.H.: Approximate calculation of multiple integrals. Prentice-Hall, Englewood Cliffs (1971)
2. Mysovskikh, I.P.: Interpolatory Cubature Formulas. Izdat. 'Nauka', Moscow-Leningrad (1981)
3. Cools, R., Haegemans, A.: Construction of minimal cubature formulae for the square and the triangle using invariant theory, Report TW 96, Dept. of Computer Science, K.U. Leuven (1987)
4. Kim, K., Song, M.: Symmetric quadrature formulas over a unit disk. Korean J. Comput. Appl. Math. 4 (1997)

5. Wandzura, S., Xiao, H.: Symmetric quadrature rules on a triangle. Computers and Mathematics with Applications 45 (2003)
6. Cools, R.: An encyclopaedia of cubature formulas. J. Complex 19, 3 (2003)
7. Cools, R., Schmid, H.: On the (non)-existence of some cubature formulas: gaps between theory and its applications. Journal of Complexity (2002)
8. Stańdo, J.: The cubature schemes for triangle, tetrahedron and 4-simplex exact for polynomials of fourth degree. Journal of Applied Computer Science (2003)
9. Stańdo, J., Rudnicki, M., Krawczyk-Stańdo, D.: Novel numerical integration scheme for the solution of inverse magnetic problems. In: Proceedings, 10th International IGTE Symposium, Graz, Austria (2002)
10. Kusztelak, G., Stańdo, J., Rudnicki, M.: Difference Selection Operator in GA. In: Rutkowski, L., Tadeusiewicz, R., Zadeh, L.A., Zurada, J.M. (eds.) Computational Intelligence: Methods and Applications, pp. 223–231. Academic Publishing House EXIT, Warsaw (2008)

Affymetrix Chip Definition Files Construction Based on Custom Probe Set Annotation Database

Michał Marczyk[1], Roman Jaksik[1], Andrzej Polański[2], and Joanna Polańska[1]

[1] Institute of Automatic Control, Silesian University of Technology, Akademicka 16,
44-100 Gliwice, Poland
[2] Institute of Informatics, Silesian University of Technology, Akademicka 16,
44-100 Gliwice, Poland
```
{Michal.Marczyk,Roman.Jaksik,Andrzej.Polanski,
              Joanna.Polanska}@polsl.pl
```

Abstract. DNA microarrays are used to characterize transcription profiles and identify changes in the cellular processes induced by various chemical and physical factors. In this paper we propose a new tool for the validation of expression measurements, based on the current genome and transcriptome knowledge. It incorporates Affymetrix Chip Definition Files (CDFs) that encode the physical design of the microarray, linking oligonucleotide probes to the respective transcripts, allowing construction of custom CDF libraries based on any probe-transcript relation table. The presented application was used to create PLANdbAffy probe annotation database CDF files which were evaluated by the analysis of two distinct datasets containing expression profiles of breast cancer patients and a technical replicates dataset. Performed re-annotation allowed to identify specific set of genes with increased probe set validity compared to CDF given from Brainarray repository. Obtained signature significantly differentiates patients showing high or low radio-sensitivity, which wasn't possible with standard Affymetrix libraries. Additionally the observed expression levels were more accurately and precisely associated with genes and their corresponding transcripts.

Keywords: Microarray preprocessing, re-annotation, CDF.

1 Introduction

Microarray technology allows measuring transcript levels of thousands of genes in a single experiment using hundreds of thousands 25-mer oligonucleotide probes grouped into gene specific sets. It has many advantages over other methods of gene expression evaluation but the results provided by this technique require multiple, complex statistical methods needed to separate changes induced by experimental factors from artifacts and effects of inaccurate measurement. These methods, developed and improved by many researchers, concern several levels of expression signals processing, such as probe definitions, normalization, gene grouping based on annotations, and many others [1-4].

R. Katarzyniak et al. (Eds.): Semantic Methods, SCI 381, pp. 135–144.

Affymetrix CEL files created for each sample contain only the information about intensity level of each probe. The definition of probe sets incorporating probes, spread across the entire arrays surface, which represent expression level of a given transcript are stored in a Chip Definition File (CDF). This file is unique for each microarray design and unlike annotation files it is not updated by the manufacturer in order to preserve data similarity making the results much more comparable across various experiments. This leads to various discrepancies between the original probe-transcript association and current genomic databases which are updated very frequently.

The problem of probe annotations was highlighted in literature many times [5-8] showing the need for custom CDF files creation in order to separate badly designed unspecific probes which target transcripts of multiple various genes or no genes at all therefore providing false expression level intensities. Remaining probes which show perfect match similarity to the respective target transcript are assigned into new gene or transcript specific sets. There are many approaches into the problem of probes reannotation with various probe selecting criteria and different definitions of minimal amount of probes needed to create a set.

In this paper we present analyses concerning CDF files constructions based on annotation databases. We focused on the probe quality described in the PLANdbAffy database [9] comparing it with official CDF file provided by Affymetrix, and the most popular custom CDF file repository Brainarray [5]. Brainarray was chosen for the comparison study as the main competitive custom methodology, not only because of its high popularity, but also due to the fact that other introduced approaches [6,7] failed to provide up-to-date versions of CDF files for both microarray platforms described in this article.

Later in the text only names of the databases were used to describe probe annotations created with the use of these databases. Our goals were to prove (i) increased accuracy and precision of observed expression levels compared to Affymetrix annotation and (ii) increased validity of probe sets in the obtained gene signature compared to Brainarray annotation. Proofs are given in the further analyses. Our results can be very useful in analyses of datasets with low- or medium levels of signal to noise ratio. The application of the software, which we have elaborated is not limited to the Affymetrix technology, but can work with any custom probe set definition. This increases its possible further applications.

2 Materials and Methods

In this work we used DNA microarray data from two different studies. First consists technical replicates of two samples repeated in five different laboratories and second includes breast cancer patients. Three different probe set annotations were compared: standard Affymetrix and two updated probe set definitions: Brainarray [5] and our custom CDF based on PLANdbAffy database. Both datasets were preprocessed with the robust multi array analysis (RMA) [10].

2.1 Gene Expression Data

Technical replicate data comes from five different laboratories [11]. Each lab was given two identical aliquots from two different samples containing variable amounts of four individual PEX mutant cell lines (samples noted as A and B and aliquots as A1,A2 and B1,B2). Relative expression between pairs of samples was quantified with log2 fold change. Duplicate RNA samples were provided to each expert lab in order to evaluate the performance of technical replicates in terms of labeling and hybridization similarity level. To obtain a realistic expectation of expression level fold changes real-time polymerase chain reaction (RT-PCR) studies were used for the four altered (PEX) genes, as well as 12 additional target genes.

Patients dataset originates from the GENEPI low-RT project and consist of lymphocyte cells gene expression profiles derived from breast cancer patients undergoing radiotherapy. Based on clinical examination of the late responses to irradiation patients were assigned RS (radiation sensitive) or RR (radiation resistant) status. Examined samples were exposed to low dose radiation (200 mGy). All patients were divided into three sets from which in this work we used pilot study set that includes 20 samples, 10 RRs and 10 RSs.

2.2 Annotation Based on PLANdbAffy Database

PLANdbAffy [9] assigns probes to four distinct groups marked with different colours. Green probes are the best ones in terms of transcript specificity level while black ones represent the worst. Green probes have to meet three conditions. First, the probe is aligned to the target gene without mismatches. Second, there are no matches of the probe to other genes. Third, there are no perfect alignments of the probe to noncoding regions. Yellow probe has a perfect match to uncoding region, unlike the green probe although it still shows a perfect match to the target gene and no matches to other genes. Red probes are the probes that have a perfect match to the target gene and at least one alignment to other genes with no more than one mismatch. Black probe is aligned to the target gene with at least one mismatch.

Only green probes were selected to create a new CDF library using the presented application since they show the highest quality of design. A minimum of five such probes per single target transcript was required in order to create a new set, instead of three used in Brainarray, which makes probe sets statistically more meaningful.

2.3 Software

Developed custom CDF File Creation Tool allows to make various Affymetrix microarray CDFs for many popular microarray designs by selecting probes that will be used in the further study and assigning them to gene- or transcript-specific sets. To prepare the basic CDF version only a list of probe sets with individual probe coordinates is required. Optionally the user can provide a list of probe sequences that will be used to fill the middle probe nucleotide field in the CDF file required by some preprocessing algorithms implemented in Bioconductor. Additional option of the application allows to create library files in PGF format used by Affymetrix Expression Console software. The presented application is equipped with a very easy to use graphical interface and does not require any additional software to operate.

2.4 Discriminative Gene Selection and Classification

Differences of gene expression values between two groups (RR and RS) were evaluated with the standard t-test since the assumption of normality of expression distributions and equal variances of genes expression levels for both samples was confirmed. This method was also used for gene selection. For all tests performed in this paper we set significance level to 0.05.

Support vector machines were applied for classification task. SVM tries to find an optimal separating hyperplane between the classes [12]. When the classes are linearly separable there is a maximal distance between the hyperplane and the nearest point of any of the classes. When the data are not separable we still try to maximize the margin but allow some classification errors subject to the constraint that the total error is less than a constant (soft margin parameter).

3 Results

3.1 Re-definition of Probe Sets

Entire PLANdbAffy microarray probe annotation database [9] was downloaded from the authors website and used to create custom CDF library for two Affymetrix microarray designs (HuGene 1.0 ST and HG-U133A) based only on the highest quality "green" probes. Probe quality criteria proposed in PLANdbAffy are based on additional data available in the UniGene [13] database compared to the EntrezGene only Brainarray approach [9,5]. Additionally PLANdbAffy gene and transcript assignments are based on USCS [14] genome sequence and transcript annotation data while Brainarray focuses only on NCBI repository [13]. This results in only 502788 identical probe assignments between the two approaches for HuGene 1.0 ST array and 153416 for HG-U133A. In the presented method at least five probes, instead of three in Brainarray are required to create a new gene specific set which leads to a higher average probe number as shown in Table 1.

Table 1. Comparison of the most popular CDF libraries with the custom PLANdbAffy based

Human Gene 1.0 ST	Affymetrix	Brainarray	PLANdbAffy
Unique probes	788012	552076	573760
Probe sets	32321	19628	19532
Average probes # per set	26.1	28.1	29.4
% of excluded probes	-	30	27.3
Probe assignment as in Affy	-	542459	522612
Human Genome U133A	Affymetrix	Brainarray	PLANdbAffy
Unique probes	247965	174284	174878
Probe sets	22215	12065	11755
Average probes # per set	11.2	14.4	14.9
% of excluded probes	-	29.7	29.5
Probe assignment as in Affy	-	171508	166056

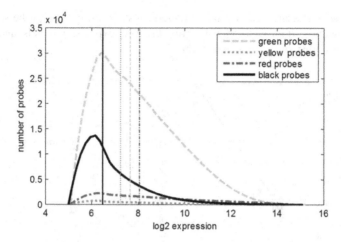

Fig. 1. Average raw probe observed signal intensity distribution across 20 samples from the pilot Genepi-lowRT data separately for each PLANdbAffy probe quality group. Horizontal lines represent median values for each of the distributions.

Figure 1 shows the average raw probe signal intensity distribution based on all 20 samples from the pilot Genepi-lowRT data. Distributions are plotted separately for each PLANdbAffy quality group incorporating various amounts of probes. Horizontal lines represent median values for each of the distributions. Green probes, which are the most frequent ones, show a relatively high median intensity level proving their high number of specific bindings with a potential to measure transcript levels of genes expressed on variable scale level. Yellow probes intensity median is lower than the median of green probes since they can target non coding gene regions. Red probes show the highest median which most likely is a result of overlapping signal intensities of multiple target genes caused by low quality level of probes in this category. Finally, black probes are the worst ones due to fact that this group incorporates large amount of probes which do not target any known transcript therefore providing noise-level signal intensities.

3.2   Technical Replicates Data

Following the methodology proposed in [11] we analyzed the effect of different probe set definitions on expression levels using data created in five different laboratories with use of two RNA samples replicated twice within each lab. Strictly speaking, we estimated precision and accuracy of expression levels by analyzing log2 expression fold changes between the samples.

Precision, which measures reproducibility and variability of the data, was defined as a Pearson correlation coefficient between log2 ratios of replicates of samples (A1/B1 and A2/B2) and calculated as a mean result value from each lab and different probe set definition files. Also standard deviation of the log2 fold change difference between replicates for each lab was calculated and its mean value describing variability is presented in table 2. Increase of precision with comparison to Affymetrix

annotation was measured by two-tailed paired t-test. Both updated probe sets definitions significantly improved precision of observed expression levels and reduced
technical variability of the replicates (table 2).

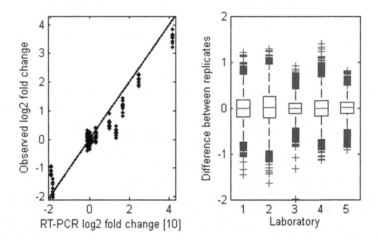

Fig. 2. Precision and accuracy of log2 fold change of expression level for PLANdbAffy based
probe set definition. Observed log2 fold changes from five laboratories versus calculated from
RT-PCR experiment for 16 genes with solid line as identity function representing perfect accu-
racy(left) and box plot of the difference in log2 fold change between replicates for each laboratory (right).

Accuracy, which measures how close observed expression level are to the real expression values estimated using RT-PCR method, was defined as the slope given from
linear regression between previously mentioned values for 16 genes. Increase of accuracy with comparison to Affymetrix annotation was measured by two-tailed paired t-
test and this time also Brainarray and PLANdbAffy based annotation significantly
improved the accuracy (table 2). Comparing updated definitions between them, no
statistical differences can be observed.

Table 2. Summary of accuracy (closer to 1 is better) and precision (higher is better)
measurements on technical replicates data for three different probe set definitions. P-values are
result of statistical comparison of obtained measures between updated probe set definitions and
Affymetrix annotation.

	Affymetrix	Brainarray	PLANdbAffy
Precision	0,54	0,61	0,60
	-	p=9,73E-05	p=1,74E-04
Variability	0,271	0,254	0,256
	-	p=6,10E-03	p=3,00E-03
Accuracy	0,759	0,780	0,779
	-	p=1,94E-02	p=1,29E-02

3.3 Pilot Genepi-LowRT Data

Despite the large amount of genes encoded in the microarray gene expression data, only a very small fraction of them is informative for a certain task. In analysis of cancer patients radiosensitivity, differences between analyzed groups are hardly detectible. To avoid gene filtering, results of discriminative gene selection are presented without multiple testing correction. Figure 1 presents histogram of p-values from a two sample t-test for all expression levels for three probe set definitions. Proportion of genes with $p<0.05$ comparing to Affymetrix annotation is significantly increased when updated annotations are used according to the z-test for two proportions (table 3).

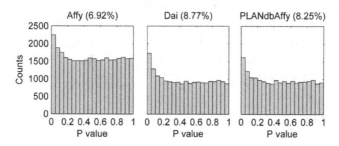

Fig. 3. The histograms of p-values from radiosensitivity data for three different probe set definition. In brackets percent of probe sets with $p<0.05$ is presented.

In order to measure the quantity of radiosensitivity status in expression profiles, principal variance components analysis (PVCA) was performed [15]. It integrates two methods commonly used in data analysis: principal components analysis (PCA) and variance components analysis. First is used to efficiently reduce data dimension with maintaining the majority of the variability in the data, next fits a mixed linear model using factors of interest as random effects to estimate and partition the total variability. Model was built on the first three principal components since they represent more than a half of data variability and as a factor radiosensitivity status was set. Both updated probe set definitions improved averaged proportion of expression levels variance explained by status of radiosensitivity (table 3).

Pilot Genepi-lowRT data contains only 20 samples, which for low informative signal create large difficulties in the classification task, imposing split-sample accuracy estimate (splitting a data set into training set used to construct classifier and test set used for estimate predictive accuracy). Leave-one-out cross-validation (LOOCV) in which the number of splits (iterations) equals the number of cases was used. Each test set includes single case while the rest of the samples is used for classifier training. Such procedure introduces high correlation among models created in each split, however when dealing with a small number of samples (lower than 40) accuracy bias is smaller than in a single split-sample. In all experiments predictive performance of the classifier was measured by the average error rate, defined as the number of all misclassification rates divided by number of all validation trials. Validation error was decreased when updated probe set annotations were used (table 3). Using probe set definitions based on PLANdbAffy database we were able to obtain the smallest

LOOCV error for only 9 probe sets with increased validity compared to 26 probe sets for the Brainarray approach. Both signatures have 7 common probe sets with comparable number of probes. In Brainarray signature 12 probe sets have at least one probe marked as not green according to PLANdbAffy database.

Table 3. Summary of discriminative gene discovery, PVCA and LOOCV classification on pilot Genepi-lowRT data.

	Affymetrix	Brainarray	PLANdbAffy
All probe sets	32321	19628	19532
Discr. probe sets	2238	1722	1611
	6,92%	8,77%	8,25%
	-	p=1,38E-14	p=2,52E-08
PVCA	11,70%	15,20%	15,10%
LOOCV error	45%	30%	30%
Signature size	30	26	9

4 Conclusions

Our research suggests that standard, out-of-date Affymetrix probe set annotation does not reveal the growing opportunities in microarray studies. The application presented in this study was proven to create custom CDF libraries of high quality evaluated by the analysis of two independent datasets based on microarrays of distinct designs separated by over five years of microarray technology development.

The use of new probe set definitions increased relative amount of discriminative genes, which resulted in more accurate differentiation between the evaluated treatment groups compared to the standard Affymetrix approach. Increased validity of defined probe sets allows the construction of classifiers with reduced dimensionality and better separation of analysed data. The usage of technical replicates which improve reproducibility as well as accuracy of the observed expression levels for updated probe sets was verified.

Data preprocessing based on redefined probe sets described in the PLANdbAffy database and the library derived from Brainarray repository results in similar precision and accuracy of expression values. Differences were noticed in classification task of low excitation signal where error bias should be minimized. Increased probe sets validity allowed less complex signature to achieve a similar discriminatory power.

The need of redefined CDF libraries usage for microarray data analysis becomes more common every year which is being reflected by the citation index of articles describing various approaches into this problem. The presented application allows to create custom CDF files without advanced bioinformatic knowledge based on available probe definition databases [9,16] or other CDF libraries allowing to select probes or probe sets of interest for further study. This work shows that by incorporating the current transcriptomic and genomic information databases with expression profiling studies it is possible to significantly increase the outcome of the analysis even for microarrays with over 10 years old designs (HG-U133A). Redefinition of microarray probes is a very difficult task driven by the need of keeping expression profiling studies up to date with current genomic and transcriptomic repositories which become more accurate with every new release.

Acknowledgments. Microarray experiments from radiosensitivity data were performed by institution partners in GENEPI low RT project, University Tuebingen, Germany and LUMC (Leiden University Medical Centers, Netherlands). Examination of late responses to irradiation was done by clinicians from ICR (Institute of Cancer Research), the Royal Marsden Hospital, UK.

Funding. This work was supported by the European Program FP6 – 036452, GENEPI-lowRT and Ministry of Science and Higher education BK/274/2011-young scientists grant.

References

1. Bolstad, B.M., Irizarry, R.A., Astrand, M., Speed, T.P.: A comparison of normalization methods for high density oligonucleotide array data based on variance and bias. Bioinformatics 19(2), 185–193 (2003)
2. Falcon, S., Gentleman, R.: Using GOstats to test gene lists for GO term association. Bioinformatics 23(2), 257–258 (2007)
3. Gautier, L., Cope, L., Bolstad, B.M., Irizarry, R.A.: affy–analysis of Affymetrix GeneChip data at the probe level. Bioinformatics 20(3), 307–315 (2004)
4. Smyth, G.K.: Linear models and empirical bayes methods for assessing differential expression in microarray experiments. Stat. Appl. Genet. Mol. Biol. 3, Article3 (2004)
5. Dai, M., Wang, P., Boyd, A.D., Kostov, G., Athey, B., Jones, E.G., Bunney, W.E., Myers, R.M., Speed, T.P., Akil, H., Watson, S.J., Meng, F.: Evolving gene/transcript definitions significantly alter the interpretation of GeneChip data. Nucleic Acids Res. 33 (20), e175 (2005)
6. Lu, J., Lee, J.C., Salit, M.L., Cam, M.C.: Transcript-based redefinition of grouped oligonucleotide probe sets using AceView: high-resolution annotation for microarrays. BMC Bioinformatics 8, 108 (2007)
7. Ferrari, F., Bortoluzzi, S., Coppe, A., Sirota, A., Safran, M., Shmoish, M., Ferrari, S., Lancet, D., Danieli, G.A., Bicciato, S.: Novel definition files for human GeneChips based on GeneAnnot. BMC Bioinformatics 8, 446 (2007)
8. Sandberg, R., Larsson, O.: Improved precision and accuracy for microarrays using updated probe set definitions. BMC Bioinformatics 8, 48 (2007)
9. Nurtdinov, R.N., Vasiliev, M.O., Ershova, A.S., Lossev, I.S., Karyagina, A.S.: PLANdbAffy: probe-level annotation database for Affymetrix expression microarrays. Nucleic Acids Res. 38(Database issue), D726–D730 (2010)
10. Irizarry, R.A., Hobbs, B., Collin, F., Beazer-Barclay, Y.D., Antonellis, K.J., Scherf, U., Speed, T.P.: Exploration, normalization, and summaries of high density oligonucleotide array probe level data. Biostatistics 4(2), 249–264 (2003)
11. Irizarry, R.A., Warren, D., Spencer, F., Kim, I.F., Biswal, S., Frank, B.C., Gabrielson, E., Garcia, J.G., Geoghegan, J., Germino, G., Griffin, C., Hilmer, S.C., Hoffman, E., Jedlicka, A.E., Kawasaki, E., Martinez-Murillo, F., Morsberger, L., Lee, H., Petersen, D., Quackenbush, J., Scott, A., Wilson, M., Yang, Y., Ye, S.Q., Yu, W.: Multiple-laboratory comparison of microarray platforms. Nat. Methods 2(5), 345–350 (2005)

12. Vapnik, V.: Nature of Statistical Learning Theory (1995)
13. Sayers, E.W., Barrett, T., Benson, D.A., Bolton, E., Bryant, S.H., Canese, K., Chetvernin, V., Church, D.M., DiCuccio, M., Federhen, S., Feolo, M., Fingerman, I.M., Geer, L.Y., Helmberg, W., Kapustin, Y., Landsman, D., Lipman, D.J., Lu, Z., Madden, T.L., Madej, T., Maglott, D.R., Marchler-Bauer, A., Miller, V., Mizrachi, I., Ostell, J., Panchenko, A., Phan, L., Pruitt, K.D., Schuler, G.D., Sequeira, E., Sherry, S.T., Shumway, M., Sirotkin, K., Slotta, D., Souvorov, A., Starchenko, G., Tatusova, T.A., Wagner, L., Wang, Y., Wilbur, W.J., Yaschenko, E., Ye, J.: Database resources of the National Center for Biotechnology Information. Nucleic Acids Res. 39(Database issue), D38–D51 (2011)
14. Fujita, P.A., Rhead, B., Zweig, A.S., Hinrichs, A.S., Karolchik, D., Cline, M.S., Goldman, M., Barber, G.P., Clawson, H., Coelho, A., Diekhans, M., Dreszer, T.R., Giardine, B.M., Harte, R.A., Hillman-Jackson, J., Hsu, F., Kirkup, V., Kuhn, R.M., Learned, K., Li, C.H., Meyer, L.R., Pohl, A., Raney, B.J., Rosenbloom, K.R., Smith, K.E., Haussler, D., Kent, W.J.: The UCSC Genome Browser database: update 2011. Nucleic Acids Res. 39(Database issue), D876–D882 (2011)
15. Scherer, A.: Batch Effect and Experimental Noise in Microarray Studies: Sources and Solutions. John Wiley & Sons, Chichester (2009)
16. Chalifa-Caspi, V., Yanai, I., Ophir, R., Rosen, N., Shmoish, M., Benjamin-Rodrig, H., Shklar, M., Stein, T.I., Shmueli, O., Safran, M., Lancet, D.: GeneAnnot: comprehensive two-way linking between oligonucleotide array probesets and GeneCards genes. Bioinformatics 20(9), 1457–1458 (2004)

Part III
Models for Collectives
of Intelligent Agents

Advanced Methods for Computational Collective Intelligence

Ngoc Thanh Nguyen, Radosław P. Katarzyniak, and Janusz Sobecki

Division of Knowledge Management Systems, Institute of Informatics,
Wroclaw University of Technology
Wyb.Wyspianskiego 27, 50-370 Wrocław, Poland
{Ngoc-Thanh.Nguyen,Radoslaw.Katarzyniak,
Janusz.Sobecki}@pwr.wroc.pl

Abstract. In this paper the authors present some recent results regarding advanced methods for Computational Collective Intelligence, which have been worked out in the Division of Knowledge Management Systems, Institute of Informatics, Wroclaw University of Technology. Three aspects have been focused: Knowledge Integration, Agent-based Systems and Recommendation Systems.

Keywords: Computational Collective Intelligence, Knowledge Integration, Agent-based Systems and Recommendation Systems.

1 Introduction

What is the intelligence of a collective? According to Newell [24] an intelligent collective is a social system, which is capable to act, even approximately, as a single rational agent. Lévy [22] and Russell [39], on the other hand, define collective intelligence as an intelligence that emerges from the collaboration and competition of many individuals, an intelligence that seemingly has a mind of its own.

Computational Collective Intelligence, in general, deals with soft methods for processing intelligences of members of a collective to make an intelligences of whole collective, or to realize tasks addressed to the collective members.

One of the most important aspects of this process is managing knowledge of collective members. It turned out that the knowledge of a collective as a whole is not the normal "sum" of members' knowledge. For example, if a collective consists of two members (say A and B), and if A knows that $x > y$ and B knows that $y > z$ for some real variables x, y and z, then they together know not only $x > y$ and $y > z$, but also $x > z$. If A knows that $x \geq y$ and B knows that $y \geq x$, then they together also know that $x = y$. Thus it seems that the knowledge of a collective is greater than the "sum" of its members' knowledge. However, this is not always true. Let's consider a situation in which A knows that $x > y$ and B knows that $y > x$, then they together know "nothing" because of the inconsistency of their knowledge.

The above examples are motivating for the integration processes for knowledge of collectives. It is needed to determine the relationship between the integrated

R. Katarzyniak et al. (Eds.): Semantic Methods, SCI 381, pp. 147–158.
springerlink.com   © Springer-Verlag Berlin Heidelberg 2011

knowledge (of a collective as a whole) and the knowledge states of its members. More generally, we investigate methods for processing knowledge from different autonomous sources (such as agents in multiagent environments). These methods can be useful in many fields of Artificial Intelligence, for example in recommendation systems where the content of recommendation for a user may come from an analysis of different knowledge bases of other systems or other users.

This subject of Computational Collective Intelligence has been dealt with in the Division of Knowledge Management Systems for several years. For this period we have published more than 100 works (books, chapters, journal and conference papers). In order to organize an international forum for exchanging scientific research and technological achievements accomplished by the international community in this field we founded the conference series "International Conference on Computational Collective Intelligence", for which the first event (ICCCI 2009) was held in Poland, and the second (ICCCI 2010) is being organized in Taiwan. From the beginning all these conferences got proceedings published by Springer in series LNCS/LNAI. We also founded and edit a new international journal "Transactions on Computational Collective Intelligence" which is published by Springer.

The remaining part of this paper is organized as follows: In the next section several aspects of knowledge integration are presented. Section 3 includes the description of some recent results on processing knowledge of collectives of agents. In the last section several approaches for collective knowledge processing in recommender systems are presented.

2   Knowledge Integration and Collective Intelligence

2.1 Service Discovery Problem

In our research we have investigated the problem of determining the knowledge state of a collective where the states of knowledge of its members are given. For this aim we assume that the knowledge of collective members refers to the same real world, but not necessarily in a complete and certain manner. We assume that there exists the *proper state* of knowledge of the considered real world, and the collective members' knowledge states reflect this state correctly, but only to some degrees because of their incompleteness and uncertainty.

We solved the following two problems [28, 30, 32]:

o   Given the states of knowledge of the collective members, how to determine the knowledge of the collective as a whole?
o   How to evaluate the quality of the knowledge of the collective?

For this aim we have built a mathematical model, in which the states of collective members' knowledge are placed in a distance space. The distance function is a half-metric, that it is reflexive and symmetrical. For such originally defined model we used Consensus Theory (see Sec. 2.2) to determine the consensus of the members' states. We proved the following theorems [28, 29]:

o The distance between the proper state and the consensus is smaller than the maximal distance between the proper state and the states of collective members. This means that using consensus guarantees the credibility of the procedure.

o If all distances values from the proper state to collective members' knowledge states are identical then the distance value between the consensus and the proper state is the smallest. This means that if all collective members reflect the real world correctly to the same degree, then the consensus reflects the knowledge about the real world at the best.

o The greater the inconsistency degree of the collective members' knowledge states, the smaller the distance between the consensus and the proper state. This means that inconsistency of the collective members' knowledge states does not worsen, but should improve the credibility of the collective.

These theorems prove that in general the knowledge of a collective as a whole is more credible than the knowledge of its members as individuals. Thus with some restrictions one can claim that the hypothesis "*A collective is more intelligent than one single member*" is true. This statement should be very useful for distributed knowledge processing. From this statement it follows that in many cases using different and inconsistent sources of knowledge should improve the equality of knowledge management.

2.2 Consensus Theory

A methodology for consensus choice and its applications to solving conflicts and knowledge integration in distributed environments have been worked out. Two levels of consensus choice have been investigated. On the first level general consensus methods which may effectively serve to solving multi-valued conflicts have been worked out. For this aim a consensus system, which enables describing multi-valued and multi-attribute conflicts has been defined and analyzed. Next the structures of tuples representing the contents of conflicts have been defined as distance functions between these tuples. Finally, the consensus and postulates for its choice have been defined and analyzed. For such defined structures algorithms for consensus determination have been worked out. Besides, the problems related to the susceptibility to consensus and the possibility of consensus modification, have been also investigated [6, 28, 46].

On the second level various applications of consensus methods in solving of different kinds of conflicts and knowledge integration have been analyzed. The following conflict solutions are presented: reconciling inconsistent temporal data; solving conflicts of the states of agents' knowledge about the same real world; determining the representation of expert information; creating a uniform version of a faulty situation in a distributed system; resolving the consistency of replicated data and determining optimal interface for user interaction in universal access systems [7-9, 18-20].

Consensus methods have been proved to be very useful for processing collective knowledge. As shown in Sec. 2.1, they have been used to determine the collective knowledge state which can be treated as the representative of collective members' knowledge states. Consensus methods enable making group decisions and processing knowledge among autonomous units acting in distributed environments. Web-based systems, social networks and multi-agent systems very often need these tools for working out consistent knowledge states, resolving conflicts and making decisions [23-27].

2.3 Processing Inconsistency of Knowledge

Processing inconsistency of knowledge is very important in knowledge integration processes.

Generally there are two aspects of knowledge inconsistency: *Centralization aspect* and *Distribution aspect*. The first of them is based on inconsistency of different elements of the same knowledge base, and the second is related to inconsistency of different knowledge bases referring to a common real world. The focused point of our works was the second aspect. Processing inconsistency is a very important aspect of Computational Collective Intelligence [36].

We have investigated this problem on two levels of inconsistency regarding the distribution aspect: Syntactic level and Semantic level [37, 47, 48].

Syntactic level refers mainly to logic-based bases of knowledge. Here one has to deal with a set of logic formulae, which are contradictory. We proposed the adjective *syntactic* for this level because the condition for contradiction, which means that the set of formulae is contradictory in each interpretation of the formulae. On syntactic level the approaches for inconsistency resolution can be grouped into three groups. The first group includes approaches based on knowledge base revision. The second group consists of paraconsistent logic-based methods. Paraconsistent logics are useful in processing inconsistent information. The third group includes methods which enable "monitoring" the inconsistency by measuring its degree.

On the semantic level logic formulae are considered as referring to a concrete real world, which is their interpretation. Inconsistency arises if in this world a fact and its negation may be inferred. For setting inconsistency on this level three parameters need to be determined: inconsistency subject, set of knowledge elements being in inconsistency and the definition of contradiction.

We have worked out a set of consensus-based algorithms for processing inconsistency of knowledge on these two levels.

2.4 Ontology Integration

The methods for knowledge inconsistency processing described in the previous sections have been used for ontology integration. In general, one of the basic problems of ontology integration can be formulated as follows: *For given several ontologies one should determine one ontology which best represents them.* The integration process for ontologies has been investigated on the following three levels [28, 33, 37]:

o *Integration on concept level:* The structure of concepts consists of a set of attributes and the domains of their values. The same concept can have different structures in different ontologies. Thus integration should take the inconsistency into account.

o *Integration on instance level*: An instance is understood as a physical occurrence of a concept. An instance has then attributes with their concrete values. The same instance belonging to different ontologies can have different attributes or event different values of the same attributes. Integration procedure on this level should refer to these aspects and should use appropriate methods.

o *Integration on relation level*: Between the same two concepts there may be defined different relations in different ontologies. Integration on this level requires consideration of different kinds of relations, for example, equivalence, generalization etc.

Consensus methods have been proved to be useful in ontology integration tasks. The consensus-based methods for ontology integration have been used, among others, for WordNet-based ontologies [1-5]. It turned out that these methods are very effective for this kind of ontology. Fuzzy aspect of ontology integration has also been investigated [38].

3 Processing Knowledge in Agent Collectives

3.1 Semantic Communication for Agent Collectives

Processing knowledge in agent collectives requires dedicated languages of communication. Such languages need to fulfill various commonsense requirements. In particular, they have to be interpreted and this interpretation has to be commonly accepted by all agents of collective intending to use this language for knowledge transfer. Our research into knowledge communication and transfer languages conducted for the latest years has concentrated on the question how a semantic communication language has to be related to the world by particular agents of collective to represent same knowledge for all members of collective. This fundamental problem is known as *the language grounding problem* [11]. The starting point of our research was to develop a precise model for grounding of a relatively simple communication language, choose adequate methodologies for implementations of this model in individual agents and then define further extensions to this original approach. This work is still going on.

3.2 Modal Languages in Knowledge Communication and Transfer

We have started our research into the basic of processing knowledge by agent collectives from a relatively simple modal language consisting the following classes of statements: "I (an agent) *know* that X is(is not) P", "I (an agent) *believe* that X is(is not) P" and "For all what I (agent) know *it is possible* that X is(is not) P" [12]. In human collectives statements of such language are used in situations of knowledge incompleteness. To model the process of autonomous extraction of the above classes of statements by individual agents we have assumed in our research that agents develop their own views on external worlds and their subjective empirical experience influences the content of uttered language statements, as well as their interpretation of incoming language massages. This influence is particularly important in subjective states of knowledge incompleteness.

The above mentioned work into modeling language generation for representation of incomplete knowledge states was further extended and simple descriptions were enriched with conjunctions of properties [13, 14] and disjunctions of properties [10].

At this moment intensive research is continued into the way in which autonomous agents should extract from their own knowledge bases various types of conditionals e.g. "I believe that if X is P then X is Q.", "If X is P then it is possible that X is Q", etc. Although a lot of time and effort has already been devoted by many authors to modeling of the semantics of these knowledge statements, many related research questions are still open [40, 41]. Some of them need to be reformulated again to be further considered in the context of artificial agent collectives.

It has already been proved that our original approach to modeling modal language grounding, processing and usage has opened new interesting and promising lines of research into practically important and theoretically interesting subfield of artificial collective intelligence and cognition. At this point the major theoretical result can be summarized as follows: for modal languages which we have already considered in our work, it has been proved possible to design artificial agents fulfilling various commonsense requirements known from natural language processing in human collectives. This important feature has been shown in an analytic way in the above mentioned works e.g. [12-14].

3.3 Hybridization of Agent Languages

The next interesting line of our research into the fundamentals of knowledge processing in agent collectives has concentrated on grounding a hybrid fuzzy and modal language. Our effort is to develop artificial agent collectives capable of properly using a hybrid language for communication and transfer of both incomplete and fuzzy knowledge. The need to effectively solve related theoretical and design problems follows from the fact that various integration of fuzzy and modal elements in human collective communication is a natural and common phenomenon. Unfortunately, the nature of fuzziness and the nature of modality are so different, that they require a special organization of internal knowledge bases of agents. Introductory results from research into this issue have already been presented in [15]. This line of research is intensively conducted by PhD students affiliated the Division of Knowledge Management Systems.

3.4 Methodologies for Implementing of Knowledge Processing in Agent Collectives

Another research line related to knowledge processing in agent collectives are strategies and methodologies for implementing our original models for modal and fuzzy language processing. The major target of our efforts is purely practical. Namely, in order to implement strong theoretical achievements that show the possibility to create rational communicative agents fulfilling commonsense requirements known from human collectives, we mainly use multiple soft approaches suited to processing inconsistent collections of data. A brief overview of at least some strategies for implementing our original models of knowledge representation languages has been presented in [16, 17]. These strategies are based both on the theory of consensus, intensively elaborated by members of our group, as well as on other commonly known models for data analysis.

4 Using Collective Intelligence Methods in Recommendation

The main aim of recommendation systems is to deliver customized information to a very differentiated users. They may be applied in a great variety of domains, such as [43]: net-news filtering, web recommender, personalized newspaper, sharing news, movie recommender, document recommender, information recommender, e-commerce, purchase, travel and store recommender, e-mail filtering, music recommender, student courses recommender [45], user interface recommendation [42] and negotiation systems [21].

We can consider two dimensions of the recommender systems, user modeling and user model exploitation. The former considers user profile representation & maintenance and profile learning technique. The later contains information filtering method, matching techniques and profile adaptation technique.

In this section we will present methods that we developed for dealing with these two dimensions of recommender systems. In the first subsection we will show the rough classification method used for user modeling [31, 35] and the second subsection will present two filtering methods applied to user interface recommendation [34, 42].

4.1 Rough Classification Method Applied In User Modeling

User modeling together with user model exploitation are two issues of the recommender systems. User modeling deals mainly with user profile: content, representation, initialization, maintenance and learning techniques.

In general user profile contains user data and usage data [34]. User data consists for example user demographic data and also data on his or her interests or preferences. Usage data, from the other hand, contains all the data considering different user activities using the system. The user profile may be represented in many different ways, but feature vector is one of the most often used types of representation. The feature vector is defined as $v = (f_1, f_2,... f_n)$ where f_i is the value (real number) of the feature i for $i \in \{1,2,..,n\}$. A modification of the feature vector is weighted feature vector, which uses some vector of weights $w = (w_1, w_2,... w_n)$ that alters the importance of the specified features in the whole vector in the following way: $v = (w_1*f_1, w_2*f_2,... ,w_n*f_n)$. In many of our works we used a relation structure, or in other words – a tuple, as a representation of the user profile [34, 43, 44]. The tuple t is a function $t: A \rightarrow V$, where $(\forall a \in A)(t(a) \in V_a)$ and the set A be the finite set of the profile attributes and the set V that contains attribute values, where: $V = U_{a \in A} V_a$ (V_a is the domain of attribute a).

One of the most important issue of the user modeling is selection of the most important attributes or features that describe the user in the task of the item recommendation. We must know that dealing with too much redundant data is very inefficient and may also be considered as violating users privacy. From the other hand our set of attributes may be insufficient to deliver information for the item recommendation and we should have tools for optimal set determination. In the work [35] we proposed to apply rough classification to solve this task. The main idea of the rough classification is following: the recommendation, i.e. item ratings define the partition U of the set of users, and we would like to find the subset B of the set of attributes A that define the partition P that distance to the partition U is the smallest. We proposed and proofed several theorems that enables us to construct the efficient algorithm for optimal set of attributes determination. We also verified this algorithm with the *Adult* dataset and the experiments have shown that even for relatively big data (more than 30 thousands of records) it is possible to find reduced set of attributes in reasonable time and the solution is the exactly the same as the exact solution, however the time for the obtaining the exact solution with the brute-force method is in most cases about 40 times longer.

4.2 Filtering Methods in Recommendation Systems

We can distinguish the following filtering methods which was discussed in our works: demographic filtering (DF) [34], collaborative filtering (CF) [45], item-based collaborative filtering (IBCF), content-based filtering (CBF) [43], hybrid approach (HA) [34, 42, 43].

Here we would like to outline two methods of user interface recommendation. First method [34] is a HA that switches application of DF and CF. In this method in the early stages of the system work, when we do not have sufficient number of registered users we apply simple DF to recommend user interfaces according to the predefined stereotypes that are connected with specified demographic data, i.e. age, gender and education. When the number of users of the system grows we gather more information about user interface preferences we can apply CF to determine the user interface for the new user. We used consensus-based method for user interface attributes determination. The method first using demographic data determines the several groups of users by means of k-means method, and then for each group determines consensus among all the specified user interface attributes.

The second user interface recommendation method we would like to present is using Ant Colony Metaphor [42]. The method is using graph-based structure to record user interface personalization preferences. In the graph user interface attribute values are represented as a graph nodes and the edges represent the number of users that selected this value. The path of this graph represents the specified user interface implementation. Ant Colony Metaphor basically uses two basic notion: trail and visibility in the selection rule. In our method trail is determined by means of CF (i.e. other user preferences) and the visibility is determined by CBF (i.e. the specified user interaction preferences) The standard Ant Colony Metaphor selection rule is applied in the process of traversing the graph for selection of each following edge from the start node to the finish mode. We also applied this method to movie recommendation using *MovieLens* data set [42] and in student courses recommendation [45] and in both cases we received very good prediction accuracy from 85 to 92%.

5 Conclusions

In this position paper we have presented some recent results regarding advanced methods for Computational Collective Intelligence, which have been worked out in the Division of Knowledge Management Systems, Institute of Informatics, Wroclaw University of Technology. Three aspects have been focused: Knowledge Integration, Agent-based Systems and Recommendation Systems. As it has been already stated Computational Collective Intelligence deals with soft methods for processing intelligences of members of a collective to create an intelligence of whole collective, or to realize tasks addressed to the collective members. In our work we concentrate both on developing strong theoretical foundations for dealing with Computational Collective Intelligence as well as on example realizations of this concept in actual computational systems. Theoretical foundations developed in our group locate at different levels of abstractions. At first, we deal with basic elements of computational methodologies. An example is the theory of consensus which is intensively studied and originally developed in our group. This theory has been mentioned in Section 2. It defines an effective approach to possible implementations of various examples of Collective Intelligence Phenomenon. At second, we work on higher level models for various aspects of Computational Collective Intelligence. For instance, Section s 2 and 3 name at least two references to such contribution. The first one is an original model for knowledge integration and the second - a model for modal language grounding. At third, in our research projects we work on various applications among which the following should be underlined: ontologies management tasks by agent collectives (see Section 2), extraction of fuzzy and modal linguistic summarizations from data bases (see Section 3), and generally understood recommendation being an important aspect of many agent interactions within collectives (see Section 4). To conclude we would like to stress that in many of our works, multiple lines of new interesting and inspiring research can be still defined due to the natural complexity and richness of the concept of Collective Intelligence.

Acknowledgements. This chapter is an extended version of a communication presented at the 6th International Conference on Information Technology for Education and Research 2010, Ho Chi Minh City - Phan Thiet, August 18-20, 2010 Viet Nam and was partially supported by Polish Ministry of Science and Higher Education under grant no. N N519 407437 (2009-2012).

References

[1] Duong, T.H., Nguyen, N.T., Jo, G.S.: A Method for Integration of WordNet-based Ontologies Using Distance Measures. In: Lovrek, I., Howlett, R.J., Jain, L.C. (eds.) KES 2008, Part I. LNCS (LNAI), vol. 5177, pp. 210–219. Springer, Heidelberg (2008)

[2] Duong, T.H., Nguyen, N.T., Jo, G.S.: A Method for Integrating Multiple Ontologies. Cybernetics and Systems 40(2), 123–145 (2009)

[3] Duong, T.H., Jo, G.S., Jung, J.J., Nguyen, N.T.: Complexity Analysis of Ontology Integration Methodologies: A Comparative Study. Journal of Universal Computer Science 15(4), 877–897 (2009)

[4] Duong, T.H., Nguyen, N.T., Jo, G.S.: Effective Backbone Techniques for Ontology Integration. In: Nguyen, N.T., Szczerbicki, E. (eds.) Intelligent Systems for Knowledge Management. Studies in Computational Intelligence, vol. 252, pp. 197–227. Springer, Heidelberg (2009)

[5] Duong, T.H., Nguyen, N.T., Jo, G.S.: Constructing and Mining: A Semantic-Based Academic Social Network. Journal of Intelligent & Fuzzy Systems 21(3), 197–207 (2010)

[6] Hernes, M., Nguyen, N.T.: Deriving Consensus for Hierarchical Incomplete Ordered Partitions and Coverings. Journal of Universal Computer Science 13(2), 317–328 (2007)

[7] Hwang, D., Nguyen, N.T., et al.: A Semantic Wiki Framework for Reconciling Conflict Collaborations Based on Selecting Consensus Choice. Journal of Universal Computer Science 16(7), 1024–1035 (2010)

[8] Jain, L.C., Lim, C.P., Nguyen, N.T.: Innovations in Knowledge Processing and Decision Making in Agent-Based Systems. In: Jain, L.C., Nguyen, N.T. (eds.) Knowledge Processing and Decision Making in Agent-Based Systems, pp. 1–12. Springer, Heidelberg (2009)

[9] Jung, J.J., Nguyen, N.T.: Consensus Choice for Reconciling Social Collaborations on Semantic Wikis. In: Nguyen, N.T., Kowalczyk, R., Chen, S.-M. (eds.) ICCCI 2009. LNCS, vol. 5796, pp. 472–480. Springer, Heidelberg (2009)

[10] Katarzyniak, R., Prusiewicz, A.: Grounding and Extracting Modal Responses in Cognitive Agents: 'and' Query and States of Incomplete Knowledge. International Journal of Applied Mathematics and Computer Science 14(2), 249–263 (2004)

[11] Katarzyniak, R.: The Language Grounding Problem and its Relation to the Internal Structure of Cognitive Agents. Journal of Universal Computer Science 11(2), 357–374 (2005)

[12] Katarzyniak, R.: On Some Properties of Grounding Simple Modalities. Systems Science 31(3), 59–86 (2005)

[13] Katarzyniak, R.: On Some Properties of Grounding Non-uniform Sets of Modal Conjunctions. International Journal of Applied Mathematics and Computer Science 16(3), 399–412 (2006)

[14] Katarzyniak, R.: On Some Properties of Grounding Uniform Sets of Modal Conjunctions. Journal of Intelligent 17(3), 209–218 (2006)

[15] Katarzyniak, R.: Grounding Crisp and Fuzzy Ontological Concepts in Artificial Cognitive Agents. In: Gabrys, B., Howlett, R.J., Jain, L.C. (eds.) KES 2006. LNCS (LNAI), vol. 4253, pp. 1027–1034. Springer, Heidelberg (2006)

[16] Katarzyniak, R., Nguyen, N.T., Jain, L.C.: Soft Computing Approach to Contextual Determination of Grounding Sets for Simple Modalities. In: Apolloni, B., Howlett, R.J., Jain, L. (eds.) KES 2007, Part I. LNCS (LNAI), vol. 4692, pp. 230–237. Springer, Heidelberg (2007)

[17] Katarzyniak, R.P., Nguyen, N.T., Jain, L.C.: A Model for Fuzzy Grounding of Modal Conjunctions in Artificial Cognitive Agents. In: Nguyen, N.T., Jo, G.-S., Howlett, R.J., Jain, L.C. (eds.) KES-AMSTA 2008. LNCS (LNAI), vol. 4953, pp. 341–350. Springer, Heidelberg (2008)

[18] Kiewra, M., Nguyen, N.T.: A Fuzzy-based Method for Improving Recall Values in Recommender Systems. Journal of Intelligent & Fuzzy Systems 20(1-2), 89–104 (2009)

[19] Kozierkiewicz-Hetmańska, A., Nguyen, N.T.: A Method for Scenario Modification in Intelligent E-Learning Systems Using Graph-Based Structure of Knowledge. In: Nguyen, N.T., Katarzyniak, R., Chen, S.-M., et al. (eds.) Advances in Intelligent Information and Database Systems. Studies in Computational Intelligence, vol. 283, pp. 169–179. Springer, Heidelberg (2010)

[20] Krol, D., Nguyen, N.T. (eds.): Intelligence Integration in Distributed Knowledge Management. IGI Global (2008)

[21] Lenar, M., Sobecki, J.: Using Recommendation to Improve Negotiations in Agent-based Systems. Journal of Universal Computer Science 13(2), 267–285 (2007)

[22] Lévy, P.: Collective Intelligence: Mankind's emerging world in cyberspace. Basic Books (1997)

[23] Maleszka, M., Mianowska, B., Nguyen, N.T.: Agent Technology for Information Retrieval in Internet. In: Håkansson, A., Nguyen, N.T., Hartung, R.L., Howlett, R.J., Jain, L.C. (eds.) KES-AMSTA 2009. LNCS, vol. 5559, pp. 151–162. Springer, Heidelberg (2009)

[24] Newell, A.: Unified theories of cognition. Harward University Press, Cambridge (1990)

[25] Nguyen, N.T., Szczerbicki, E. (eds.): Intelligent Systems for Knowledge Management. Springer, Heidelberg (2010)

[26] Nguyen, N.T., Katarzyniak, R., Janiak, A. (eds.): New Challenges in Computational Collective Intelligence. Springer, Heidelberg (2009)

[27] Nguyen, N.T., Jain, L.C. (eds.): Intelligent Agents in the Evolution of Web and Applications. Springer, Heidelberg (2009)

[28] Nguyen, N.T.: Advanced Methods for Inconsistent Knowledge Management. Springer, London (2008)

[29] Nguyen, N.T.: Using Consensus Methodology in Processing Inconsistency of Knowledge. In: Last, M., et al. (eds.) Advances in Web Intelligence and Data Mining. Studies in Computational Intelligence, pp. 161–170. Springer, Heidelberg (2006)

[30] Nguyen, N.T.: Processing Inconsistency of Knowledge in Determining Knowledge of a Collective. Cybernetics and Systems 40(8), 670–688 (2009)

[31] Nguyen, N.T.: Rough Classification Method – New Approach and Applications. Journal of Universal Computer Science 15(13), 2622–2628 (2009)

[32] Nguyen, N.T.: Inconsistency of Knowledge and Collective Intelligence. Cybernetics and Systems 39(6), 542–562 (2008)

[33] Nguyen, N.T.: A Method for Ontology Conflict Resolution and Integration on Relation Level. Cybernetics and Systems 38(8), 781–797 (2007)

[34] Nguyen, N.T., Sobecki, J.: "Using Consensus Methods to Construct Adaptive Interfaces in Multimodal Web-based Systems. J. of UAI 2(4), 342–358 (2003)

[35] Nguyen, N.T., Sobecki, J.: Determination of User Interfaces in Adaptive Systems Using a Rough Classification based Method. New Generation Computing 24(4), 377–402 (2006)

[36] Nguyen, N.T.: Methods for Achieving Susceptibility to Consensus for Conflict Profiles. Journal of Intelligent & Fuzzy Systems 17(3), 219–229 (2006)

[37] Nguyen, N.T.: Keynote Speech: Computational Collective Intelligence and Knowledge Inconsistency in Multi-agent Environments. In: Bui, T.D., Ho, T.V., Ha, Q.T. (eds.) PRIMA 2008. LNCS (LNAI), vol. 5357, pp. 2–3. Springer, Heidelberg (2008)

[38] Nguyen, N.T., Truong, H.B.: A Consensus-Based Method for Fuzzy Ontology Integration. In: Pan, J.-S., Chen, S.-M., Nguyen, N.T. (eds.) ICCCI 2010. LNCS, vol. 6422, pp. 480–489. Springer, Heidelberg (2010)

[39] Russell, P.: The Global Brain Awakens: Our Next Evolutionary Leap. Global Brain, Inc. (1995)

[40] Skorupa, G., Katarzyniak, R.: Applying Possibility and Belief Operators to Conditional Statements. In: Setchi, R., Jordanov, I., Howlett, R.J., Jain, L.C. (eds.) KES 2010. LNCS, vol. 6276, pp. 271–280. Springer, Heidelberg (2010)

[41] Skorupa, G., Katarzyniak, R.: Conditional Statements Grounded in Past, Present and Future. In: Pan, J.-S., Chen, S.-M., Nguyen, N.T. (eds.) ICCCI 2010. LNCS, vol. 6423, pp. 112–121. Springer, Heidelberg (2010)

[42] Sobecki, J.: Ant Colony Metaphor Applied in User Interface Recommendation. New Generation Computing 26(3), 277–293 (2008)

[43] Sobecki, J.: Implementations of Web-based Recommender Systems Using Hybrid Methods. International Journal of Computer Science & Applications 3(3), 52–64 (2006)

[44] Sobecki, J.: Consensus-based Hybrid Adaptation of Web Systems User Interfaces. Journal of Universal Computer Science 11(2), 250–270 (2005)

[45] Sobecki, J., Tomczak, J.M.: Student Courses Recommendation Using Ant Colony Optimization. In: Nguyen, N.T., Le, M.T., Świątek, J. (eds.) Intelligent Information and Database Systems. LNCS, vol. 5991, pp. 124–133. Springer, Heidelberg (2010)

[46] Śliwko, L., Nguyen, N.T.: Using Multi-agent Systems and Consensus Methods for Information Retrieval in Internet. International Journal of Intelligent Information and Database Systems 1(2), 181–198 (2007)

[47] Tran, T.H., Nguyen, N.T.: A Consensus-based Integration Method for Security Rules. In: Velásquez, J.D., Ríos, S.A., Howlett, R.J., Jain, L.C. (eds.) KES 2009. LNCS, vol. 5711, pp. 54–61. Springer, Heidelberg (2009)

[48] Tran, T.H., Nguyen, N.T.: Security Policy Integration Method for Information Systems. In: Nguyen, N.T., et al. (eds.) Proceedings of ACIIDS 2009, pp. 223–225. IEEE Computer Society Press, Los Alamitos (2009)

Identity Criterion for Living Objects Based on the Entanglement Measure

Mariusz Nowostawski[1] and Andrzej Gecow[2]

[1] Information Science Department
University of Otago
New Zealand
mariusz@nowostawski.org1
[2] Centre for Ecological Research
Polish Academy of Science
Lomianki, Poland
gecow@op.pl

Abstract. We extend the original Gecow's computational theory of life and introduce a formal specification language that provides structural and operational semantics for further experimentation on complex evolving systems. The formalisation is based on Milner's process calculus called π-calculus in conjunction with the notion of *reactive systems* and formal process algebra.

In addition, we provide formal object boundary and identity criterion that can be used to isolate an object in a complex computational dynamical system from its environment. To achieve that, we use a formal notion of entanglement, borrowed from graph theory, to identify objects within a complex and changeable computational environment characterised by multiple scales of abstraction. This allows the model to be re-implemented and used for investigations on scale-free, self-adapting, self-evolving computational systems without an explicit notions such as object, organism, fitness, or purpose.

1 Introduction

There were many attempts to formally define, model, and investigate the processes of life. Most of the models are based on the colloquial notions of biological life and simply list the properties which fit into the common understanding of the process of "life" [1]. Some are strictly biology-centric and apply only to biological life as we know it from Earth [18,14] and therefore have very limited utility. Other models are oriented towards the physical sciences and they build on the notion of living objects as a special case of dissipative structures, where material and energetic aspects are emphasized [25,23]. There are also formal and systematic attempts to define life, organism, organism identity, organism boundaries and information content based on information-centric, discrete and cybernetic models [19,4,6,7,13]. Those models have high utility and they are the most adequate and applicable in the computational realm. They also offer framework for experimentation and insights into the process of life.

R. Katarzyniak et al. (Eds.): Semantic Methods, SCI 381, pp. 159–169.
springerlink.com   © Springer-Verlag Berlin Heidelberg 2011

This work extends previous work on modelling life [6,7,8,22] and self-evolving structures [20,21]. The contribution of this work is in formalisation of the previously developed computational system through process algebra and in providing an effective way to recalculate the boundaries of objects[1] through graph entanglement analysis.

In our previous research we have used two models. One to investigate structural tendencies of evolving objects [9] based on multi-input, multi-output and multi-value Kauffman networks and spontanous self-adaptation. The second model was based on computational ensembles that use purposefully build stack-based evolvable virtual machine (EVM) [20]. Based on these two models, we have investigated and observed the process of spontaneous self-adaptation and self-improvement. Our models are more generic as compared to other contemporary theories of life. They can be used to generate an instance of an evolutionary process that does not call upon the notion of "purpose" or "utility" and does not require an explicitly defined fitness function. Our system differs from other biologically inspired evolutionary computational models in the fact that the fitness function is implicit and derived from a generic condition: "to exists". However our original approach has a number of limitations. One limitation was the lack of a well-defined effective method of calculating the boundary of object (organism) in the context of its environment. That was one of the main reasons to use the extended Kauffman networks so that we could track a process of a particular computational ensemble evolution. To address this limitation, we introduce a computational method of calculating object-environment, and object-object boundaries based on a novel measure of entanglement.

In the following sections we unify and formalise our computational model, and later provide a formal object boundary criterion. The formalisation is based on Milner's calculus of communicating systems (CCS) [15] in conjunction with the notion of *reactive systems* and process algebra. We use a formal notion of entanglement [3], borrowed from graph theory, to identify objects within an environment, on multiple scales of abstraction. The entanglement measure (`ent`) allows our system to be unified and re-implemented. The new model will allow us more detailed investigations on self-adapting, self-evolving computational systems without an explicit notions such as object, organism, fitness, or purpose.

2   Information and Communication

We use two generic terms when discussing information: *Information*, a measure based on a selection from a set of available choices; *Algorithmic information*, a compression rate or information encoding relative to the given computing machine[4]. The concept of information is used when discussing communication complexity, whereas algorithmic information is used when discussing complexity of a computational processes.

---

[1] We use the terms *object*, *ensemble* and *organism* interchangeably, with preference to object and ensemble as they have less biological bias.

The amount of information is based on the ability to make a correct selection from a given set. Let us consider a unique code $k \subset X \times Y$, i.e. $y = k(x)$, where $x \in X, y \in Y$, x represents a given condition, and y represents the correct selection. Let us use index g to indicate a particular selection of x for a given y (goal). We can express now $x_g = k^{-1}(y_g)$ as a process that "decodes" code k and guessing the cause (condition). Selection of a single unique condition x_g gives us all necessary classical Shannon information to obtain the output y_g: $I(x_g) = -log\,p(x_g)$, where $p(x_g)$ is the probability of picking a correct condition x, and I is the information content of a particular x_g (Shannon Theory of information).

Unique selection and subset selection. Any process of constraining the choices from X to $A \subset X$ is treated as information. Thus a process of elimination of any causes which do not lead to the effect or lead to the effect with insufficient probability is to be treated as information. The information content in such a case is equal to: $I_X(g; A) = -log\,p_X(A) = -log \sum_A p_X(x)$. This information is called *set information* in contrast to *individual information* which is based on selecting a unique individual cause x. If the set A can be derived from some set of conditions, the process of selecting these causes will be equivalent to $I_x(g; A)$.

Communication is a mechanism for passing signals from an information source to an information sink. Most contemporary life models assume of what constitutes an object and what constitutes the environment. This assumption makes modelling easier. There are also important models based on autonomous

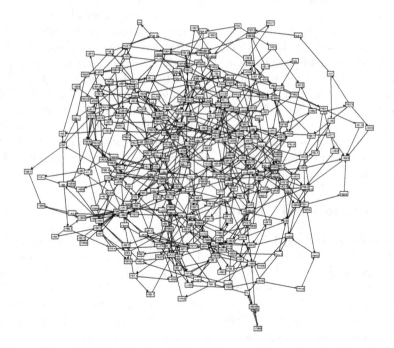

Fig. 1. Example of highly entangled computational graph with 100 nodes.

networks, such as Random Boolean Networks [11] that only investigate object properties, and thus avoid to tackle the problem of object boundery and identity. However, this also leads to certain bias, and restricts the model to a single scale (or abstraction level). In addition, the communication between objects is assumed to be direct and performed always on a single fixed hierarchy level. This is a simplistic representation and rather crude approximation, that does not capture properties observed in the natural phenomenon of life. Within natural life not only is communication performed on different levels simultaneously, but also it is never direct. Signals have to pass from the source through multiple encoding-decoding phases so be eventually received and interpreted by the receiver.

Communication is not something external to life, it is not an external utility, which living forms simply use or exploit. It is an integral part of the process of life. It operates on different scales and carries different information for different levels of structural organisation. As Wittgenstein once pointed out [27], even if we knew the language of lions, we would not understand them. Wittgenstein's point is that because humans and lions do not share the same inter-dependencies and scales, they cannot share language or understanding. The language (communication channel) is being shaped by all the outward criteria and dependencies, and as such, it is constrained to a particular scale or level or class of entities only.

Scale-free property. As indicated in Section 1, there are many formal approaches to define life. Some of the proposals are very narrow and precise (with strong biological grounds) [26]; some are broader and more abstract [5]. As diverse as they may appear, the models have some common characteristics. One such shared feature is the lack of scale. Most the models can naturally be applied to systems and objects regardless of their respective scales and topology of their interactions with the environment. This can be indicative of an inherent property of the phenomenon of life itself: scale-free property. Our own model of life is also inherently scale-free, and it is designed so it can span multiple levels of organisation. It is important to support a hierarchy of organisational levels. This property is quite universal, and we believe it is the crucial element of any modern theory of evolution, or life in general. Our formalism presented below supports this scale-free approach.

3   Computational Model and Its Structure

In classical computational models, such as those based on λ-calculus or Turing machines, the computation is modelled as a partial function that maps input signals into outputs. We depart from such conceptions because we are primarily interested in non-terminating computations. Our model is based on a notion of computing systems that do not terminate, that interact with their surroundings and exchanging information. These systems react to stimuli from their computing surrounding, change their state and send signals (information) to their outputs and by doing so, in turn, influence their surroundings. To model such computational systems we use the notion of *reactive systems* [10].

Reactive systems are by their nature parallel systems that exhibit an algebraic structure. Due to this reason, a number of formal models in process algebra has evolved [2]. In our work, we base our formalism on π-calculus [17] which is extended version of Calculus for Communicating Systems (CCS) [15,16]. CCS on the other hand, is mappable and semantically equivalent to a labelled transition system (LTS) [16].

In our model both, active and passive entities are modelled as computational processes that exchange information through communication channels composed of *input* and *output ports*. One can think of the process as a single "black box" that transforms incoming input signals into it's internal state and some output activations. Processes are arranged into a complex dependency network due to the nature of their appropriate input-output port interconnections. We refer to our model as Evolvable Virtual Machine (EVM), following the formalism presented in [22]. The entire system forms a large, complex graph of interdependencies (see Figure 1).

To recall, TLS is a triple $(P, A, \xrightarrow{a} |a \in A)$, where:

- P is a set of states, ranged over by s;
- A is a set of actions, ranged over by a;
- $\xrightarrow{a} \subseteq P \times P$ is a *transition relation*, for every $a \in A$. For readability we use an equivalent notation $s \xrightarrow{a} s'$ in lieu of $(s, s') \in \xrightarrow{a}$.

For the EVM specification we assume countably infinite collection $\mathcal{A}$ of channel names, and the set $\overline{\mathcal{A}} = \{\overline{a}|a \in \mathcal{A}\}$ as the set of complementary names.

$$\mathcal{L} = \mathcal{A} \cup \overline{\mathcal{A}} \tag{1}$$

is the set of all labels, and

$$A = \mathcal{L} \cup \tau \tag{2}$$

is the set of all actions. To ensure that we never run out of names for processes, we assume a countably infinite collection $\mathcal{K}$ of *process names*. We also use a set of indexes I.

The EVM system provides the following process constructs:

1. the nil process; 0, is a process whose execution is complete and has stopped. Example: $P \overset{\text{def}}{=} checkmate.0$ might represent a final stage of a chess game.
2. action prefixing; if a is an action ($a \in A$), and P and P_1 are process names ($P, P_1 \in \mathcal{K}$), then $P \overset{\text{def}}{=} a.P_1$ is a process such that action a is followed by process P_1. Using LTS notation $P \xrightarrow{a} P_1$. Note, given a communication channel name a, we use a for depicting the input port and $\overline{a}$ for depicting output port. We refer to label a as an *action*. Example: $Clock \overset{\text{def}}{=} tick.Clock$ represents a recursive definition of a *Clock* process, that *ticks* indefinitely, i.e. action *tick* is followed by an action *tick* ab infinitum.
3. channel name restriction; if P is a process and A is a set of port names, then operation $P' \overset{\text{def}}{=} P \setminus A$ is a process, with the scope of the port names in A restricted to P. This works like normal variable name scoping rule in

any programming language. $P \setminus a$ restricts the port named a to process P. Example: $Mariusz \stackrel{\text{def}}{=} birth.P \setminus birth$ restricts the action $birth$ exclusively to process $Mariusz$.

4. relabelling; if P is a process and f is a function from labels to labels, $f : A \times A$ (A from Equation 2) satisfying Equation 6, then $P[f]$ is a process. Example: consider a generic vending machine VM that accepts a coin and dispenses a generic item $VM \stackrel{\text{def}}{=} coin.\overline{item}.VM$. We can instantiate a concrete vending machine that dispenses coffee $CoffeVM \stackrel{\text{def}}{=} VM[coffee/item]$.

5. concurrency; if P and Q are processes, then $P|Q$ is a process in which P and Q are executed concurrently. Axioms for parallel composition:
   - $P|Q \equiv Q|P$
   - $(P|Q)|R \equiv P|(Q|R)$
   - $P|0 \equiv P$

6. non-deterministic choice operator; if P and Q are processes, then $P + Q$ means that either P or Q executes, but not both. Consider a simple example of a vending machine with two possible output choices, tea and coffee: $VM \stackrel{\text{def}}{=} coin.(\overline{tea}.VM + \overline{coffee}.VM)$

7. simple communication; *input prefixing* $c(x).P$ is a process waiting for a message that was sent on a communication channel named c before proceeding as P, binding the name received to the name x. *output prefixing* $\overline{c}\langle y \rangle.P$ describes that the name y is emitted on channel c before proceeding as P.

8. polyadic communication; processes can communicate by an arbitrary number of names in a single action through polyadic input $x(z_1, ...z_n).P$ and polyadic output $\overline{x} < z_1, ...z_n > .P$.

9. match operator; $[x = y]P$ can proceed as P if and only if x and y are the same name.

10. replicated input process; $!x(y).P \equiv x(y).P|!x(y).P$. Replicated input process such as $!x(y).P$ awaits on channel x to be invoked by clients. Invocation spawns a new copy of the process $P[a/y]$, where a is the name passed by during the latter's invocation.

11. full process replication; $!P$, a process which can always create a new copy of P. The axiom:
    - $!P \equiv P|!P$

12. creation of a new name; $(\nu x)P$, which may be seen as a process allocating a new constant x within P. The constants of π-calculus are defined by their names only and are always reference to communication channels.

Note, that the actual implementation in addition to the above allows also: higher-order processes, i.e. process can be passed through communication channel, and asynchronous communication. It has been shown [24] that these can be formally reduced to the above constructs hence we limit our formal discussion only to the subset of EVM specification language.

4 Structural Operational Semantics

To formally capture the semantics of the specification language we use the following rules, which follow from CCS SOS rules [17]. The rules have the format:

premise (above the bar), conditions (on the right-hand-side of the bar) and conclusion (above the bar).

$$\frac{P_j \xrightarrow{\alpha} P_j'}{\Sigma_{i \in I} P_i \xrightarrow{\alpha} P_j'} j \in I \qquad \frac{}{\alpha.P \xrightarrow{\alpha} P} \qquad \frac{P \xrightarrow{\alpha} P'}{Q \xrightarrow{\alpha} P'} Q \overset{\text{def}}{=} P \qquad (3)$$

The first rule specifies the non-deterministic choice operation: given the fact that a given process P_j can afford a transition α for $j \in I$, we conclude that for the non-deterministic choice over all P_i with $i \in I$ the transition α is affordable. The second rule (ACTION rule) provides the basis for actions. The third rule specifies the equivalence relationship.

$$\frac{P \xrightarrow{\alpha} P'}{P|Q \xrightarrow{\alpha} P'|Q} \qquad \frac{Q \xrightarrow{\alpha} Q'}{P|Q \xrightarrow{\alpha} P|Q'} \qquad \frac{P \xrightarrow{\alpha} P' \quad Q \xrightarrow{\bar{\alpha}} Q'}{P|Q \xrightarrow{\tau} P'|Q'} \qquad (4)$$

The equations above provide the basis for parallel composition rules. The first two describe the fact that individual transitions in a parallel composition afford a transition α if each of them affords it separately. The final third rule specifies the special synchronised coordination, which is depicted by the τ transition (one process produces and the other consumes a given action).

$$\frac{P \xrightarrow{\alpha} P'}{P[f] \xrightarrow{f(\alpha)} P'[f]} \qquad \frac{P \xrightarrow{\alpha} P'}{P \setminus L \xrightarrow{\alpha} P' \setminus L} \alpha, \bar{\alpha} \notin L \qquad (5)$$

The first rule provides the basis for relabelling operations. Given a process that affords transition α we conclude that the relabelled process will afford relabelled transition. The second rule provides the basis for restriction operation.

Note, f is a relabelling function satisfying the following constraints:

$$f(\tau) = \tau \ and \ f(\bar{a}) = \overline{f(a)} \ \text{for each label } a \in \mathcal{L} \qquad (6)$$

5   Entanglement as the Object Identity Measure

In our system, the information sources, information sinks, and computational ensembles work in a large complex and changing interconnected graph of dependencies (see e.g. Figure 1). The system evolves (in a physical sense) from one global state into the next state through the number of local concurrent transitions τ (see Equation 4). Up to now, all our experiments were based on isolated ensembles that we could track over time and analyse for their structural and dynamical characteristics (for more details please consult [6,7]). This was due the effective inability to isolate an individual computational ensemble (object) from its environment during runtime. Now we have solved this limitation thanks to entanglement measure.

In 2005, Dietmar Berwanger and Erich Grädel proposed a novel measure of complexity of finite directed graphs [3]. Let $G = (V, E)$ be a finite directed graph.

The entanglement of G, denoted $ent(G)$, measures to what extent the cycles of G are entangled (or intertwined). $ent(G)$ is defined by way of a game, played by a thief against k detectives on G according to the following rules. Initially the thief selects an arbitrary position v_0 of G and the detectives are outside the graph. In any move the detectives may either stay where they are, or place one of them on the current position v of the thief. The thief, in turn, has to move to a successor $w \in vE$ that is not occupied by a detective. If no such position exists, the thief is caught and the detectives have won. Note that the thief sees the move of the detectives before she decides on her own move, and that she has to leave her current position no matter whether the detectives stay where they are or not. The entanglement of G is the minimal number $k \in \mathcal{N}$ such that k detectives have a strategy to catch the thief on G.

With the entanglement measure, it is now possible, at each step of the simulation, to isolate individual objects from their environment. This allows us to establish the *object boundary*. In addition, we can keep track of how the system evolves into a stable self-sustaining ensembles. This provides us with the effective *object identity* measure. The graph in step $t+1$ that shares the most of its structure with the state of the system at time t. Consider a simple example on Figure 2. In the first case two computational ensembles exchanging signals with surrounding environment have entanglement value 1. In the second case, the joined entanglement is 2. In our previous models we had to pre-design the ensembles and artificially keep track of how their structure changes under evolutionary pressures of a complex dynamical process. However, with the entanglement measure, we can recalculate and identify ensembles (sub-graphs) with varying degree of entanglement, and not only observe the evolution of a single ensemble, but also monitor all spontaneously created hypercycles and we will be able to observe

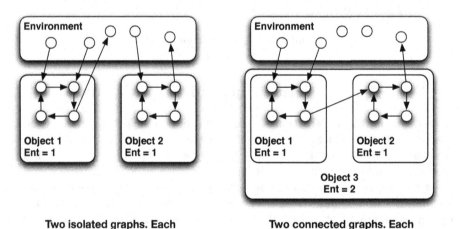

Two isolated graphs. Each object with entanglement = 1

Two connected graphs. Each object with entanglement = 1, joined entanglement = 2

Fig. 2. Example of two ensembles with their entanglement measure.

the process of symbiogenesis in which lower-order ensembles (with small entanglement measure) form larger, and more entangled, self-sustaining and adopting clusters.

6 Implications, Summary and Future Work

In our modelling of life we have decided to introduce information as a primary category, which therefore needs no explanation or proof. The information incorporates both material and ideal. Information never exists without the material carrier on one hand, but it is not limited to the carrier's physical properties. In such a model fields, particles, biological organisms and other concepts become the derivatives of the information concept. Any interaction is in fact an exchange of information. We treat laws of physics as an encoding process which translates causes into effects, then the reversed, decoding process, i.e. fitting causes into an assumed effect. In such a model one can use a formalism based on the notion of reactive systems and calculus for communicating systems, to describe the processes and interactions to a desirable level of accuracy. Process of life in our model is limited exclusively to one generic criterion: to exist. Any process that is capable of sustaining itself will continue to exist. This we define as life. Objects (or organisms) that participate in the process of life contain specific information encoded in their structure that allows them to perpetuate their existence through interactions with their environment.

In this article we have presented formal description of our life simulation framework and provided a necessary component that allows us to compute and re-compute object hierarchy in each of the simulation steps dynamically during runtime. Our system does not use predefined conception of a living entity (organism) and the entanglement measure is now used to establish object and environment boundaries dynamically. Despite the fact that sound formal foundations have been laid out, and effective measures and methodology have been established, the computational model requires efficient algorithmic implementation due to the nature of the problem: large complex graphs and large state spaces. We plan to employ latest techniques in system modelling and verification [12] and work on efficient implementation capable for large scale experiments of complex system modelling and analysis. With an efficient algorithm we will continue to investigate the details of the evolving ensembles and how they self-organise spontaneously.

References

1. Adami, C.: Introduction to Artificial Life. Springer, Heidelberg (1997)
2. Baeten, J.: A brief history of process algebra. Theoretical Computer Science 335(2-3), 131–146 (2005)
3. Berwanger, D., Grädel, E.: Entanglement–a measure for the complexity of directed graphs with applications to logic and games. In: Logic for programming, artificial intelligence, and reasoning, pp. 209–223. Springer, Heidelberg (2005)

4. Chaitin, G.J.: Toward a mathematical definition of "life". In: Levine, R.D., Tribus, M. (eds.) The Maximum Entropy Formalism, pp. 477–498. MIT Press, Cambridge (1979)
5. Emmeche, C.: Life is an abstract phenomenon: is artificial life possible? In: Varela, F.J., Bourgine, P. (eds.) Proceedings of the First European Conference on Artificial Life, pp. 466–474. MIT Press, Cambridge (1992)
6. Gecow, A., Nowostawski, M., Purvis, M.: Structural tendencies in complex systems development and their implication for software systems. Journal of Universal Computer Science 11(2), 327–356 (2005)
7. Gecow, A.: Structural Tendencies – effects of adaptive evolution of complex (chaotic) systems. Int. J. Mod. Phys. C 19(4), 647–664 (2008)
8. Gecow, A.: Emergence of Growth and Structural Tendencies During Adaptive Evolution of System. In: Aziz-Alaoui, M., Bertelle, C. (eds.) From System Complexity to Emergent Properties, pp. 211–241. Springer, Heidelberg (2009)
9. Gecow, A.: The differences between natural and artificial life towards a definition of life. arXiv (1012.2889) (2010)
10. Harel, D., Pnueli, A.: On the development of reactive systems. In: Logics and Models of Concurrent Systems, vol. 13, pp. 471–498. NATO Advanced Study Institute (1985)
11. Kauffman, S.A.: The origins of order: self-organization and selection in evolution. Oxford Press, New York (1993)
12. Knottenbelt, W.J., Bradley, J.T.: Tackling large state spaces in performance modelling. In: Bernardo, M., Hillston, J. (eds.) SFM 2007. LNCS, vol. 4486, pp. 318–370. Springer, Heidelberg (2007)
13. Korzeniewski, B.: Cybernetic formulation of the definition of life. Journal of Theoretical Biology 209, 275–286 (2001)
14. Margulis, L., Sagan, D.: What is life? Simon and Schuster, New York (1995)
15. Milner, R.: A Calculus of Communicating Systems. In: Jones, N.D. (ed.) Semantics-Directed Compiler Generation. LNCS, vol. 94, Springer, Heidelberg (1980)
16. Milner, R.: Communication and concurrency. Prentice-Hall, Inc., Upper Saddle River (1989)
17. Milner, R.: Communicating and mobile systems: the π-calculus. Cambridge University Press, Cambridge (1999)
18. Muller, H.J.: The gene material as the initiator and organizing basis of life. American Naturalist 100, 493–517 (1966)
19. von Neumann, J.L., Burks, A.W.: Theory of self-reproducing automata (1966)
20. Nowostawski, M., Epiney, L., Purvis, M.: Self-Adaptation and Dynamic Environment Experiments with Evolvable Virtual Machines. In: Brueckner, S.A., Di Marzo Serugendo, G., Hales, D., Zambonelli, F. (eds.) ESOA 2005. LNCS (LNAI), vol. 3910, pp. 46–60. Springer, Heidelberg (2006)
21. Nowostawski, M., Purvis, M.K.: Evolution and Hypercomputing in Global Distributed Evolvable Virtual Machines Environment. In: Brueckner, S.A., Hassas, S., Jelasity, M., Yamins, D. (eds.) ESOA 2006. LNCS (LNAI), vol. 4335, pp. 176–191. Springer, Heidelberg (2007)
22. Nowostawski, M., Purvis, M.K., Cranefield, S.: An architecture for self-organising evolvable virtual machines. In: Brueckner, S.A., Di Marzo Serugendo, G., Karageorgos, A., Nagpal, R. (eds.) ESOA 2005. LNCS (LNAI), vol. 3464, pp. 100–122. Springer, Heidelberg (2005)
23. Prigogine, I., Stengers, I.: Order out of chaos: man's new dialog with nature. Bantam Books (1984)

24. Sangiorgi, D., Walker, D.: The π-calculus: A Theory of Mobile Processes. Cambridge University Press, Cambridge (2001)
25. Schrödinger, E.: What is life?: the physical aspect of the living cell. University Press, Cambridge (1945)
26. Smith, J.M., Szathmary, E.: The major transitions in evolution. W.H. Freeman, Oxford (1995)
27. Wittgenstein, L.J.J.: Philosophical Investigations. Basil Blackwell, Malden (1953)

Remedial English e-Learning Study in Chance Building Model

Chia-Ling Hsu

No.151, Yingzhuan Rd., Danshui Dist., New Taipei City 25137, Taiwan (R.O.C.)
clhsu@mail.tku.edu.tw

Abstract. Society and learning change with the development of technology. In order to keep up the latest development of technology in education, this study focused on a remedial English e-learning course in a university. Hopefully, by using a chance building model, rare and important elements that are so called 'chances' would be acquired. The chance building model was based on the text mining, KeyGraph technology and Grounded theory to present visualized scenarios. The process of the grounded theory was applied in chance building model to detect the chances. The participants were graduate school students who took the e-learning course. A questionnaire is composed of the ARCS (attention, relevance, confidence, satisfaction) model and preference selection with material topics and types. The results indicated that abstract learning style and business school learner characteristic would be the factors to improve the course instruction. More studies in chance building model and in empirical experiment are needed.

Keywords: Instructional Design, Chance Discovery, e-learning, Remediation Instruction, Course Evaluation.

1 Introduction

Using technologies in education is not only for gaining students' attention but also for providing remedial instruction. Studies indicated that applying multimedia to courses would increase students' learning motivation especially when using computers as educational media [1][2]. The computer assisted instruction (CAI) is a good example. Learning with CAI, students would learn by themselves with their own learning speed and level. Many studies, therefore, proposed that CAI could be used for remedial education and for individual learning [3][4].

Since Web 2.0 made the technology change drastically, the instructional design in using educational technology appeared different. E-learning is a new trend in education. For example, blogs, wiki systems, and game-based environments have been developing rapidly. Hence, how to adapt these technologies into instructional design will be a new and important issue.

R. Katarzyniak et al. (Eds.): Semantic Methods, SCI 381, pp. 171–182.
springerlink.com   © Springer-Verlag Berlin Heidelberg 2011

1.1 Purpose of the Research

This study is not to develop new platform but to promote a way of evaluation and analyzing a remedial English e-learning course. In other words, this study proposed a model of reanalyzing the data collected from an e-learning course, and of representing a chance building map. Therefore, the previous studies were to evaluate the remedial English e-learning course by a motivation questionnaire, to analyze the data to present a chance building map, and to provide instructional design strategies in remedial e-learning courses[5][6]. However, the results suggest that the chances were the import but not rare elements.

Based on the results of the previous studies, the main factors for remedial English e-learning instruction were the concrete-sequential learning style, the Engineering school, Science school and Management school. However, one important element in ARCS model was missing: Confidence. What factors would influence the remedial English e-learning instruction? In other words, what would be the rare and important factors to raise students' confidence? Therefore, this study re-examined the remedial English e-learning course in order to find the factors which affected the students learning. Those rare and important factors were so called the "chances" for instructors to improve their teaching.

2 Relative Literature

The instructional design and motivation models were applied in this study. In addition, the approach of text mining and KeyGraph technologies were used in this study, too. A further illustration of the models and technologies mentioned above is offered as follows.

2.1 Instructional Design

Instructional design is to provide teaching blueprints, and to examine teaching and offer solutions. Accordingly, the practice of instructional design is to target specific learners, select specific approaches, contents, and strategies, and make an effective teaching policy [7]. Instructional design is often presented and explained through models [8].

2.2 ARCS Model

Keller proposed the attention (A), relevance (R), confidence (C) and satisfaction (S) motivation model in order to solve motivational problems for instructional designer [9][10]. There are two main issues facing with the ARCS model: one is attitude, and the other is evaluation. The evaluation for the motivation is not to evaluate the learning efficiency but to evaluate the motivation character of the learning motivation in the instructional design model. The ARCS Model identifies four essential components for motivation instruction:

Attention: strategies for arousing and sustaining curiosity and interest.

Relevance: strategies that link to learners' needs, interests and motives.

Confidence: strategies that help students develop a positive expectation for successful achievement.

Satisfaction: strategies that provide extrinsic and intrinsic reinforcement for efforts.

The ARCS was also applied in adult learning [11]. Hence, this study used the ARCS model to evaluate the graduate students' motivation in a remedial e-learning English course in a university.

2.3 Text Mining and Chance Building Model

Weiss, Induskhya, Zhang, and Damerau [12] indicated that data mining technology would find out the pattern in structure data base but not in none or semi structure data base. Hence, Hearst [13] pointed out that data mining would not satisfy the human needs of pursuing information and knowledge. Fortunately, text mining is created to apply the language and statistics to analyze text data in order to attain new information [14]. Therefore, text mining became one of the important issues.

Ohsawa in 1998 proposed the KeyGrpah technology as a kind of data visualization tool in order to discover the chance [15]. The KeyGraph technology brought the text mining research into a new era. Montero and Araki, in 2005, showed that a text could be divided into some different subgroups and there was an association link to each subgroup [16]. Some phases would be connected to each other but some were not. At the same time, 2005, Sakakithara and Ohsawa sorted out different subgroups and used KeyGraph format to present these subgroups [17]. They also defined the high frequency element as a "black node", and the number of baskets which contain two elements and the high frequency co–occurrence as a "black link". In this model, the KeyGraph technology becomes a very powerful tool in many areas. Oshawa in 2002 pointed out that the value of KeyGraph technology as an extractor of causalities from an event - sequence, and as a words abstractor from a document [18]. Moreover, the main point of the KeyGraph technology provided some chances which would reverse the bad situation into a better one, especially in a feeble industry.

Wang, Hong, Sung, and Hsu applied this method to get the validity of KeyGraph [19]. The results indicated that although the statistics data showed no significant difference, the KeyGraph technology provided more information. Hsu and other educators also applied the KeyGraph technology in education setting. The results pointed out that the learners' scenario map would tell more information than the traditional statistics results. Huang, Tsai, & Hsu also applied the KeyGraph technology to exploring the learners' thinking [20]. Tsai, Huang, Hong, Wang, Sung, and Hsu [21] used Keygraph technology and tried to find the chances in instructional activity.

What are chances? According to chance discovery theory, the definition of chances is that chances are the rare but import events or factors [22][23]. However, Watts emphasizes the importance of linking [24]. Hsu has been applying chance discovery model to education especially in class activities. The findings were novel and would provide teachers with chances to design the activities in different points of view [25] [26]. The model we proposed would help instructional designers evaluate the course development as well as find chances to improve the quality of e-learning course. The data were analyzed in text format and represented in a chance building map. In this study we took students' motivation as an indicator to evaluate the course. The KeyGraph was another indicator to improve the course. With evaluation and reflection, the quality of the remedial e-learning English course would be better. Hopefully, few but important elements would be found.

3 Research Model

This study was to analyze the students' motivation and preference in a remedial English e-learning course. Although the traditional ARCS questionnaire was applied, this study employed a new model, chance building, with a view to acquiring more information in learners' motivation. In addition, this study intends to combine the data of ARCS motivation with interesting study materials to form a possible teaching plan. The participants, instrument, and chance building model were elaborated as follows.

There are two main core ideas in ground theory approach. One is that the ground theory leads researcher to examine all possible theoretical explanations for the empirical findings. The other is that the iterative process of moving back and forth between empirical data and emerging analysis makes the collected data progressively more focused. First, the researchers checked one possible theoretical explanation, learner factors. In this stage, the open coding in the ground theory method was to conceptualize the learner factors into gender, learning style and school. In axial coding stage, the main factors were determined in each factor (the gender, learning style, and school factors).

Then after having found the core variables, the selective coding stage was conducted with the research objectives, that is, the learners' evaluation, ARCS. The impact factors within the learner factors (the gender, learning style, and school) were interacted with the learners' evaluation and appeared the impact factors in the remedial English e-learning course.

Finial the theoretical codes integrate the impact factors in learner factors and the learners' evaluation to examine the third possible theoretical explanation, the teaching strategies. The result was indicated that all the possible theoretical explanation in this study, the learner factors, learners' evaluation, and teaching strategies were integrated together.

The main reason of this study applied the Grounded Theory is that the instructional design contains main different theoretical parts. Most study would look at one part or look at the relationship with two parts. However, using ground theory makes this study can look at three parts of the instructional design theory. Moreover, the ground theory approach would not manipulate the variables thought the data were the survey data. In other words, with the technology the ground theory may use the survey data as well as the interview data.

3.1 Participants

A remedial English e-learning in a university for graduate students who did not pass the graduation threshold on English tests such as TOEIC, TOEFL and GEPT. 186

students enrolled in this class. At the end of the course, students filled out online the questionnaire voluntarily. Hence, 179 questionnaires were collected. Table 1 showed population of the students.

Table 1. Population of the students

School	Number of students
Engineer	104
Science	32
Management	21
Business	9
Education	5
Liberal Arts	3
others	5
Total	179

3.2 Instrument

The instrument was contained the learning style testing, the questions for instructional design and the ARCS questionnaire for course evaluate. The ARCS questionnaire is composed of 34 items. The Attention factor includes items 1, 4*, 10, 15, 21, 24, 26*, and 29. The items marked with * mean the inverse items. The Relevance factor contains items 2, 5, 8, 13, 20, 22, 23, 25, and 28. The Confidence factor covers items 3, 6*, 9, 11*, 17*, 27, 30, and 34. The Satisfaction factor contains items 7, 12, 14, 16, 18, 19, 31*, 32, and 33. The score is calculated for each item by 5 scale points, from non-agree to very agree.

In addition, two more questions were asked. One was a multiple-choice question about the material types. The other was an open question about the material topics.

According to the information procession model, the learning style testing would classify the learner into one of the 4 learning styles, concrete-sequential, concrete-random, abstract-sequential, and abstract-random.

In this case study, the revised items in ARCS questionnaire were encoded into statistics consistent numbers. Then, the numerical data were transferred by a scoring system into a text data format in order to find the relationship between the students' character and the teaching' character. The numerical value of material types was returned to the original question items, short-essay, conversation, and both. The material topics collected were the response from students, including travel, health, technology, business, international, entertainment, education, and campus.

3.3 Chance Building Model

Using KeyGraph technology would provide scenarios for possible chances. The KeyGraph was a kind of data-driven technology which meant that the scenarios were not driven by objectives. With the objectives and the interesting topics we added to the original data-driven information, a more meaningful scenario would be formed. Figure 1 presented the procedure of chance building model. Each step was explained as follows.

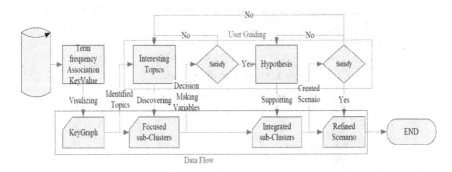

Fig. 1. The chance building model

4 Data Analysis – Chance Detection

The procedure of the research model was described below:

Step 1: Term frequency and Association Key value in KeyGraph

The data of questionnaires were collected from the WebCT platform. Numerical data were transferred by the scoring system into text data form in order to find the relationship between students' character and teaching' character. Using the KeyGraph technology, the thresholds of key frequency (51) and Jaccard relation (.333) were determined (figure 2 and figure 3).

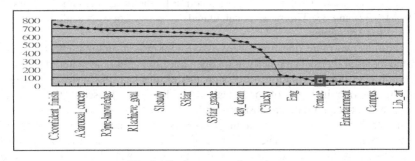

Fig. 2. The frequency of word term

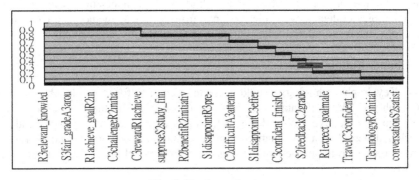

Fig. 3. The Jaccard value

Step 2: The sub-clusters in Value focus system.

Using the Jaccard association rule, the results of sub-clusters maps were shown in figure 4, 5, and 6. Figure 5, school cluster, appeared that the engineer school, science school, and management school were associated with other factors. Figure 6, learning style indicated that the concrete-sequential learning style were associated with other factors.

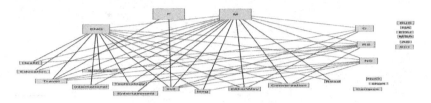

Fig. 4. The association of the course evaluation, ARCS, and the gender factor

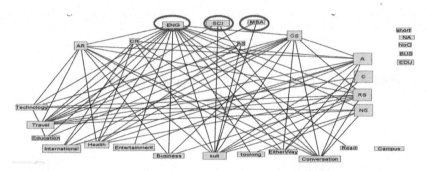

Fig. 5. The association of course evaluation, ARCS, and the learning style factor

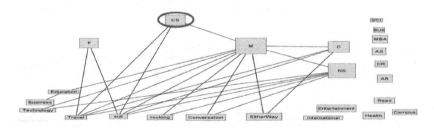

Fig. 6. The association of course evaluation, ARCS, and the school factor

Step 3: The integrated sub-cluster graph was generated and would propose an innovated scenario.

The integrated sub-cluster graph indicated that the impact factors in learner characteristics affected the course evaluation with ARCS by the factors in teaching perspective in figure 7. In other words, learners who were in management school, engineer school, science school, or with concrete-sequential learning style felt that the course would draw their attention, be related to their needs, and satisfy their efforts.

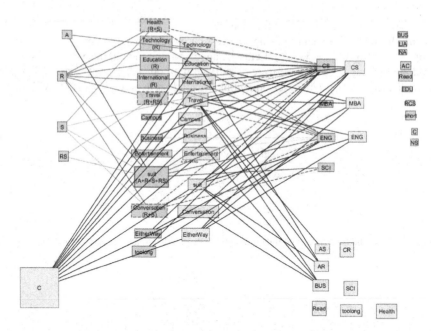

Fig. 7. A scenario of the impact factors in a remedial English e-learning course

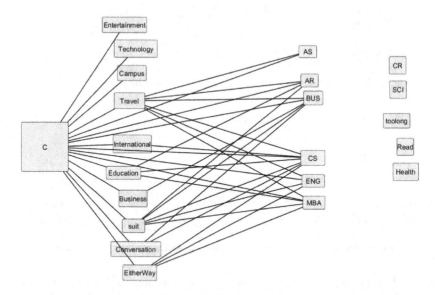

Fig. 8. The confidence element with main learner factors

The internal factors were the topics of health, technology, education, international, and travel. In addition, the topics of teaching materials and the suitable length of articles were also very important factors when learners evaluated the course. The teaching method was by means of conversation.

Step 4: Refined the scenario to detecting the rare and important factors

Although the important learners' factors affect the instruction through the material types and topics, the 'confidence' element in ARCS model did not appear in the graph. The factors that affected the students' motivation in confidence aspect would be the rare and important factors for instructors to improve the remedial English e-learning course. Hence, according to the grounded theory, the select code was determined. Then, the chance graph was shown in figure 8.

Comparing the ARCS elements (Attention, Relevance, Confidence, and Satisfaction) in figure 7 and figure 8, the rare and important factors were determined. Figure 9 indicated that the material types remained the same situation, and the material topics also remained the same topics except for the health topics which were missing. Therefore, the rare and important factors were 'abstract' learning style and business school.

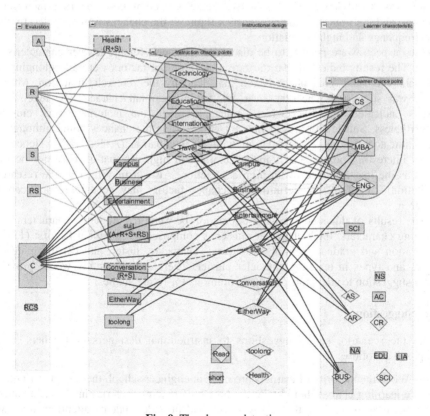

Fig. 9. The chances detection

5 Conclusion and Suggestion

This study applied the chance building model to detecting the rare and important factors as chances for instructors to improve the remedial English e-learning course. The results indicated that the rare and important factors were "abstract learning style," "business school," and "health material topics".

5.1 Discussion and Conclusion

The research method was based on the process of Grounded theory to reveal the impact factors in a remedial English e-learning course. Although this study aimed at detecting the chances factors, the principle would not yet be generated because of the research limitation. The processes of Grounded theory in chance building model would detect the chances for the remedial English e-learning course.

In the chance build model, the chances are defined as the rare and important factors. In this case study, we find the important factors in figure 7 both in the learner characteristic level and the instructional characteristic level. However, those factors reflected the high frequency and association. The rare part was missing. The rare factor was 'Confidence' in this study. Therefore, combining the rare factor, C, scenario and the important factor scenario, figure 9, the chances were detected by the low frequency and high association.

Two aspects were needed to be discussed. One was for the learner characteristic level. The results indicated that the rare and important factors were all belonging to the school cluster. This phenomenon may lead by the research limitation. Moore study was suggested. The other was for the instructional characteristic level. The results indicated that the chances points were in the material topics cluster. Nevertheless, Suit length article was not appeared as the chances point, although it was indicated the highly association with the evaluation (A+R+S+RS). Hence, it raises interesting points. The import factors with high association may not be the same as the low frequency with high association. Also, it may lead by the research limitation. Therefore, the definition of chance factors may need more application research.

The results of this study provide interaction between the students' characters and teaching characters. Each character cluster is composed of two main factors: (1) the motivation and graduate school in students' character cluster, and (2) the material topics and types in teaching character cluster. Therefore, this study would provide some suggestion to educators and some opinions for future studies.

5.2 Suggestion

From the scenario, some suggestions to instructional designers or teachers were proposed as follows.

●   When the majority of learners are male in engineer school, the remedial English e-learning course should take the length of reading material into consideration because in figure 7 the suit length would provide students attention, relevant and satisfaction.

- Although the topics of reading passages will affect the participants' course evaluation, the length of reading passages is a more important factor.
- The rare and important factors, so-called "chances", might be the critical factors for instructional designers.

This study finds such impact factors as engineer school, science school, management school, and concrete-sequential learning style in the aspect of learner characteristics. Based on these learner characteristics, a scenario of instructional strategies is proposed. According to the chances detection, the remedial English e-learning course may focus on the confidence of students. The material may need to adjust for the abstract learning style learner.

For the future studies, more various phases should be considered together to conduct the analysis. The intelligent thinking can be applied to the chance building model to provide intelligent scenarios. More research is needed for future study.

Acknowledgments. We would like to thank Dr. Hong, C. F. and Dr. Chiu, T. F. from Aletheia University for their support of computer system. The e-learning course was supported by Tamkang University.

References

1. Hsu, C.L., Chang, Y.F.: Study of the Relationship with the Media Material and the Students' Learning motivation. J. Educational Study 116, 64–76 (2003)
2. Taylor, R.P.: The computer in the school: Tutor, tool and tutee. Teacher College Press, New York (1980)
3. Hsu, C.L.: E-CAI Case Study. Educational Technology and Media 33, 28–35 (1997)
4. Hsu, C.L., Kuo, C.H.: Study of e-Learning Material Technology. In: 2000 e-Learning Theory and Practice Conference, pp. 61–65. National Chiao Tung University, Shin-Chu (2000)
5. Hsu, C.-l., Wang, A.-l., Lin, Y.-c.: Instructional design for remedial english e-learning. In: Pan, J.-S., Chen, S.-M., Nguyen, N.T. (eds.) ICCCI 2010. LNCS, vol. 6422, pp. 363–372. Springer, Heidelberg (2010)
6. Hsu, C.L., Lin, Y.C.: The Impact Factors in Remedial English e-Learning Instruction. In: ISSC 2010 (2010)
7. Smith, P.L., Ragan, T.J.: Instructional design. Macmillan, New York (1993)
8. Michael, T., Marlon, M., Roberto, J.: The third dimension of ADDIE: A cultural embrace. TechTrends 46(12), 40–45 (2002)
9. Keller, J.M.: Motivation and Instructional Design. A Theoretical perspective. J. Instructional Development, 26– 34 (1979)
10. Keller, J.M.: Motivational design of instruction. In: Reigeluth, C.M. (ed.) Instructional design theories and models: An overview of their current status. Erlbaum, Hillsdale (1983)
11. Visser, J., Keller, J.M.: The Clinical Use of Motivational Messages: an Inquiry into the Validity of the ARCS Model of Motivational Design. J. Instructional Science 12, 467–450 (1990)
12. Weiss, S.M., Indurkhya, N., Zhang, T., Damerau, F.: Text mining predictive methods for analyzing unstructured information. Spring Science-Business Media, Inc., New York (2005)

13. Hearst, M.A.: Untangling text data mining. In: The 37th Annual Meeting of the Association for Computational Linguistics, University of Maryland (1999)
14. Grimes, S.: Mining text can boost research. Information Outlook 9(11), 20–21 (2005)
15. Ohsawa, Y., Benson, N.E., Yachida, M.: KeyGraph: Automatic Indexing by Co-occurrence Graph based on Building Construction Metaphor. In: Proc. Advanced Digital Library Conference (IEEE ADL 1998), pp. 12–18. IEEE Press, New York (1998)
16. Montero, C.A.S., Araki, K.: Discovering Critically Self-Organized Chat. In: Proc. the fourth IEEE International Workshop WSTS, pp. 532–542. IEEE Press, New York (2005)
17. Sakakibara, T., Ohsawa, Y.: Knowledge, Discovery Method by Gradual Increase of Target Baskets form Sparse Dataset. In: Proc. the Fourth IEEE International Workshop WSTS, pp. 480–489. IEEE Press, New York (2005)
18. Ohsawa, Y.: KeyGraph as Risk Explorer in Earthquake-Sequence. J. of Contingencies and Crisis Management. 10, 119–128 (2002)
19. Wang, L.H., Hong, C.F., Hsu, C.L.: Closed - ended Questionnaire Data Analysis. In: 10th International Conference on Knowledge-Based & Intelligent Information & Engineering Systems, KES 2006, United Kingdom, pp. 1–7 (2006)
20. Huang, C.J., Tsai, P.H., Hsu, C.L.: Exploring Cognitive Difference in Instructional Outcomes Using Text Mining Technology. In: Proc. of 2006 IEEE International Conference on Systems, Man, and Cybernetics, pp. 2116–2120. IEEE Press, New York (2006)
21. Tsai, P.H., Huang, C.J., Hong, C.F., Wang, L.H., Sung, M.Y., Hsu, C.L.: Discover Learner Comprehension and Potential Chances from Documents. In: The 11th IPMU International Conference, Paris, France (2006)
22. Ohsawa, Y., McBurney, P.: Chance Discovery. Springer, Germany (2003)
23. Ohsawa, Y., Tsumoto, S.: Chance Discoveries in Real World Decision Making. Springer, New York (2006)
24. Watts, D.: Small Worlds: the Dynamics of networks begween order and randowness. Princeton, New York (1999)
25. Hsu, C.C., Wang, L.H., Hong, C.F., Sung, M.Y., Tasi, P.H.: The KeyGraph Perspective in ARCS motivation model. In: the 6th IEEE International Conference on Advanced Learning Technologies, pp. 970–974. IEEE Press, New York (2006)
26. Hsu, C.C., Wang, L.H., Hong, C.F.: Understanding students' Conceptions and providing Scaffold Teaching Activities. In: International Conference of Teaching and Learning for Excellence, Tamsui, pp. 166–175 (2007)

Using IPC-Based Clustering and Link Analysis to Observe the Technological Directions

Tzu-Fu Chiu[1], Chao-Fu Hong[2], and Yu-Ting Chiu[3]

[1] Department of Industrial Management and Enterprise Information,
Aletheia University, Taiwan, R.O.C.
[2] Department of Information Management, Aletheia University, Taiwan, R.O.C.
[3] Department of Information Management, National Central University, Taiwan, R.O.C.
{Chiu,cfhong}@mail.au.edu.tw, gloria@mgt.ncu.edu.tw

Abstract. To explore the technological directions of an industry is essential for companies and stakeholders to anticipate the future situations and R&D activities. Patent data contains plentiful technical information, which is appropriate to be used in technological analysis in order to find out the technical topics and possible directions. Due to the complex nature of patent data, two data mining methods: IPC-based clustering and link analysis, are used to figure out the potential tendencies on thin-film solar cell. An IPC-based clustering algorithm will be proposed and utilized to generate the significant categories via the IPC and Abstract fields, while the link analysis will be adopted to draw a link diagram for the whole dataset via the Abstract, Issue Date, and Assignee Country fields. During experiment, the technical categories will be identified using the IPC-based clustering, and the technological directions will be found through the link analysis. Finally, the recognized technical categories and technological directions will be provided to the managers and stakeholders for assisting their decision making.

Keywords: technological direction, IPC-based clustering, link analysis, thin-film solar cell, patent data.

1 Introduction

Solar cell (especially thin-film solar cell), one of green energies, is growing at a fast pace with its long-lasting and non-polluting natures. Meanwhile, up to 80% of the technological information disclosed in patents is never published in any other form [1]. It is worthwhile for researchers and practitioners to devote their efforts to explore the directions of this technology via patent data. According to the textual characters of patent documents (in Abstract, Claim, and Description fields), a clustering method, IPC-based clustering is proposed to manipulate the homogeneity and heterogeneity of patent data so as to increase the cohesiveness (similarity) within a category and the dispersion (dissimilarity) among categories. In patent database, the IPC codes are provided by the examiners of patent office [2] and contain the professional knowledge of the experienced examiners. It would be reasonable for a research to base on the IPC code and the term vectors of Abstract to classify the patents into a certain number

R. Katarzyniak et al. (Eds.): Semantic Methods, SCI 381, pp. 183–197.
springerlink.com   © Springer-Verlag Berlin Heidelberg 2011

of categories. Afterward, association analysis is employed to find out the subcategories in each category, while link analysis is adopted to discover the relations between year (/country) and category. Consequently, the technical categories and subcategories will be obtained and the technological directions of thin-film solar cell will be observed for assisting the decision making of managers and stakeholders.

2   Related Work

As this study is attempted to observe the technological directions for companies and stakeholders via patent data, a research framework is required and can be built via a proposition of an IPC-based clustering and an adoption of association analysis and link analysis. In order to manipulate the homogeneity and heterogeneity of patent data, an IPC-based clustering is proposed for dividing the patents into different categories. For handling the textual nature of patent data (mainly the Abstract, Claim, and Description fields), association analysis is used to find out the technical subcategories in every category. Due to the collected data spreading over ten years (2000 to 2009), link analysis is employed to discover the relations between year (/country) and technical category. Subsequently, the research framework will be applied to recognize the technical trends on thin-film solar cell. Therefore, the related areas of this study would be technological direction, thin-film solar cell, IPC-based clustering, association analysis, and link analysis, which will be described briefly in the following subsections.

2.1   Technological Direction

Technology strategy represents managers' efforts to think systematically about the role of technology in decisions affecting the long-term success of the organization [3]. Strategic planning is an organization's process of defining its strategy, or direction, and making decisions on allocating its resources to pursue this strategy, including its capital and people [4]. Strategic planning processes are: mission definition, objectives setting, external analysis, internal analysis, strategic choice, strategy implementation, and competitive advantages [5]. Technological direction is an issue in the technology strategic planning which is to find out strategic directions of technology so as to allow the organization to make use of its strengths and grasp the possible opportunities. In order to draw the data mining techniques for observing the technological directions via patent data, a research framework of IPC-based clustering and link analysis will be formed to perform patent analysis on thin-film solar cell in this study.

2.2   Patent Data and Thin-Film Solar Cell

A patent document is similar to a general document, but includes rich and varied technical information as well as important research results [6]. Patent data, among the better structured and monitored data sources, is the official filings of inventions [7]. Patent documents can be gathered from a variety of sources, such as the United States Patent and Trademark Office [8], the European Patent Office [9], the Intellectual Property Office in Taiwan [10], and so on. A patent document includes numerous fields [8], such as: Patent Number, Title, Abstract, Issue Date, Application Date,

Application Type, Assignee Name, Assignee Country, International Classification (IPC), Current US References, Claims, Description, etc.

Solar cell, a sort of green energy, is clean, renewable, and good for protecting our environment. It can be mainly divided into two categories (according to the light absorbing material): crystalline silicon (in a wafer form) and thin films (of other materials) [11]. A thin-film solar cell (TFSC), also called a thin-film photovoltaic cell (TFPV), is made by depositing one or more thin layers (i.e., thin film) of photovoltaic material on a substrate [12]. The most common materials of TFSC are amorphous silicon and polycrystalline materials (such as: CdTe, CIS, and CIGS) [11]. In 2009, the photovoltaic industry production increased by more than 50% (average for last decade: more than 40%) and reached a world-wide production volume of 11.5 GWp of photovoltaic modules, whereas the thin film segment grew faster than the overall PV market [13]. Therefore, thin film is the most potential segment with the highest production growth rate in the solar cell industry, and it would be appropriate for academic and practical researchers to contribute efforts to explore the technological directions of this segment.

2.3 IPC-Based Clustering

An IPC (International Patent Classification) is a classification derived from the International Patent Classification System (supported by WIPO) which provides a hierarchical system of symbols for the classification of patents according to the different areas of technology to which they pertain [14]. IPC classifies technological fields into five hierarchical levels: section, class, subclass, main group and sub-group, containing about 70,000 categories [15]. As stated by the Intellectual Property Office of UK [2], each patent document published will have at least one IPC code applied to it; and the EPO and other patent offices worldwide also use it to classify their own patent documents. The IPC codes of every patent are assigned by the examiners of the national patent office and contain the professional knowledge of the experienced examiners [16]. Therefore, it would be reasonable for a research to base on the IPC code and the term vectors of Abstract to cluster the patents into a number of categories. The IPC codes have been applied for assisting patent retrieval in some researches [16, 17].

In this study, the mean vector of the patents comprised in the same IPC code group will be utilized as a centroid for clustering to divide the whole dataset of patents into a certain number of categories.

2.4 Association Analysis

Association analysis is a useful method for discovering interesting relationships hidden in large data sets. The uncovered relationships can be represented in the form of association rules or co-occurrence graphs [18]. An event map, a sort of co-occurrence graphs, is a two-dimension undirected graph, which consists of event clusters, visible events, and chances [19]. An event cluster is a group of frequent and strongly related events. The occurrence frequency of events and co-occurrence between events within an event cluster are both high. The co-occurrence between two events is measured by the Jaccard coefficient as in Equation (1), where e_i is the ith event in a data record (of the data set *D*). The event map is also called as an association diagram in this study.

$$Ja(e_i, e_j) = \frac{Freq(e_i \cap e_j)}{Freq(e_i \cup e_j)} \tag{1}$$

In this study, the association diagram will be adopted to generate the subcategories of each IPC category for direction observation.

2.5 Link Analysis

Link analysis is a collection of techniques that operate on data that can be represented as nodes and links [20]. A node represents an entity such as a person, a document, or a bank account. A link represents a relationship between two entities such as a reference relationship between two documents, or a transaction between two bank accounts. The focus of link analysis is to analyze the relationships among entities. The areas related to link analysis are: social network analysis, search engines, viral marketing, law enforcement, and fraud detection [20]. Additionally, in social network analysis, the degree centrality of a node can be measured as in Equation (2), where $a(P_i, P_k) = 1$ if and only if P_i and P_k are connected by a link (0 otherwise) and n is the number of all nodes [21]. In data mining, the strength of a relationship (i.e., the similarity between nodes) can be measured as in Equation (3), where Q and D are the term frequency vectors and n is the vocabulary size [22].

$$C_D(P_k) = \sum_{i=1}^{n} a(P_i, P_k) / (n-1) \tag{2}$$

$$Sim(Q, D) = \sum_{i=1}^{n} (Q_i \times D_i) / \sqrt{\sum_{i=1}^{n} (Q_i)^2} \times \sqrt{\sum_{i=1}^{n} (D_i)^2} \tag{3}$$

In this study, link analysis will be employed to generate the linkages between IPC category and year (as well as country) for direction observation.

3   A Research Framework for Technological Direction Observation

A research framework for technological direction observation, based on IPC-based clustering and link analysis, has been constructed as shown in Fig. 1. It consists of five phases: data preparation, procedure of IPC-based clustering, execution of IPC-based clustering, association analysis and link analysis, and new findings; and will be described in the following subsections.

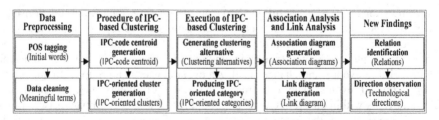

Fig. 1. A research framework for technological direction observation

3.1 Data Preprocessing

In first phase, the patent data of thin-film solar cell (during a certain period of time) will be downloaded from the USPTO [8]. For considering an essential part to represent a patent document, the Abstract, Issue Date, and Assignee Country fields are selected as the objects for this study. Afterward, two processes, POS tagging and data cleaning, will be executed to clean up the textual data of the abstract field.

(1) POS Tagging: An English POS tagger (i.e., a Part-Of-Speech tagger for English) from the Stanford Natural Language Processing Group [23] will be employed to perform word segmenting and labeling on the patents (i.e., the abstract field). Then, a list of proper morphological features of words needs to be decided for sifting out the initial words.

(2) Data Cleaning: Upon these initial words, files of n-grams, synonyms, and stop words will be built so as to combine relevant words into compound terms, to aggregate synonymous words, and to eliminate less meaningful words. Consequently, the meaningful terms will be obtained from this process.

3.2 Procedure of IPC-Based Clustering

Second phase is used to describe the IPC-based clustering proposed by this research, including IPC-code centroid generation and IPC-oriented cluster generation. IPC-code centroid generation is used to generate the centroid for every IPC code group. IPC-oriented cluster generation is designed to produce the clusters of patents according to the IPC-coded centroids and term vectors.

(1) IPC-coded Centroid Generation: As IPC codes of patent are provided by the examiners and contain abundant professional knowledge, they will be exploited to direct the clustering process. The patents with the same IPC code will be put together to form an IPC code group, if a patent which has more than one IPC code will be assigned to multiple groups. Patents in an IPC code group will then be applied to calculate the group centroid M_i via the term vector x using Equation (4) [24] where C_i is the ith group.

$$M_i = \frac{1}{|C_i|}\sum_{x \in C_i} x \tag{4}$$

The centroids obtained will be called as IPC-coded centroids and used for forming the IPC-oriented clusters in the next process.

(2) IPC-oriented Cluster Generation: According to the IPC-coded centroids and term vectors, the whole dataset of patents will be distributed into a certain number of clusters using the Euclidean distance measure as in Equation (5) [24] where X_i is a term vector.

$$D(X_i, M_j) = \sqrt{\sum_{k=1}^{n}(x_{ik} - m_{jk})^2} \tag{5}$$

A patent will be assigned to a specific IPC code cluster, while the shortest distance $D(X_i, X_j)$ exists between that patent and the IPC code centroid M_j. The number of clusters will be determined by the research requirements and the domain knowledge. The clusters derived will be called as IPC-oriented clusters and utilized for execution and analysis in the following phases.

3.3 Execution of IPC-Based Clustering

Third phase is utilized to execute the IPC-based clustering according to the above-proposed procedure. The alternatives of IPC-oriented clusters will be created successively for consisting of a certain number of clusters. An appropriate alternative will then be selected based on the F score measure for creating the IPC-oriented categories for the whole patent dataset.

(1) Generating Clustering Alternative: At the beginning, the IPC code groups are sorted descendingly according to the number of comprising patents. Secondly, the term vectors of comprising patents are applied to calculate the centroid for every IPC code group. Thirdly, a clustering alternative is constructed via the IPC-coded centroids and the term vectors of Abstract data by distributing patents of the whole dataset into the code groups (e.g., 4 groups in the first alternative) according to the shortest distance via Equation (5). Fourthly, the patents distributed into a code group form an IPC-oriented cluster; and the including IPC-oriented clusters together compose that alternative. Lastly, the other alternatives (from 5 to 31) are then constructed successively.

(2) Producing IPC-oriented Category: Among the clustering alternatives, F score (in Equation (6)) is employed to evaluate the accuracy of the clustering results, where the Precision and Recall are in Equation (7) and (8) [25].

$$F = \frac{2}{(1/\mathrm{Re}\,call) + (1/\mathrm{Pr}\,ecision)} \tag{6}$$

$$\mathrm{Pr}\,ecision = \frac{|\{relevant \cap retrieved\}|}{|\{retreieved\}|} \tag{7}$$

$$\mathrm{Re}\,call = \frac{|\{relevant \cap retrieved\}|}{|\{relevant\}|} \tag{8}$$

An original or adjusted alternative with the higher F score will be selected as the appropriate clustering result. In the selected alternative, every cluster is regarded as an IPC-oriented category and will be utilized for further analyzing in the following association analysis and link analysis.

3.4 Association Analysis and Link Analysis

Fourth phase is designed to perform the association analysis for generating the association diagrams and to implement the link analysis for producing the link diagram so as to obtain the technical subcategories and the relations between years (/countries) and technical categories.

(1) Association Diagram Generation: An association diagram will be drawn via the meaningful terms of the Abstract data for every IPC-oriented category, so that a number of subcategories can be generated through the threshold setting of frequency and co-occurrence. These subcategories will be named using the domain knowledge and regarded as technical subtopics to represent the IPC-oriented category.

(2) Link Diagram Generation: In order to generate the link diagram, the node centrality and relationship strength of the entities (i.e., IPC-oriented categories, years, and countries) need to be calculated via Equation (2) and Equation (3). Based on the node centrality and relationship strength, a link diagram will be drawn, so that the linkages between year (/countries) and IPC-oriented categories can be constructed through the threshold setting of node centrality and relationship strength. These linkages will be utilized to identify the relations between technical categories and Issue Year (as well as Assignee Country) and then to explore the technological directions in the next phase.

3.5 New Findings

Last phase is intended to identify the relations between technical categories and years (/countries) and to recognize the technological directions, based on the link diagram.

(1) Relation Identification: According to the association diagrams and link diagram, the technical subtopics of each IPC-oriented category and the relations between the technical categories and years (/countries) will be identified. Based on the relations between categories and years, four relation types are likely found: a category existing in the full period of time (i.e., not less than 5) (type A1), existing in the first half (type A2), existing in the second half (type A3), and existing randomly in the period of time (type A4). Based on the relations between the categories and countries, three relation types are likely found: a category spreading in various countries (i.e., not less than 5) (type B1), spreading in the dominant countries (i.e., JP with 70 patents and US with 52 patents) (type B2), and spreading in the non-dominant countries (type B3).

(2) Trend Recognition: In accordance with the relation types of technical category, the directions of thin-film solar cell will be recognized and then provided to the managers and stakeholders for assisting their decision making.

4 Experimental Results and Explanation

The experiment has been implemented according to the research framework. The experimental results will be explained in the following five subsections: result of data preprocessing, result of IPC code group and IPC-coded centroid, result of clustering alternatives and IPC-oriented categories, result of association analysis and link analysis, and result of new findings.

4.1 Result of Data Preprocessing

As the aim of this study is to explore the trends of thin-film solar cell, the patent documents are the target data for the experiment. Mainly, the Abstract, Issue Date, and Country fields were used in this study. The issued patents (160 records) during year 2000 to 2009 were collected from USPTO, using key words: "'thin film' and ('solar cell' or 'solar cells' or 'photovoltaic cell' or 'photovoltaic cells' or 'PV cell' or 'PV cells')" on "title field or abstract field". Afterward, the POS tagger was triggered and the data cleaning process was executed to do the data preprocessing upon Abstract data. Consequently, the abstract data during year 2000 to 2009 were cleaned up and the meaningful terms were obtained.

4.2   Result of IPC Code Group and IPC-Coded Centroid

According to the IPC field, the number of IPC code groups (down to the fifth level) in 160 patents were 190, as many patents contained more than one IPC code, for example, Patent 06420643 even contained 14 codes. But there were up to 115 groups consisting of only one patent. If the threshold of the number of comprising patents was set to 5, there were 31 leading groups including the first group H01L031/18 (consisting of 48 patents), the second group H01L021/02 (27 patents), and down to the 31st group H01L031/0236 (5 patents), as in Fig. 2.

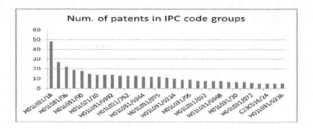

Fig. 2. The number of patents in IPC code groups

The patents contained in each IPC code group were used to generate the IPC-coded centroid for that group via Equation (3). These IPC-coded centroids would be utilized to produce the IPC-oriented clusters afterward.

4.3   Result of Clustering Alternatives and IPC-Oriented Categories

The clustering alternatives contained the ones from including 4 leading groups, to 5 leading groups, ..., till 31 leading groups. The first alternative was constructed by distributing patents of the whole dataset into the 4 leading groups (i.e., H01L031/18, H01L021/02, H01L031/06, and H01L031/036) via Equation (4), using the IPC-coded centroids of 4 groups and the term vectors of Abstract data. The number of patents distributed into 4 groups was: 118 in H01L031/18, 11 in H01L021/02, 19 in H01L031/06, and 12 in H01L031/036. The accuracy (i.e., average F score) of this alternative was 0.4514. Each IPC code group with its distributed patents was regarded as an IPC-oriented cluster. The other alternatives (from including 5 to 31 groups) were then constructed successively. Some of the clustering alternatives with their including clusters and accuracies were calculated and summarized as in Table 1.

Table 1. A summary of clustering alternatives with their including clusters and accuracies

Alternative	Num. of clusters	Num. of patents in cluster	Accuracy
1	4	118, 11, 19, 12	0.4514
2	5	93, 11, 15, 10, 31	0.5109
3	6	87, 11, 15, 10, 29, 8	0.5213
4	7	84, 5, 15, 8, 29, 8, 11	0.5140
5	8	77, 5, 12, 8, 26, 8, 11, 13	0.5055
6	9	77, 5, 12, 1, 26, 8, 11, 13, 7	0.5052
7	10	73, 5, 12, 1, 24, 8, 11, 10, 7, 9	0.4988
8	15	55, 2, 2, 1, 24, 8, 0, 7, 10, 9, 10, 0, 10, 10, 12	0.5466
9	20	48, 2, 2, 0, 16, 8, 0, 5, 8, 5, 9, 0, 10, 10, 11, 8, 1, 6, 8, 3	0.5251
10	25	47, 2, 2, 0, 16, 7, 0, 5, 6, 5, 5, 0, 5, 10, 1, 8, 0, 6, 7, 3, 10, 2, 2, 5, 6	0.4992
11	31	42, 2, 0, 0, 16, 7, 0, 2, 6, 5, 5, 0, 4, 7, 1, 8, 0, 3, 6, 3, 10, 2, 0, 0, 6, 5, 4, 5, 5, 2, 4	0.5192
adjusted	9	70, 23, 15, 12, 10, 10, 7, 7, 6	0.5617

According to Table 1, the accuracies of most alternatives varied from 0.50 to 0.53. An adjusted method was suggested which selected the leading clusters from the alternative 11 (with 31 clusters) by setting the threshold of the number of comprising patents to 6, so as to increase the accuracy to 0.5617. After the IPC-oriented cluster generation, the adjusted alternative, including 9 clusters: H01L031/18, H01L031/00, H01L021/00, H01L021/20, H01L031/052, H01L031/048, H01L031/04, H01L031/0336, and H01L031/20, were shown in Table 2 (with the containing patents and IPC code description). These nine clusters were regarded as the IPC-oriented categories and used in the following association analysis and link analysis.

Table 2. The adjusted alternative with including clusters and IPC code description

Cluster	Num. of patents	IPC code description
H01L031/18	70	Processes or apparatus specially adapted for the manufacture or treatment of these devices or of parts thereof
H01L031/00	23	Semiconductor devices sensitive to infra-red radiation, light, electromagnetic radiation of shorter wavelength, or corpuscular radiation and specially adapted either for the conversion of the energy of such radiation into electrical energy or for the control of electrical energy by such radiation; Processes or apparatus specially adapted for the manufacture or treatment thereof or of parts thereof; Details thereof
H01L031/048	15	encapsulated or with housing
H01L031/052	12	with cooling, light-reflecting or light- concentrating means
H01L021/20	10	Deposition of semiconductor materials on a substrate, e.g. epitaxial growth
H01L031/0336	10	in different semiconductor regions, e.g. Cu_2X/CdX hetero-junctions, X being an element of the sixth group of the Periodic System
H01L021/00	7	Processes or apparatus specially adapted for the manufacture or treatment of semiconductor or solid state devices or of parts thereof
H01L031/04	7	adapted as conversion devices
H01L031/20	6	such devices or parts thereof comprising amorphous semiconductor material

4.4   Result of Association Analysis and Link Analysis

Using association analysis, the association diagrams of 9 IPC-oriented categories were drawn via the meaningful terms of Abstract data of the comprising patents so as to find out their including sub-clusters. Taking the diagram of H01L031/18 as an example, 15 sub-clusters were found while the number of consisting nodes of a sub-cluster was set to no less than five. According to the domain knowledge, the sub-clusters were named as follows: a1-photoelectron-spectroscopy, a2-layer-deposition, a3-organic-light-emitting, a4-semiconductor-layer, a5-selenium-indium-layer, a6-substrate-support-member, a7-heating-jig, a8-electrode-layer, a9-Zn-compound, a10-conductive-layer, a11-electromagnetic-radiation, a12-aluminum-substrate, a13-coating-chamber, a14-pin-junction, and a15-power-density. These named sub-clusters were regarded as technical subcategories in the following analysis. Fig. 3 is the association diagram of the H01L031/18-oriented category. The other eight association diagrams were then drawn successively for the remaining IPC-oriented categories and used in link analysis.

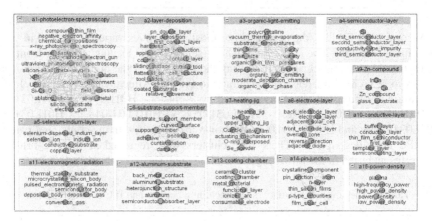

Fig. 3. An association diagram of H01L031/18-oriented category

Using link analysis, the node centrality and relationship strength of the IPC-oriented categories, years, and countries were calculated and summarized in Table 3 and 4, where the items in italic face were the ones over the threshold setting 0.05.

Table 3. The strength of relationship between IPC-oriented categories and years

Category	2000	2001	2002	2003	2004	2005	2006	2007	2008	2009
H01L031/18	*0.0588*	*0.1325*	*0.0824*	*0.1646*	*0.0741*	*0.0274*	*0.0921*	*0.1579*	*0.0411*	*0.0506*
H01L031/00	0.0238	0.0444	*0.0976*	0.0465	*0.0526*	0	0.0286	*0.1081*	*0.0741*	*0.1613*
H01L031/048	*0.0606*	*0.1143*	*0.1563*	0.0278	*0.0667*	0	0.0370	0	0	0
H01L031/052	0	*0.1250*	*0.0968*	*0.0625*	0.0357	0	0.0417	0.0345	0	0
H01L021/20	*0.1538*	0.0303	0.0323	*0.0667*	0.0385	*0.0714*	0	0	0	0
H01L031/0336	*0.2000*	0	*0.0667*	0	0.0385	*0.0714*	0	0	0	0.0455
H01L021/00	0	0	0	0	0.0435	*0.0909*	*0.0526*	0.0417	*0.0833*	*0.1111*
H01L031/04	0.0385	0	0	0.0357	*0.1429*	0	*0.0526*	0	0	*0.0526*
H01L031/20	*0.0833*	*0.0714*	0	0.0370	0	0	*0.0556*	0	0	0

Table 4. The strength of relationship between IPC-oriented categories and countries

Category	JP	US	DE	NL	FR	KR	AU	CA	BE	IT	CH	TH	FI	TW
H01L031/18	0.2500	0.2323	0.1467	0.0411	0.0139	0	0	0.0286	0.0141	0	0	0.0143	0	0
H01L031/00	0.0690	0.2500	0.0263	0	0	0	0	0	0	0	0.0435	0	0	0
H01L031/048	0.1333	0.0308	0	0.0500	0.0588	0	0	0	0	0	0	0	0.0667	0
H01L031/052	0.0513	0.0492	0.0769	0	0	0	0.1667	0	0.0769	0	0	0	0	0
H01L021/20	0.1111	0.0333	0	0	0	0	0	0	0	0	0	0	0	0
H01L031/0336	0.0667	0.0333	0.0400	0.0667	0.0833	0	0	0	0	0	0	0	0	0
H01L021/00	0.0132	0.0727	0	0	0	0.1250	0	0	0	0.1429	0	0	0	0
H01L031/04	0.0694	0	0	0	0	0.1250	0	0	0	0	0	0	0	0.1429
H01L031/20	0.0411	0.0175	0.0476	0.0909	0	0	0	0	0	0	0	0	0	0

Based on Table 3 and 4, a link diagram for 9 IPC-oriented categories was drawn via the Abstract, Issue Date, and Assignee Country fields of the comprising patents in order to show up the relations between the categories and the years (/countries), which was shown in Fig. 4. In the diagram, the digits under the year and country nodes were the node centrality (e.g., year 2000: 0.37; JP: 0.39); the digits on the link lines were the relationship strength (e.g., between H01L031/18 and 2000: 0.05; between H01L031/18 and JP: 0.25).

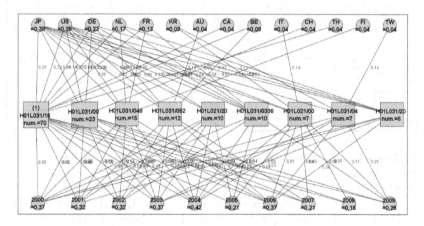

Fig. 4. A link diagram for 9 IPC-oriented categories (H01L031/18 to H01L031/20)

4.5 Result of New Findings

The association diagrams and link diagram will be utilized to identify the technical subcategories and relation types. Afterward, the technical subcategories and relation types will be applied to observe the technological directions.

(1) Relation Type Identification: According to the link diagram (Fig. 4) and the summarized table (Table 3), the relation type A1 (between technical categories and years) were: H01L031/18 and H01L031/00. The relation type A2 were: H01L031/048, H01L031/052, H01L021/20, and H01L031/0336. The relation type A3 were: H01L021/00 and H01L031/04. The relation type A4 was: H01L031/20. In addition, according to the link diagram (Fig. 4) and the summarized table (Table 4), the relation type B1 (between technical categories and countries) were: H01L031/18, H01L031/048, H01L031/052, and H01L031/0336. The relation type B2 were: H01L031/00, H01L021/20, and H01L031/20. The relation type B3 were: H01L021/00 and H01L031/04. Subsequently, the relation types between technical categories and years as well as between technical categories and countries were summarized below in Table 5 and then used to recognize the technological directions.

Table 5. A summary of technical categories and relation types

Category	Subcategory	Focused year	Type	Related country	Type
H01L031/18	a1-layer-deposition, a2-photoelectron-spectroscopy, a3-organic-light-emitting, a4-electrode-layer, a5-coating-chamber, a6-semiconductor-layer, a7-Zn-compound, a8-aluminum-substrate, a9-power-density, a10-heating-jig, a11-pin-junction, a12-porous-layer, a13-conductive-layer, a14-selenium-indium-layer, a15-electromagnetic-radiation, a16-substrate-support-member	2000, 2001, 2002, 2003, 2004, 2006, 2007, 2009	A1	JP, US, DE, NL, FR, CA, BE, TH	B1
H01L031/00	b1-roll-to-roll-processing, b2-conductive-oxide-layer, b3-polycrystalline-film, b4-single-crystal-thin-film, b5-light-transmissive-substrate	2002, 2004, 2007, 2008, 2009	A1	JP, US, DE, CH	B2
H01L031/048	c1-electrode-layer, c2-porous-structure, c3-crystal-semiconductor, c4-thermal-emissive-coating	2000, 2001, 2002, 2004	A2	JP, US, NL, FR, FI	B1
H01L031/052	d1-anti-reflection-coating, d2-composite-structure, d3-glass-substrate, d4-reflective-film	2001, 2002, 2003	A2	JP, US, DE, AU, BE	B1
H01L021/20	e1-amorphous-annealing, e2-CVD-method, e3-annealing-process, e4-separated-substrate, e5-SOI-substrate	2000, 2003, 2005	A2	JP, US	B2
H01L031/0336	f1-roll-to-roll, f2-light-absorbing-layer, f3-carrier-concentration	2000, 2002, 2005	A2	JP, US, DE, NL, FR	B1
H01L021/00	g1-absorbing-layer, g2-thermal-budget, g3-transparent-substrate	2005, 2006, 2008, 2009	A3	JP, US, KR, IT	B3
H01L031/04	h1-amorphous-silicon-film, h2-aromatic-enediyne	2004, 2006, 2009	A3	JP, KR, TW	B3
H01L031/20	i1-optical-band-gap, i2-CVD-plasma, i3-accumulated-charge, i4-deposition-rate	2000, 2001, 2006	A4	JP, US, DE, NL	B2

(2) Technological Direction Observation: According to the link diagram (Fig. 4) and the above summarized table (Table 4), the technical categories and subcategories had been recognized, and the technological directions of thin-film solar cell could be observed. The technical categories were: H01L031/18, H01L031/00, H01L021/00, H01L021/20, H01L031/052, H01L031/048, H01L031/04, H01L031/0336, and H01L031/20. The technical subcategories of H01L031/18-oriented category, as an example, were: a1-photoelectron-spectroscopy, a2-layer-deposition, a3-organic-light-emitting, a4-semiconductor-layer, a5-selenium-indium-layer, a6-substrate-support-member, a7-heating-jig, a8-electrode-layer, a9-Zn-compound, a10-conductive-layer, a11-electromagnetic-radiation, a12-aluminum-substrate, a13-coating-chamber, a14-pin-junction, and a15-power-density. The technological directions of each IPC-oriented category would be observed and described as follows.

(a) H01L031/18-oriented category: Referring to Table 5, this technical topic was continuously developing in the full period of time from 2000 to 2009 (relation type A1) and widely spreading in eight countries (relation type B1). It seemed to be an essential topic of the industry, as it is related to the manufacturing processes or devices.

(b) H01L031/00-oriented category: From the above Table 5, this topic existed in the whole period of time from 2002 to 2009 (type A1) and spread in the dominant countries (type B2). It seemed that the topic was growing constantly and participated by the technologically advanced countries. It is related to "semiconductor devices sensitive to infra-red radiation" and "the conversion of the energy".

(c) H01L031/048-oriented category: According to Table 5, this topic existed in the first half of the period of time (type A2) and spread in the various countries (type B1). It was likely that the topic had been active during 2000 to 2004 and was out of the technical mainstream afterward. It was emphasized by several countries. It is about the "encapsulated or with housing".

(d) H01L031/052-oriented category: Referring to Table 5, this topic existed in the first half of the period of time (type A2) and spread in the various countries (type B1). It seemed that the topic had been popular during 2001 to 2003 and declined gradually. It is relating to the "cooling, light-reflecting or light- concentrating means".

(e) H01L021/20-oriented category: From the above Table 5, this topic existed in the first half of the period of time (type A2) and spread in the dominant countries (type B2). It was likely that the topic had been common during 2000 to 2005 and became minor afterward. It was focused mainly by the dominant countries JP (Japan) and US (United States). It is regarding the "deposition of semiconductor materials on a substrate".

(f) H01L031/0336-oriented category: According to Table 5, this topic existed in the first half of the period of time (type A2) and spread in the various countries (type B1). It was plausible that the topic had been active during 2000 to 2005 and became unimportant eventually. It was stressed by several countries. It is concerning the "different semiconductor regions, e.g. Cu2X/CdX hetero-junctions".

(g) H01L021/00-oriented category: Referring to Table 5, this topic existed in the second half of the period of time (type A3) and spread in the non-dominant countries (type B3). It seemed that the topic became popular lately from 2005 to 2009 and was contributed by the non-dominant countries as KR (Korea) and IT (Italy). It is relating to the manufacturing processes or devices of semiconductor.

(h) H01L031/04-oriented category: From the above Table 5, this topic existed in the second half of the period of time (type A3) and spread in the non-dominant countries (type B3). It was likely that the topic gained emphasis slowly from 2004 to 2009 and was participated by the non-dominant countries like KR (Korea) and TW (Taiwan). It is concerning the "adapted as conversion devices".

(i) H01L031/20-oriented category: According to Table 5, this topic existed randomly in the period of time (type A4) and spread in the dominant countries (type B2). It seemed that the topic was not in the technical mainstream and supported randomly the dominant countries as JP, US, DE (Germany) and NL (Netherlands). It is regarding the "devices comprising amorphous semiconductor material".

In addition, the significant H01L031/18-oriented category possesses 70 patents (about 44%), which reflects that a big portion of patents put efforts in the manufacturing process and device aspects. The dominant countries JP and US possess 70 and 52 patents (about 44% and 32%) respectively, which shows that these two countries held the powerful innovative ability and resources in this industry and can affect the technological directions strongly.

5 Conclusions

The research framework, based on IPC-based clustering and link analysis, for observing the technological directions on thin-film solar cell has been formed. The experiment was performed and the experimental results were obtained. The association diagrams and link diagram were generated. The basic technological directions were: H01L031/18, H01L031/00, H01L021/00, H01L021/20, H01L031/052, H01L031/048, H01L031/04, H01L031/0336, and H01L031/20. The specific topics which existed in the full period of time and developed continuously were: H01L031/18 and H01L031/00 categories. The specific topics which existed in the first half of the period of time and became active earlier were: H01L021/00, H01L021/20, H01L031/052, and H01L031/048 categories. The specific topics which existed in the second half of the period of time and became common lately were: H01L031/04 and H01L031/0336 categories. The dominant countries which possessed the powerful innovative ability and resources in the thin-film solar cell industry were Japan and United States. The above experimental results and findings would be helpful to the managers and stakeholders for their decision making on R&D aspects.

In the future work, the other aspects of company information (e.g., the public announcement, open product information, and financial reports) can be included so as to enhance the validity of research result. Additionally, the patent database can be expanded from USPTO to WIPO or TIPO in order to perform the technological direction observation on thin-film solar cell widely.

Acknowledgments. This research was supported by the National Science Council of the Republic of China under the Grants NSC 99-2410-H-156-014.

References

1. Blackman, M.: Provision of Patent Information: A National Patent Office Perspective. World Patent Information 17(2), 115–123 (1995)
2. Intellectual Property Office, Patent classifications, http://www.ipo.gov.uk/pro-types/pro-patent/p-class.htm (2011/03/15)
3. Floyd, S.W., Wolf, C.: Technology Strategy. In: Narayanan, V.K., O'Connor (eds.) Encyclopedia of Technology and Innovation Management, pp. 37–45. John Wiley & Sons, Chichester (2010)
4. Wikipedia, Strategic planning, http://en.wikipedia.org/wiki/Strategic_planning (2010/10/15)
5. Barney, J.B., Hesterly, W.S.: Strategic Management and Competitive Advantage: Concepts and Cases. Prentice-Hall, Englewood Cliffs (2010)
6. Tseng, Y., Lin, C., Lin, Y.: Text Mining Techniques for Patent Analysis. Information Processing and Management 43, 1216–1247 (2007)
7. Russell, S.: Technology Forecasting. In: Narayanan, V.K., O'Connor (eds.) Encyclopedia of Technology and Innovation Management, pp. 37–45. John Wiley & Sons, Chichester (2010)
8. USPTO (2010) USPTO: the United States Patent and Trademark Office, http://www.uspto.gov/ [2010/07/14]
9. EPO (2010) EPO: the European Patent Office, http://www.epo.org/ [2010/07/14]

10. TIPO (2010) TIPO: the Intellectual Property Office, http://www.tipo.gov.tw/ [2010/07/14]
11. Solarbuzz, Solar Cell Technologies, http://www.solarbuzz.com/technologies.htm (2010/10/20)
12. Wikipedia, Thin film solar cell, http://en.wikipedia.org/wiki/Thin_film_solar_cell (2010/10/20)
13. Jager-Waldau, A.: PV Status Report 2008: Research, Solar Cell Production and Market Implementation of Photovoltaics, JRC Scientific and Technical Reports (2010)
14. WIPO, Preface to the International Patent Classification (IPC), http://www.wipo.int/classifications/ipc/en/general/ preface.html (2010/10/30)
15. Sakata, J., Suzuki, K., Hosoya, J.: The analysis of research and development efficiency in Japanese companies in the field of fuel cells using patent data. R&D Management 39(3), 291–304 (2009)
16. Kang, I.S., Na, S.H., Kim, J., Lee, J.H.: Cluster-based Patent Retrieval. Information Processing & Management 43(5), 1173–1182 (2007)
17. Chen, Y.L. and Chiu, Y.T.: An IPC-based Vector Space Model for Patent Retrieval, Available online June (2010) (in press)
18. Tan, P.N., Steinbach, M., Kumar, V.: Introduction to Data Mining. Pearson Addison-Wesley (2006)
19. Ohsawa, Y., Benson, N.E., Yachida, M.: KeyGraph: Automatic Indexing by Co-Occurrence Graph Based on Building Construction Metaphor. In: Proceedings of the Advanced Digital Library Conference (IEEE ADL 1998), pp. 12–18 (1998)
20. Maimon, O., Rokach, L. (eds.): Data Mining and Knowledge Discovery Handbook, 2nd edn. Springer, Heidelberg (2010)
21. Freeman, L.C.: Centrality in Social Networks: Conceptual Clarification. Social Networks 1, 215–239 (1979)
22. Lee, D.L., Chuang, H., Seamons, K.: Document Ranking and the Vector-Space Model. IEEE Software 14(2), 67–75 (1997)
23. Stanford Natural Language Processing Group, Stanford Log-linear Part-Of-Speech Tagger, http://nlp.stanford.edu/software/tagger.shtml (2009/10/15)
24. Feldman, R., Sanger, J.: The Text Mining Handbook: Advanced Approaches in Analyzing Unstructured Data. Cambridge University Press, Cambridge (2007)
25. Hotho, A., Nürnberger, A., Paaß, G.: A brief survey of text mining. LDV Forum - GLDV Journal for Language Technology and Computational Linguistics 20(1), 19–62 (2005)

Using the Advertisement of Early Adopters' Innovativeness to Investigate the Majority Acceptance

Chao-Fu Hong[1], Tzu-Fu Chiu[1], Yuh-Chang Lin[1], Jer-Haur Lee[2], and Mu-Hua Lin[3]

[1] Aletheia University, Tamsui, New Taipei City, Taiwan
au4076@au.edu.tw, chiu@email.au.edu.tw, au1258@mail.au.edu.tw
[2] Yuan Ze University, Taoyuan, Taiwan
jerhaur.lee@gmail.com
[3] National Chengchi University, Taipei City, Taiwan
95356503@nccu.edu.tw

Abstract. It is important to find a way to across the gap of innovation and ease the process of innovation diffusion. A method for innovation diffusion is also important to launch a new product successfully to the market. According to the innovation diffusion model proposed by Rogers, early adopters hold the key points and features to across the innovation gap. Thus, this research intends to collect the information of usage values of innovative products written by early adopters via HCCS. A questionnaire is designed to investigate whether freshmen of university (i.e., the majority) would be convinced by the information from early adopters and have higher willingness to use the new product. This experimental result reveals that 14% of additional freshmen in the experimental group (comparing to those in the control group) agree with the values of technological features of netbook computers in entertainment. The result indicates that using the proposed method to assist a new product to market has a high possibility to across the innovation gap. This research also proved a new way to predict the acceptance rate of the majority to a new technology in the early stage of joining in the market. The prediction can be used to assist a company to enter the market earlier and to consolidate its position. It is a managerial contribution of this research.

Keywords: Innovativeness, Word-of-Mouth, Innovative Diffusion Model (IDM).

1 Introduction

According to Rogers, innovation diffusion means the diffusion of an innovative product through early adopters, early majority, and late majority to laggards. The acceptance scenario for innovative products is also influenced by the characteristics of these five types of people. The early adopters take on the traits of easily accepting innovative products and creating use value out of the innovative products. Early adopters are a group of practical people. Accordingly, Moore (1999) put forward an idea that there is a chasm existing between the early adopters and early majority, the chasm which stops the innovation from diffusion.

R. Katarzyniak et al. (Eds.): Semantic Methods, SCI 381, pp. 199–213.
springerlink.com   © Springer-Verlag Berlin Heidelberg 2011

Additionally, with the emergence of the Internet blogs, friends who both use Internet tend to interact with each other on the Internet blogs, exchanging and sharing experience and thought about purchasing with each other. According to BIGresearch, in its investigation in 2005, it says that for those consumers who tend to collect pre-information on the Internet before they buy things, 87% of them think that blogs are a very important sources of information for them. Therefore, the thought about their purchasing experience devoted by the bloggers accumulates increasingly on the Internet, which contributes to the so-called word-of-mouth (WOM).

With the recent and rapid development of the Internet, which provides people with rapid and convenient lifestyle, Internet has also been an important channel for consumers to search for related information before purchasing goods. Therefore, the information about the bloggers who share their thought of using certain products and their information about products is the one to be dependent on by consumers to make the most accurate purchasing decision and to save time for purchasing (Hennig-Thurau, Thorsten and Gianfranco Walsh, 2003).

This study is to use the creative early adoption and the creative use value (characteristics) of new products deriving from early adopters (Hirschman, 1980) (Hong, 2009). We modified these wisdom and value into new reflection articles. We tried to use experiments to test whether the new wisdom and value deriving from the early adopters' employment of the innovative products is able to convey to those browsers and exercise influence on them, who look for article messages via blogs, and to a certain extent, this will contribute to significant effect of positive commercial. The consumers' intention of buying can even be reinforced. This is to test whether the extracted value can help enterprises to form a force to cross the chasm and to influence consumers with a view to achieving the goal of innovation diffusion.

2   Literature Review

Related literary review can be carried out from the following two perceptions. One from the aspect of how to perceive the use value created by early adopters takes on the power to influence the public; the other one is how to test and justify the effect of the use value created by the early adopters on the promotion of commercial.

2.1   How to Perceive the Use Value Created by Early Adopters Taking on the Power to Influence the Public

The Early Adopters
What is the definition of early adopters? Ram and Jung (1994) pointed that the early adopters had significantly higher use innovativeness and product involvement as compared to the early majority. Furthermore, early adopters may possess the specialized skills and abilities required for using the product in a wide variety of ways (Price & Ridgway, 1982). Besides, if a product is quite innovative and discontinuous, only the creative consumers may discover the additional uses, too. (Robertson, 1971; Venkatesh & Vitalari, 1986). According to the experiment of Morrison et al. (2000), we learn that there are about 26% of users who has "leading edge status" and their in-house technical capabilities, and many innovating users freely share their innovations with others.

Word-of Mouth Transmitting and Blogs
The concept of word-of-mouth communication derives from Lazarsfeld's second-level communication theory. He found that the behaviors of voters are affected more by interpersonal communication than by mass media. In other words, idea leaders make use of the power of owning information and transmitting information to influence others (Webster, 1970). In the study of marketing and consumers, Sheth (1971) also agreed that word-of-mouth communication has three research orientations: second-level communication, innovation transmission, and reference group, which corresponds to the idea that early adopters' word-of-mouth transmission has important influence on innovation diffusion.

Furthermore, with the development of network technology, blogs provide people with social interaction, informal discussion, and chance of interlocution (Kavanaugh et al., 2006). According to Marketing Week, it pointed out that in terms of the use of internet, the reliability of messages from blogs is next to newspapers. Nearly 24% of the respondents perceive that blogs are the most trustworthy sources of information (Brown, Broderick & Lee, 2007). Blogs are also considered one of the important communication media (Bickart and Schindler, 2001). Therefore, by the use of the Internet information from blogs to acquire early word-of-mouth information of the products, to realize the products value created by the early adopters, and to identify the social status of the early adopters (Lin, 2001) will become a chance to identify whether innovative products can cross the chasm.

Visualizing the Community: Data Mining and Knowledge Extract
After collecting relative data, data mining technology was employed to do analysis. For example, the algorithm of Apriori is used. It is an algorithm to utilize traditional consuming database to mine association rules. By satisfying the criteria with minimum support and confidence, we obtained rules of consuming behaviors. These rules can be employed not only to realize consuming patterns but also to infer meaningful consuming behaviors (Agrawal, Imielinski, & swami, 1993). Those low-frequency data which fail to pass the criteria tend to be ignored. However, not all low-frequency data are not without its value. For example, although some high-priced products (diamonds, for example) with low frequency to be purchased, they tend to bring business sectors high profits when they are purchased. There are still other products with low frequency to be purchased; however, when one item is purchased another one tends to be purchased accordingly (for example, purchasing a food processor and at the same time, buying a cooking pan). These data cannot be taken as noisy data. In order to prevent the data with lower frequency from being sifted out, Liu, Hsu and Ma (1999) developed a rare association analysis with the data carrying the characteristics of low support and high confidence. Although the problem with low frequency is solved, it fails to link with the diffusion route.

Grounded Theory
Grounded theory is based on three coding processes to formulate a new theory which has its own persuasion. The first step is open coding. This is a step to scrutinize the data collected, find out some phenomena, and give names to certain phenomena. This is an analysis work for the classification. The data collected are mainly texts, which are segmented, investigated, conceptualized, and domain-categorized. The second

step is axial coding. In this step, researchers are based on one coding paradigm, which is the strategies and results that are obtained from qualifications, contexts, actions, and interaction used to analyze the phenomena. Then, the researchers connect the result of various domains analyzed in the previous stage with the one acquired in this stage. The third step is selective coding. This is a step to select a core category from the categories induced in previous two stages, and then to connect it systematically with other categories and test the relationship between among them. This is a process to supplement the categories which are not completely conceptualized. The final integration stage is to conduct a theoretical sampling. Based on the new theory derived from verified and evolving theories which show relativity to do the sampling, some questions and comparison tend to emerge from the theoretical sampling and analysis. These questions and comparison help researchers to find and link relative categories, and the quality and dimensions related to its categories. If we base on grounded theory to analyze the early adopters' word of mouth and the data about the early majority behaviors, we have chance to know the feasibility of innovation diffusion. However, to complete whole process requires a huge amount of man force. Therefore, how to integrate the problems mentioned above and to develop some workable information technology will become an important research issue.

Hong (2009) proposed a qualitative chance discovery model (QCD) combining with the grounded theory. It is a model to segment research issues, to reduce the complexity brought by Chance Discovery, and to solve the problems with the Grounded Theory which is in the absence of assistance from information technology. Although it is called a qualitative analysis approach, the research development itself still tend to be algorithm-oriented, which is in the absence of defining research issues and ways beforehand.

2.2 How to Test and Justify the Effect of the Use Value Created by the Early Adopters on the Promotion of Commercial

Commercial Effect

Commercial effect, according to Rosenberg and Hovland (1960), can be analyzed through the Tricomponent Attitude Model, in which attitude is composed of three dimensions: (1) cognitive component, (2) affective component, (3) conative component. Cognitive component indicates personal experience, thought, and point of view on value and the knowledge and belief on products through the integration of information from different aspects. Affective component refers to the emotion and feeling toward a certain brand or product. For example, good/bad, with texture/without texture, availability of use/unavailability of use. Conative component means that by the integration of consumers' cognition and affection towards products, consumers are supposed to generate an orientation or intention to purchase the products. This study is to use experimental texts to be browsed by consumers with a view to adding the wisdom and value of the early adopters to the texts, and to a certain extent, the positive commercial effect can be observed among consumers.

Task-Technology Fit Model

Task-Technology Fit (TTF) is a model which was put forward by Goodhue and Thompson in 1995. From the angel of task orientation, this study is to research on the users' idea about the degree of the fit between technology use and task characteristics. Combining the performance of technology and working effect, we find that the higher the fit, the higher the use intention (Goodhue and Thompson, 1995). The theoretical framework of TTF is listed below.

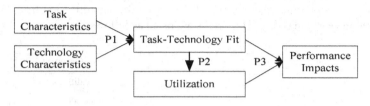

Fig. 1. A theoretical framework of TTF (Goodhue & Thompson, 1995).

Therefore, this study is to use the texts written by early adopters, including the description of the early adopters' idea about personal tasks, the technological characteristics of netbooks, and the degree of TTF. We even employ the description of the netbooks by task orientation to meet the need so as to achieve one's task. When the new value and wisdom extracted from the employment of these new products, a piece of commercial covering the early adopters tends to be obtained. We then employ TTF, which make it possible for us to see whether the value created by early adopters can promote the commercial effect.

In the next section, this study is going to adopt the analysis concept derived from QCD, combining with the sociology concepts, to construct a so-called social diffusion scenario and to define research issues. Adding the data from the cluster of the early adopters, we may get an initial word-of-mouth information network which derived from the social network diffusion of the traditional word-of-mouth marketing (Engel, Kegerreis, & Blackwell, 1969; Herr, Kardes, & Kim, 1991). This way, we may identify and extract the word-of-mouth message which we want to convey, and to a certain extent, through the expert's comparison of different word-of-mouth messages, we may see the feasibility of Internet diffusion. Additionally, by means of experiments, this study utilized TIF to test and verify whether the innovative use value created by early adopters can be conveyed to consumers via Internet texts, which contributes to the generation of influence on consumers.

3   Research Methods

This research has two purposes: (1) How to perceive the creative value deriving from early adopters, (2) To test whether early adopters' value is capable of promoting the intention of acceptance by the majority. The details are delineated as follows:

3.1   Human-Centered Computing System

Whether the innovative products can diffuse in the society is a life-or-death issue for an industry. The major problem lies in the chasm existing between the early adopters and the early majority (Moore, 1999), the chasm which prevents the innovative products from being accepted by the early majority. Therefore, how to extract the use value (word of mouth) created by the early adopters and how to make sure whether this diffusion route can be available in making the early majority accept innovative products is an important issue in assessing whether the innovative products are capable of crossing the chasm. Therefore, this study put forward a human-centered computing system to integrate the data of the word-of-mouth diffusion from the early adopters to the early majority. The details are listed as follows.

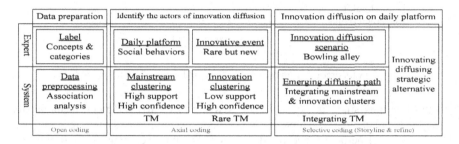

Fig. 2. A theoretical framework of TTF (Goodhue & Thompson, 1995).

Phase 1: Preparation for Data and Initial Analysis
From the point of view of word of mouth transmission, some researchers thought about how the innovative products diffuse from the cluster of the early adopters to that of the early majority (Engel, Kegerreis, & Blackwell, 1969; Herr, Kardes, & Kim, 1991). Therefore, for the purpose of creating an innovation diffusion scenario, the expert first of all collects relevant data to conduct a preprocessing on the data and then combines the Grounded Theory with text mining to process the data. The detailed process is listed as follows.

Step 1: Data preprocess
1-1-1H) The expert defines the domain and relevant key terms on which he intends to study.
1-1-1C) To sift out the data, which correspond to the key term, from the Internet, and the data which analyze the time.
1-1-2H) Based on his domain knowledge, the expert interprets the texts, and at the same time, he segments texts into terms, removes any meaningless terms, and marks the segmented terms with conceptual labels.

The system processes the first-stage open coding analysis, which contributes to the visualization by using data association analysis. This visual image helps the expert to focus on the extractable various theme values which are obtained from the data analysis.

Step 2: Terms co-occurrence analysis (open coding)
1-2-1C) To use equation 1 to calculate terms frequency and association value.

$$assoc(w_i, w_j) = \sum_{s \in D} \min(|w_i|_s, |w_j|_s) \tag{1}$$

where i and j is ith term and jth term; s is stands for sentence; and D is all textual data

1-2-2C) To visualize the analysis result, which contributes to an association diagram drawn by the co-occurrence terms.

1-2-1H) The expert identifies the concepts and categories derived from the co-occurrence association diagram, which helps the expert to preliminarily realize the various theme values presented in the data.

Phase 2: To define the various roles and linking storylines in the process of innovation diffusion (axial coding and selective coding)

Based on the first-stage data analysis in which the social association concept derived from mainstream market clusters, and the social categories by identifying the mainstream market clusters, the expert develops various types of mainstream market clusters and innovative products clusters derived from innovators (axial coding). Then the expert draws a storyline which links mainstream market clusters with innovators' innovative products clusters to make the emergence of innovation diffusion scenario possible (selective coding).

Step 1: To develop innovative products clusters (Axial coding)

In the last stage, the expert has preliminarily identified the number of mainstream products which has currently existed in the social context. In this step, the expert intends to create a single mainstream product network scenario. Based on the characteristics of "rareness" and "early period", the expert extracts the data about the innovative products to create a single Internet scenario for innovative products. The process is illustrated as follows.

2-1-1H) The expert needs to decide what innovative products ($w_{innovation}$) are.

2-1-1C) Circle the products from the first to the last one.

2-1-2C) Products ($w_{innovation}$) are used to sift out the rare but meaningful sentences.

This variable is used to confirm the research topics and to remove irrelevant sentences with a view to narrowing down the data range and to sift out valid sentences. That is, valid sentences must include terms related to products $w_{innovation}$, which is shown as equation 2.

$$if\,((w_1, w_2, \ldots)_s \cap (w_{innovation})) \neq \phi \quad then$$
$$sentence \ \in \ Valide \ \ sentences \ \ set \tag{2}$$

2-1-3C) To integrate all the valid sentences related to product terms, and create integrated association diagrams, then go to 2-1-1C.

2-1-2H) Based on his relevant domain knowledge, the expert confirms both the mainstream products scenario and the innovative products scenario.

Step 2: Linking Crossing-chasm Storyline with Innovative Products and the Innovation Diffusion Scenario Emerging from Mainstream Market Cluster (Selective Coding)

The expert applies the storyline of crossing the chasm to link two clusters in which the innovation diffusion scenario emerges. Its process is listed as follows.

2-2-1H) The expert makes use of the storyline of early adopters' innovative use values by the innovative products are defined.

2-2-1C) We cycle from the first product till the last product to develop the innovative market.

2-2-2C) Product ($w_{innovation}$) is used to collect the rare but meaningful sentences. This variable is used to confirm the research theme and to remove irrelevant sentences with a view to narrowing down the data range and to sift out effective sentences. That is, effective sentences must include terms with products $w_{innovation}$ shown as equation 3:

$$if\,((w_1, w_2, ...)_s \cap (w_{ma\,int\,ech})) \neq \phi \quad then$$
$$sentence \ \in \ Valid \ \ sentences \ \ set \ \ then \tag{3}$$
$$go \quad to \quad 2-2-1C$$

2-2-3C) To integrate the effective sentences concerning innovative products to create an integrated association diagram.

2-2-2H) Based on his domain knowledge, the expert confirms the diffusion scenario about the innovative products to cross the chasm.

3.2 A Research Framework of Using TTF to Investigate the Public's Intention of Acceptance

The study makes use of the texts written by early adopters. It includes the early adopters' description of personal task, technological characteristics of netbooks, and the degree of TTF, to a certain extent, to use task orientation to describe how netbooks satisfy their need to achieve their task. We employ HCCS mentioned above to extract the new value and wisdom of these products so as to form a piece of commercial which contains the use value created by early adopters. Then, we adopt TTF and via laboratory experimental method to test and verify the feasibility of this experimental model. Its framework is elaborated below:

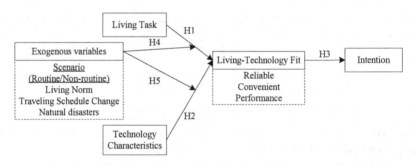

Fig. 3. A theoretical framework of TTF (Goodhue & Thompson, 1995).

Related Research Hypothesis

The premise assumption of TTF is that the users themselves tend to base on the task characteristics and technology to finish their job. Among a group of undertakers, this kind of phenomenon will lead to different degree of fit. Additionally, this study adds early adopters' value to blogs with a view to using their use wisdom to conduct a commercial operation on the public, and to a certain extent, the public's expectation toward the degree of the TTF can be elevated. Therefore, the research proposed the following hypotheses:

H1: The life characteristics have positive effect on the life TTF.
H2: The technology function has positive effect on the life TTF.
H3: The life TTF has positive effect on the intention of use.
H4: External variables have interfering effect on the life characteristics and the life TTF.
H5: External variables have interfering effect on the technology function and the life TTF.

Data Collection

After the observation of netbooks circulation messages in the market, we find that there is time difference of the transmitting speed on the Internet between products and users' utilizing information. This study focuses on the target innovative product: ASUS EeePC901. Therefore, we collected its blogs data for six months after it was introduced to the market to conduct the analysis. The time period for collecting the given data is from 1 October 2008 to 31 October 2008. Using Google blogs (http://blogsearch.google.com/blogsearch) and giving key terms, netbook + thought, to search for the data, we obtained 63 related data from blogs articles.

In addition, explain that purchase innovativeness and use innovativeness tend to affect the majority's acceptance of the new products. Therefore, after carefully reading the collected data, removing the text box without innovativeness and the distracting articles, there were 7 articles with use innovativeness. Then, value extraction was executed from these 7 articles. Based on the human-centered computing system proposed by this study, we integrate human intelligence with text mining technology.

Research Design

We extracted the co-generated new wisdom and value from these early adopters and added them to the texts, which makes them different from those in the control group (the difference between the texts with or without new wisdom and value deriving from the early adopters). At the same time, the major group of using blogs ranges in age from 15 to 24. Therefore, this study chooses university students who are supposed to be familiar with the operation of Internet as the experimental subjects. On the other hand, the research background is mainly built on the interaction of the Internet, so we employ the experimental interface for the test-takers to browse the texts and answer the questionnaire.

4 Experimental Results and Discussion

4.1 Human-Centered Computing System: Early Adopters' Use of Innovative Scenario

From Fig. 4, experts realize netbook computers have special functions of real time communication and portability. The early adopters of netbook computers have more innovative performance in mobile information, entertainment and social communication. Besides, from long distance movement, experts also see the need of real time scenario, so long distance movement is set to be the core concept and related document/sentence is selected. Some interesting netbook applications have drawn researchers' attention such as real time social talk, real time mind writing and real time web page browsing. According to the above procedure, real time axial chart is constructed shown as Fig. 4.

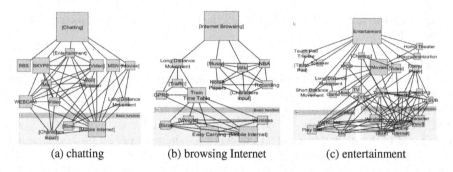

(a) chatting   (b) browsing Internet   (c) entertainment

Fig. 4. Axial coding.

Constructing Storyline

Expert recognizes how the early adopters use the netbook's portability and wireless Internet access to create their daily innovative value. After analysis, expert focuses on long distance movement, such as the travel, and then he tries to integrate some of the all innovative values into a bigger innovative value, such as storyline in real life long distance movement. Experts construct three story scenarios: long distance movement with chatting, long distance movement with web page access and long distance movement with entertainment. They are described as follows:

Scenario of Long Distance Movement with Chatting
From its outlook there is nothing different, but more tools such as Skype, BBS are added. Meanwhile, webcam and microphone are treated as standard equipment. Actually, Skype alone can provide online phone or video phone, the communication capability is enhanced largely. The users of netbook computers always can reach out and touch someone no matter where they are.

Scenario of Long Distance Movement with Web Page Access
From its outlook there is nothing different, but web page browsing capability enables the users gets real time information no matter where on earth they travel. They never have to worry about maintaining their normal life.

Scenario of Long Distance Movement with Entertainment
From its outlook there is nothing different, but as far as the entertainment is concerned, TV programs, online music, online films and online games are added. This kind of function makes the life of long-distance traveler easier and more colorful.

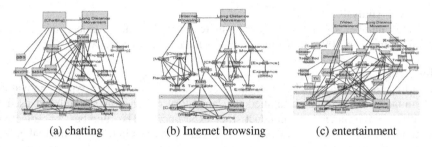

| (a) chatting | (b) Internet browsing | (c) entertainment |

Fig. 5. Three kinds of long-distance real-time movement.

4.2 Modifying TTF Questionnaire

Previous summary discussing results represents that our HCCS is a useful system. For even during travelling, the customers can communicate real time with their family and friends to share their travel life just as they do at home. Of course, he can enjoy the videos and audios entertainment to forget lonely travel, too. Finally, when meeting the unexpected event, they can real time and easily change their travel program by accessing the train time table.

As the experiment results above, an early adopter brings numerous creativities of using a netbook computer during his/her traveling or long-commuting time. A freshman might have a more colorful lifestyle comparing to the regular lifestyle in high school while he/she enters university. While there is a big change for a freshman in his/her life, it is interesting to observe whether he/she is affected by an early adopter of netbook more easily to be willing to use a netbook. This research collected and analyzed the articles in blogs related to creative usages of a netbook computer written by early adopters to discover possible strategies for encouraging a freshman to use a net book computer. It intends to evaluate if the colorful life of a university freshman would increase the effectiveness of using a netbook computer. This research modified the questions in the questionnaire of TTF to fit the experiment of investigating the fitness of a freshman to a netbook computer.

4.3 Experiment Process of TTF

In the experiment, it was planned to use the convenient sampling method to find freshmen in a university in northern Taiwan, and then randomly divided these experimenters into experimental group and control group evenly.

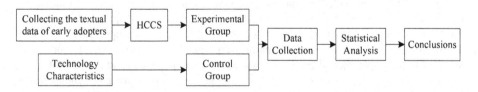

Fig. 6. Three kinds of long-distance real-time movement.

Statistical Analysis
The total number of returned questionnaires was 145, the number of effective questionnaires was 109, and the rest 36 questionnaires are ineffective.

Clustering Analysis
According to the Bass model, consumers can be divided into two types: innovators and imitators. This research applied the laboratory experiment method and used a controlled environment to conduct the experiment. The number of effective questionnaires was 109; 54 of them from the experimenters in the experimental group and 55 of them from the experimenters in the control group. Applying the two-step clustering algorithm on the effective questionnaire data, it suggested using two clusters for grouping data. Accordingly, the K-means clustering algorithm was then applying to the effective questionnaire data from the experimental group and control group, and k was set to be two. The result is shown as below:

Table 1. The clustering results

	Experimental Group	Control Group
Accepted	36 (67%)	29 (53%)
Not Accepted	18 (33%)	26 (47%)

The result of the separating clusters from the experimental group and control group fits the result (i.e., grouping data into two clusters) obtained from the Bass model. Using the blogs with traditional functions and regular introductions to stimulate experimenters in control group, 29 experimenters (53%) were accepted and 26 experimenters (47%) were not accepted. The accepted group can be viewed as innovators or early majority, and the non-accepted group can be viewed as late majority or laggards in Rogers' innovation diffusion model. On the other hand, the blogs with innovative usages written by early adopters were used to stimulate experimenters in experimental group. The accepted group contains 36 experimenters (67%) and the non-accepted group consists of 18 experimenters (33%). This result reveals that using early adopters' innovative usage values of netbook computer to stimulate participants brings 14% of additional affected majority. Explaining this result on the basis of Rogers' innovation diffusion model, half of the late majority was able to be convinced by the extra values created from the innovative usage.

Difference Analysis
The major difference between the experimental group and the control group is the contents of blogs provided by the system. One contains the information innovative usages of netbook computers, but one doesn't. This research applied Independent Samples T Test to examine the difference between these two groups. Only six out of 23 questions (26%) in the questionnaire were significant. However, 18 out of 54 samples in the experimental group are non-accepted population. The answers of these 18 participants should not be included in the analysis. Conducting the Independent Samples T Test using 36 accepted samples in experimental group and the samples in the control group, 14 out of 23 questions (60%) in the questionnaire are significant. The details of the result are shown in the following figure.

According to the experimental result, the experimenters (i.e., freshmen) have already realized the convenient features of wireless and enduring battery for the netbook computers. They are also aware the lightness of weight and the thinness of size. They usually use netbook computers for entertainment purposes. Hence, the fitness of technology to entertainment in the experiment is the most significant. High fitness draws higher intention of purchasing.

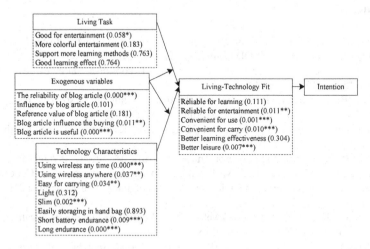

Fig. 7. Result of Independent Sample T Test.

Additionally, innovative and creative usages provided by early adopters can stimulate 14% of freshmen as the so-called late majority to agree with the values of technological features of netbook computers in entertainment. This means that only a few information of early adopters' innovative usage can bring a good result. This research also proved that if the values of innovative usage from early adopters can be discovered in the early stage of new product adoption, the acceptance of a new product or a new technology for the majority can be predicted in the early stage via the proposed method. The prediction can be used to assist a company to enter the market earlier. It is a managerial contribution of this research.

5   Conclusions

This research collected the using experiments of netbook computers written by early adopters on the internet. The collected information was then utilized to discover the values of early adopters of innovative usages of netbook computers using HCCS system. This research applied the laboratory experiment method to the collected data to investigate the affection of early adopters' ideas to the freshmen in the university. According to the experimental results, it was pointed out that 18% of additional participants in the experimental group comparing to the control group agreed the usage values of technological features of netbook computers in entertainment. It proved the usability of the proposed method. This research also proved that if the values of innovative usage from early adopters can be discovered in the early stage of new product adoption, the acceptance of a new product or a new technology for the majority can be predicted in the early stage via the proposed method. The prediction can be used to assist a company to enter the market of a new product or technology earlier. It is a managerial contribution of this research.

References

1. Agrawal, R., Imielinski, T., Swami, A.N.: Mining Association Rules between Sets of Items in Large Databases. In: SIGMOD Conference, pp. 207–216 (1993)
2. Bass, F.M.: A new product growth for model consumer durables. Management Science 15(5), 215–227 (1969)
3. Bickart, B., Schindler, R.M.: Internet forums as influential sources of consumer information. Journal of Interactive Marketing 15(3), 31–40 (2001)
4. BIGresearch. Executive Briefing – Confidence in the Economy Declines for the Third Straight Month (2010/08/02),
   http://www.bigresearch.com/news/EBJune06.htm (2006/06/16)
5. Brown, J., Broderick, A.J., Lee, N.: Extending Social Network Theory to Conceptualise On-Line Word-of-Mouth Communication. Journal of Interactive Marketing 21(3), 2–19 (2007)
6. Engel, J.F., Kegerreis, R.J., Blackwell, R.D.: Word-of-mouth communication by the innovator. Journal of Marketing 33(3), 15–19 (1969)
7. Goodhue, D.L., Thompson, R.L.: Task-Technology Fit and Individual Performance. MIS Quarterly 19(2), 213–236 (1995)
8. Hennig-Thurau, T., Walsh, G.: Electronic Word-of-Mouth: Motives for and Consequences of Reading Customer Articulations on the Internet. In: International Journal of Electronic Commerce, vol. 8(2), pp. 51–74 (Winter 2003)
9. Herr, P.M., Kardes, F.R., Kim, J.: Effects of word-of-mouth and product-attribute information on persuasion: an accessibility-diagnosticity perspective. The Journal of Consumer Research 17(4), 454–462 (1991)
10. Hirschman, E.C.: Innovativeness, Novelty Seeking, and Consumer Creativity. The Journal of Consumer Research 7(3), 283–295 (1980)
11. Holak, S.L.: Determinants of innovative durables adoption an empirical study with implications for early product screening. Journal of Product Innovation Management 5(1), 50–69 (1988)
12. Hong, C.-F.: Qualitative Chance Discovery - Extracting Competitive Advantages. Information Sciences 179(11), 1570–1583 (2009)
13. Kavanaugh, A.L., Zin, T.T., Carroll, J.M., Schmitz, J., Perez-Quinones, M., Isenhour, P.: When opinion leaders blog: new forms of citizen interaction. In: Proceedings of the 2006 International Conference on Digital Government Research, pp. 79–88 (2006)
14. Lin, N.: Social capital: a theory of social structure and action. Cambridge University Press, New York (2001)
15. Liu, B., Hsu, W., Ma, Y.: Mining Association Rules with Multiple Minimum Supports. In: Proceedings of the 1999 International Conference on Knowledge Discovery and Data Mining, pp. 337–341 (1999)
16. Moore, G.A.: Inside the tornado: marketing strategies from silicon valley's cutting edge. Harper Business, New Youk (1999)

17. Morrison, P.D., Roberts, J.H., von Hippel, E.: Determinants of User Innovation Sharing in a Local Market. Management Science 46(12), 1513–1527 (2000)
18. Price, L.L., Ridgway, N.M.: Use Innovativeness, Vicarious Exploration and Purchase Exploration: Three Facets of Consumer Varied Behavior. In: Walker, B. (ed.) AMA Educator's Conference Proceedings, pp. 56–60 (1982)
19. Ram, S., Jung, H.S.: Innovativeness in Product Usage: A Comparison of Early Adopters and Early Majority. Psychology & Marketing 11(1), 57–67 (1994)
20. Robertson, T.S.: Innovative behavior and communication. Holt, Rinehart and Winston, Inc., New York (1971)
21. Rogers, E.M.: Diffusion of innovations, 5th edn. Free Press, New York (2003)
22. Rosenberg, M.J., Hovland, C.I.: Cognitive, affective, and behavioural components of attitudes. In: Hovland, C.I., Rosenberg, M.J. (eds.) Attitude Organisation and Change: An Analysis of Consistency Among Attitude Components, pp. 1–14. Yale University Press, New Haven (1960)
23. Sheth, J.N.: Word-of-mouth in low-risk innovations. Journal of Advertising Research 11(3), 15–18 (1971)
24. Venkatesh, A., Vitalari, N.P.: Computing technology for the home: Product strategies for the next generation. Journal of Product Innovation Management 3, 171–186 (1986)
25. Webster Jr., F.E.: Informal Communication in Industrial Markets. Journal of Marketing Research 7, 186–189 (1970)

The Chance for Crossing Chasm: Constructing the Bowling Alley

Chao-Fu Hong[1], Yuh-Chang Lin[1],
Mu-Hua Lin[2], Woo-Tsong Lin[2], and Hsiao-Fang-Yang[2]

[1] Aletheia University, Tamsui, New Taipei City, Taiwan
au4076@au.edu.tw, au1258@mail.au.edu.tw
[2] National Chengchi University, Taipei City, Taiwan
95356503@nccu.edu.tw, lin@mis.nccu.edu.tw,
hfyang.wang@gmail.com

Abstract. It is widely accepted that the research in innovation diffusion can be divided into two categories. One is to cross the chasm and to carry out the diffusion for innovative products. The other one is to do the diffusion forecast for the innovative products. However, crossing the chasm is the most important issue for innovation diffusion. Accordingly, this research put forward the following logic thinking with an intention to solve the problem of crossing the chasm: When thinking about how to construct a bowling alley in the chasm, we need first of all to make sure whether the bowling pins still exist. Then we move on to observe the gap difference between the two ends of the bowling alley and try to fill the gap difference in order to make the innovation diffusion go smoothly. Under this kind of thinking pattern, this study conducts a case study by adopting a clustering method with multi kinds of clusters in it to identify the bowling pins: First of all, we divide the TAM testing collected samples into two clusters and three clusters. We can observe a newly-formed third cluster which is composed of two sub-clusters. This sub-cluster is made up of partial members from the two-cluster division. These two sub-clusters exist separately at both ends of the chasm. They can also be merged as a new cluster. Therefore, we made an assumption that these two sub-clusters may help to build a bowling alley to cross the chasm. We find that the sub-clusters (the early adopters) derived from the public are exactly the bowling pins. They are the target cluster for innovation diffusion. Moreover, we use HCCS to analyze the difference between two sub-clusters which are derived from the third cluster. We further analyze the innovative use value created by the early adopters in the blogs. Through experts' brainstorming meeting, we finish organizing a remedial scheme for the difference resulting from these two sub-clusters. At the end, we proceed again to observe the degree of acceptance by the target cluster. The experimental result confirms that the remedial scheme takes on the capacity to influence the early majority in crossing the chasm, which also proves the effectiveness of the research method.

Keywords: Chasm, Bowling alley, Innovative Diffusion Model (IDM).

R. Katarzyniak et al. (Eds.): Semantic Methods, SCI 381, pp. 215–229.
springerlink.com   © Springer-Verlag Berlin Heidelberg 2011

1 Introduction

In Rogers' innovation diffusion model, he stresses that when the innovative products emerge, there is about 20% of the early adopters who would accept new products, for they are quicker and easier to accept new products (Rogers 2003). The early adopters are more capable to develop the value of use innovativeness from the new products, so can the lead users affect the early majority of customers to accept new products (Hippel 1986; Morrison 2000). Meanwhile, Moore (1999) proposed that there was a huge gap, also called chasm, between early adopters and early majority. If we fail to distinguish early adopters (as well as their value) from the whole population, innovation diffusion can be impeded.

Furthermore, another issue is who are the members of the target cluster of innovation diffusion and which cluster is the nearest one from the early adopters. Based on the information of this nearest cluster, the chasm of innovation diffusion is likely to be overcome and resolved. Therefore, this study firstly converts the value of use innovativeness of early adopters into a TAM (Technology Acceptance Model) type questionnaire (Davis 1989) to survey the general customers and then obtains the clusters upon the majority's degree of accepting innovative products using the k-mean clustering. By observing these clusters, we will try to find out the bowling alley. In addition, the clusters are formed by the cohesiveness of behavioral features of the comprising members. According to the analysis of the behavioral features of members of a specific group, the characteristics of a cluster will be obtained and the differences between the early adopters and individual members of clusters can be derived. Consequently, the target cluster of innovation diffusion can be recognized and the characteristics of its members would also be identified, so that a possible alternative of bowling alley may be constructed to check whether it is capable to cross the chasm between early adopters and early majority.

This study employs information technology to collect the textual data related to innovative products in the virtual community. A human-centered computing system, which is based on experts' recognition and concept of the textual data, is proposed to use experts' implicit knowledge (value-driven) to activate information technology so as to integrate data (data-driven) with a view to showing the use value created by early adopters in the virtual community. Nowadays, notebook computers are popular with university students and become one of their necessary learning tools. With the formation of the network of Taipei Mass Rapid Transit System, commuting university students are increasing, and the time for commuting is gradually prolonged. To most university students, aside from learning, entertainment and social life are two indispensible parts in terms of their daily routines. Therefore, this study aims at a more convenient product - netbook, as a study object. In this study, we make use of its lightness and being durable in providing immediate use to examine whether the proposed alternative of bowling alley can cross the chasm to bring the innovative products to the early majority customers.

2 Literature Review

2.1 Innovative Diffusion Model (IDM) and Chasm

Innovation diffusion model (Rogers, 2003) illustrates that innovative products are diffused by inventors' invention, subjective adoption by early adopters and creating new service value, passing on the new value to early majority and making them to accept the products, and eventually resulting in a pressure of following majority which helps to lead the late majority and laggards to accept new products. The result resembles the chasm between early adopters and early majority proposed by Moore (1999). Moore pointed out that if innovative products fail to satisfy the criteria to cross the chasm, the product will not become a popular product accepted by the majority, which indicates that early adopters are important figures in terms of crossing the chasm. However, contrary to the aforementioned argument, Bass applied the variation of sales data and conducted a time series variable regression research. Bass (1969) focused on forecasting the number of the acceptance of innovative products in the next time point. Whether it can cross the chasm or not is not the main concern of this research.

The research on the innovation diffusion can be divided into two aspects. The first one is about how to cross the chasm and carry out the diffusion for innovative products. The second one is the diffusion forecasting for innovative products. Because this study focuses on the first aspect, in next chapter, we are going to explore the characteristics of early adopters and how early adopters select objects of diffusion.

2.2 Early Adopters

However, before we explore the issue about crossing the chasm, first of all, we need to know what is the definition of early adopters? Ram and Jung (1994) pointed out that the early adopters had significantly higher use innovativeness and product involvement when compared to the early majority.

Therefore, the early adopters are able to create innovative value to lead the popularity. When exploring these two clusters' behaviors (early adopters and early majority) of the acceptance of innovative products, aside from taking on the characteristics of the early adopters stated above, we find that they still show us another characteristic of doing word-of-mouth network transmitting which is an important factor in deciding whether crossing the chasm is possible or not. Next, we are going to explore the word-of-mouth transmitting network like the bowling alley to crossing the chasm.

2.3 The Bowling Alley Which Crossing the Chasm: A Phenomenon of Diffusion

Egmond et al. (2006) further explored the innovation diffusion from the perspectives of mainstream and early markets. Namely, innovative products first reach the early market actors who are driven by a visionary attitude. Hereafter, the products must reach the much larger mainstream market whose actors are driven by a pragmatic attitude. Many innovations, however, only appeal to the visionary early market and do not meet the more pragmatic needs of the mainstream. For that reason, innovations often fall in a "chasm" between the early market and the mainstream and fail to reach the mainstream. That is the reason why Egmond et al. (2006) try to avoid falling into chasm. They take the niche market as the early market. Then they search for the target cluster from the mainstream market, and at the same time, they inquire the target cluster for their view on niche. All the efforts they made are trying to build a bowling alley so as to confirm whether the niche market can cross the chasm.

2.4 The Predicaments Which Innovation Diffusion Is Facing

Although Moore stated that crossing the chasm should be based on the existence of the blowing alley, aside from Egmond et al.'s (2006) assessment of the relationship between niche market and mainstream market to confirm that whether the bowling alley between niche market and mainstream market may help cross the chasm, there was not any scholar who proposed methods to construct the bowling alley. This is because when we want to construct the bowling alley, we should face the following questions. For example, if early adopters do not know where the early majority is or where the early majority who is nearest to them is, it is hard to assess how to construct the blowing alley. In addition to the problems mentioned above about the confirmation of those clusters to be diffused, the correlation difference of the behavioral characteristics between early adopters and early majority is also an important factor to affect whether innovative products can step on the bowling alley and cross the chasm. However, up to now, little attention has been paid to these two problems of innovation diffusion. Therefore, this study aims at the aforementioned problems to propose a methodology as a solution.

3   Research Methods

Simon proposed a four-step decision making process to solve problems: intelligence, design, choice, and review activities (Simon, 1955). Among them, the intelligence activity which searches the environment for conditions calling for decision is the most essential one. In reality, if the experts can observe the environment and identify the problem clearly, it would be easier to get good result. This study employs grounded theory and text mining techniques to construct a Human-Centered Computing System, so as to collect and analyze the articles of early adopters to obtain their value of use innovativeness. Subsequently, the TAM type questionnaire will be modified and tested (i.e., the Design activity). Lastly, an appropriate one will be selected after analyzing the different clusters (i.e., the Choice activity). Therefore, this study will identify the diffusible target cluster and then assist the firms to design and evaluate the alternatives in order to cross the chasm. The research flowchart is shown as follows (in Fig. 1).

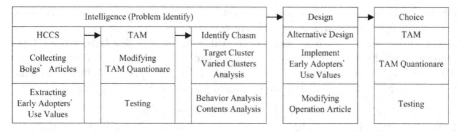

Intelligence (Problem Identify)			Design	Choice
HCCS	TAM	Identify Chasm	Alternative Design	TAM
Collecting Bolgs' Articles	Modifying TAM Quantionare	Target Cluster Varied Clusters Analysis	Implement Early Adopters' Use Values	TAM Quantionare
Extracting Early Adopters' Use Values	Testing	Behavior Analysis Contents Analysis	Modifying Operation Article	Testing

Fig. 1. Research flowchart

3.1 Intelligence

Human Centered Computing System (HCCS)

Phase 1: Preparation for Data and Initial Analysis
From the point of view of word of mouth transmission, some researchers thought about how the innovative products diffuse from the cluster of the early adopters to that of the early majority (Engel, Kegerreis, & Blackwell, 1969; Herr, Kardes, & Kim, 1991). Therefore, for the purpose of creating an innovation diffusion scenario, the expert first of all collects relevant data to conduct a preprocessing on the data and then combines the Grounded Theory with text mining to process the data. The detailed process is listed as follows.

Step 1: Data preprocess
1-1-1H) The expert defines the domain and relevant key terms on which he intends to study.
1-1-1C) To sift out the data, which correspond to the key term, from the Internet, and the data which analyze the time.
1-1-2H) Based on his domain knowledge, the expert interprets the texts, and at the same time, he segments texts into terms, removes any meaningless terms, and marks the segmented terms with conceptual labels.

The system processes the first-stage open coding analysis, which contributes to the visualization by using data association analysis. This visual image helps the expert to focus on the extractable various theme values which are obtained from the data analysis.

Step 2: Terms co-occurrence analysis (open coding)
1-2-1C) To use equation 1 to calculate terms frequency and association value.

$$assoc(w_i, w_j) = \sum_{s \in D} \min(|w_i|_s, |w_j|_s) \qquad (1)$$

where i and j is ith term and jth term; s is stands for sentence; and D is all textual data
1-2-2C) To visualize the analysis result, which contributes to an association diagram drawn by the co-occurrence terms.
1-2-1H) The expert identifies the concepts and categories derived from the co-occurrence association diagram, which helps the expert to preliminarily realize the various theme values presented in the data.

Phase 2: To define the various roles and linking storylines in the process of innovation diffusion (axial coding and selective coding)
Based on the first-stage data analysis in which the expert develops various types innovative products clusters derived from innovators (axial coding). Then the expert draws a storyline which links with innovators' innovative products clusters to make the emergence of innovation diffusion scenario possible (selective coding).

Step 1: To develop innovative products clusters (Axial coding)
In the last stage, based on the characteristics of "rareness" and "early period", the expert extracts the data about the innovative products to create a single Internet scenario for innovative products. The process is illustrated as follows.

2-1-1H) The expert needs to decide what innovative products ($w_{innovation}$) are.
2-1-1C) Circle the products from the first to the last one.
2-1-2C) Products ($w_{innovation}$) are used to sift out the rare but meaningful sentences.

This variable is used to confirm the research topics and to remove irrelevant sentences with a view to narrowing down the data range and to sift out valid sentences. That is, valid sentences must include terms related to products $w_{innovation}$, which is shown as equation 2.

$$if\,((w_1, w_2,...)_s \cap (w_{innovation})) \neq \phi \quad then$$
$$sentence \;\in\; Valide \;\; sentences \;\; set \tag{2}$$

2-1-3C) To integrate all the valid sentences related to product terms, and create integrated association diagrams, then go to 2-1-1C.

Step 2: Linking Innovative Products for Extracting Early Adopters' Innovative use values (Selective Coding)
The process is listed as follows.
2-2-1H) The expert makes use of the storyline of early adopters' innovative use values by the innovative products are defined.
2-2-1C) We cycle from the first product till the last product to develop the innovative market.
2-2-2C) Product ($w_{innovation}$) is used to collect the rare but meaningful sentences. This variable is used to confirm the research theme and to remove irrelevant sentences with a view to narrowing down the data range and to sift out effective sentences. That is, effective sentences must include terms with products $w_{innovation}$ shown as equation 3:

$$if\,((w_1, w_2,...)_s \cap (w_{maintech})) \neq \phi \quad then$$
$$sentence \;\in\; Valid \;\; sentences \;\; set \;\; then \tag{3}$$
$$go \;\; to \;\; 2-2-1C$$

2-2-3C) To integrate the effective sentences concerning innovative products to create an integrated association diagram.
2-2-2H) Based on his domain knowledge, the expert confirms the early adopters' innovative use values about the innovative products to create the diffusion scenario.

Modifying TAM Questionnaires and Testing

This study intends to extract how early adopters use innovative products in their daily life, and how regular consumers' purchasing behaviors are easily affected by different lifestyles (Anderson and Golden 1984; Assael 2003). The TAM questionnaires, which consist of two dimensions: useful and ease of use, were used to test the degree of acceptance of new soft systems which enterprises intend to adopt (Davis 1989). Different from the original TAM questionnaires, this study also plans to test the extent of the affection from the value of use innovativeness of early adopters. After the test, the model of this study can be used to analyze whether the majority's acceptance of new products will be affected by the innovative value proposed by early adopters.

Identify Chasm

Target Cluster: Varied clusters analysis to discover a Bowling Alley
If the number of clusters is set to two to fit the Bass model, the two clusters would represent innovator and imitator clusters. The difference between these two clusters can be viewed as the chasm mentioned by Moore. If the number of clusters is set to three, the three clusters would represent innovator, imitator, and a mix of innovator and imitator clusters. The third cluster (i.e., a mix of innovator and imitator cluster) is formed by a portion of innovators (early adopters) and imitators (early majority) from the innovator cluster and the imitator cluster respectively. According to the special property of clustering result, samples within the mix of innovator and imitator cluster must have some characteristics in common. This means that some of the imitators are not far from the innovators. Thus, the bridge for crossing the chasm between innovator and imitator can be built upon the mix of innovator and imitator cluster. This can also be explained using the diffusion theory that a new idea is easier to diffuse while the idea creator is closer to the idea follower. Accordingly, the mix of innovator and imitator cluster (i.e., diffusible target cluster) is valuable for business to cross the diffusion chasm.

Behavior Analysis - Contend Analysis
The Independent-Sample T Test is used to test whether the two sub-clusters in the mix of innovator and imitator cluster have the same average or centroid. After conducting the cluster analysis, we carry out a test on different combination of clusters by the employment of T Test to see whether there is a significant difference between clusters. The significant questions are then used to conduct a content analysis. The different items are converted to be the input information for HCCS to generate the visualized outcome. As mentioned in the previous section, the difference means the variations of the two clusters separated by the chasm. The way of overcoming the chasm to make up the variations will be discussed in the next section.

3.2 Alternative Design

According to the previous section, the bowling alley across the chasm can be built upon the target cluster (i.e., the mix of innovator and imitator cluster). The variations of the two sub-clusters can be figured out using the difference test. If there are no significantly different questions between the two sub-clusters of the target cluster, the imitators (early majority) already have the same characteristics of innovators (early adopters). This also indicates that the bowling alley to cross the chasm has been built. Otherwise, if there are significantly different questions between them, the imitators (early majority) and the innovators (early adopters) have nothing in common. In order to cross the chasm, it needs to investigate the differences aside the bowling alley. Using the content analysis and difference comparison diagram of significant items accomplished in the previous section, the differences of the bowling alley is clearly defined. Furthermore, experts read the diagram of the value of use innovativeness (i.e., innovator cluster), and conduct the brainstorming to create the scenario of building the bowling alley. Eventually, the scenario can be used to generate a solution to diminish the distance between early majority and early adopters.

3.3 Choice

A choice means whether a solution proposed by this study can affect the majority to accept the innovative products or not. The TAM questionnaire in Subsection 3.1.2 is used to ask the same experiment participants to investigate whether the new solution can be chose or not. The content of the participant's answers is used to conduct the clustering analysis. Since this research intends to understand the moving trend of early majority toward early adopters, the number of clusters is set to three. The names of the two sub-clusters in one of three clusters are "new A2 (nA2)" and "new B1 (nB1)" which indicate the new cluster of early adopters and the new cluster of early majority. The clustering results are compared to the results of the first questionnaire analysis. Additionally, sub-clusters nA2 and nB1 are compared to sub-clusters "old A2 (oA2)" and "old B1 (oB1)". Finally, the result of the difference test and the distribution of each cluster are used to conclude the moving trend. If nB1 is closer to oA2 than oB1, and nB1 is farther from oB1; it means the early majority move to the side of early adopters.

4   Case study

4.1   Background

In 2007, ASUS and Intel cooperatively introduced to the market a Netbook which is smaller than 10 inches in size. The purpose of the invention is to make more people to use computers by the relatively cheap price. However, with the follow-up by the big-brand industries, the efficacy of netbooks becomes greater and their price becomes dear. The original purpose of making netbooks has been obliterated. The industries once again utilize the characteristics of the netbooks: light, thin, and easy to carry, and define the netbook as a second set of computers for the consumers. However, how to make consumers realize the innovative use value created by netbooks will be an issue for industries to overcome.

4.2 Data Resources

After the observation of the circular of netbooks in the market, we find that there is a time difference on the speed transmission on the Internet between products and users' adoption information. This study aims at an innovative product - ASUS EeePC901. Therefore, we collected its 6-month blogs data from the time this product being introduced to the market to carry out the analysis. The time period for collecting the data is from 1 October 2008 to 30 December 2008. Using Google blogs (http://blogsearch.google.com/blogsearch), we use terms netbooks + though to execute a key term search. We found 63 related blog articles in total. Based on the idea of purchase innovativeness proposed by Rogers and Shoemaker (1971) and use innovativeness and vicarious innovativeness advocated by Hirschman (1980), which will influence the intention of the majority's acceptance of the new products, the researchers remove the text box which has no use innovativeness and textual data which deviates from the main theme. There are 7 articles which have the textual data concerning the use innovativeness to be used to do the value extraction.

4.3 Intelligence

Employing HCCS to Extract the Innovative Use Value Created by Early Adopters

Based on this study, we proposed HCCS. First of all, the experts followed the process of the grounded theory to do the segmentation of the text in order to acquire meaningful terms. Then, through the calculation of terms frequency and co-occurrence value, and visualizing the association result, we conducted an axial coding to categorize the concepts created by the early adopters. After the interpretation of the fig. 2, we find that innovative use value and the related concepts, namely, the innovative concept of easy-to-use and immediate mobility (light, thin, small, and delicate shape so on). Combining with wireless Internet access which takes on the feature of easy-of-use, we generate an innovative use concept shown in the following figure.

TAM

The researchers focus on the netbooks issued in May 2008. Although it has been a while since the netbook was issued in 2008, this product remains a new product to most of the freshmen in the years to come. Furthermore, the research result concerning early adopters is shown as figure 2. We may find that the functions of the new products: light, thin, small, and durable in terms of using time, which belongs to the item of ease of use. We may further find that they have defined several single-values: webpage, idle chat, entertainment, creative writing, and transportation. However, when using the netbooks, the early adopters' use innovativeness is to integrate various single-values into new value to modify the TAM questionnaires. Then we randomly chose a university located in the northern part of Taiwan as our target object and used this questionnaire to carry out the TAM testing.

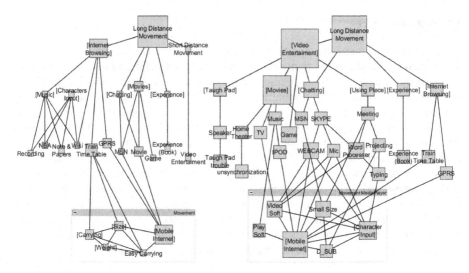

Fig. 2. Early Adopters' Purchase Innovativeness and Use Innovativeness

Identify Chasm

Target Cluster: Varied clusters analysis and o Bowling Alley
In order to test the majority's acceptance of the value created by early adopters, the researchers randomly sampled 218 freshmen who studied at a university located in northern part of Taiwan. The number of valid questionnaires is 195. We suggested that two-stage grouping can be divided into three groups. Next, the researchers applied K-means grouping to divide the subjects into three groups: two large groups and one small group. High grouping contains 97 subjects (49.7%), middle grouping includes 91 subjects (46.7%), and low grouping has 7 subjects (3.6%).

After removing 7 samples which have big difference from two groupings, we used 188 samples to run again the K-means to divide the subjects into three groupings. The percentage of the number of subjects in each group is shown in the following figure. After the examination of the members in the middle group, we found that of them, 46 subjects (49%) belong to high-score grouping, and 47 subjects (51%) belong to low-score grouping.

High (A) (49.7%)		Middle (B) (46.7%)		Low (3.6%)
High (A1) (27%)	High (A2) (49%)	Low (B1) (51%)	Low (B2) (24%)	
		Middle (49%)		

Fig. 3. A clustering illustrated for data collected

We conducted an independent sample t-test for each grouping. We obtained a table 1 shown below. In the table, each figure stands for the number of the items which show significance after conducting t-test. Take the far left column as an example. Three out of twenty two items has no significant difference. There are nineteen items bearing significant difference (Sig. <0.05).

Table 1. The number of items with *t*-test significance in each clustering

Sig.	oA1 - oA2	oA1 - oB1	oA1 - oB2	oA2 - oB1	oA2 - oB2	oB1 - oB2
-	2	2	1	8	2	5
*	0	0	1	0	1	0
**	1	0	0	4	0	2
***	19	20	20	10	19	15

From the perspective of influence on consumption, the number of the subjects in the two subgroups of the low-score group is 47 (oB1) and 44 (oB2) respectively. Aside from oB1 outnumbering oB2, the reason for the selection of oB1 as an aim to move toward high-score grouping is that: when we divide the group into two, high-score group is oA1+oA2, low-score group is oB1+oB2, when we divide the group into three, we acquire a pattern of oA1+C+oB2. Therefore, group C is composed of oA2+oB1. From here, we may realize that if we intend to move the low-score group, either oB1 or oB2, toward high-score group, the distance to move is much shorter for oB1 than for oB2. Therefore, if the aim group is oB1, we may come to a conclusion that the shorter the distance to move, the lower the cost.

If oB1 and oB2 are compared with oA1, we find that it must be very difficult to move low-score cluster to the highest score one via just one-time operation. However, when compared with oA2, we find that oB1 with more items of insignificant difference than those of oB2, so oB1 seems to us much easier to be moved. From the discussion above, we find that if we keep the focus on oB1 and oA2, then we may know the distance between these two is the nearest one. Since oA2 belongs to the innovative cluster (oA2), it is much easier to move oB1 to oA2 than to move oB2 to oA2. As we know that both oB1 and oB2 all belong to imitative clusters. Therefore, when we want to explore if oB1 or oB2 is easier to be moved toward the innovative cluster by using TAM to analyze the imitative clusters, obviously, oB1 is much easier to stand out. So, through the aforementioned analysis, we may gain our target cluster, which is oB1. Also, from the discussion mention above, we may learn that through formation of oA2+oB1, we constitute cluster C. This is to prove that if we construct a bridge connecting the difference between these two clusters, and through the connection of oB1 and oA2, accordingly, the so-called bowling alley proposed by Moore will be constructed. Naturally, this will enable the innovation to be diffused continuously.

Behavior Analysis--Content Analysis (Similarity Tests and Difference Tests)
As stated in the previous section, if we can construct a passage to connect the target cluster and the innovative cluster between oA2 and oB1, the innovation can be diffused continuously. Therefore, after t-test, we obtain the significance for each item, and use each of them as frequency. When we apply HCCS to carry out content analysis on each item, we get a result shown as follows.

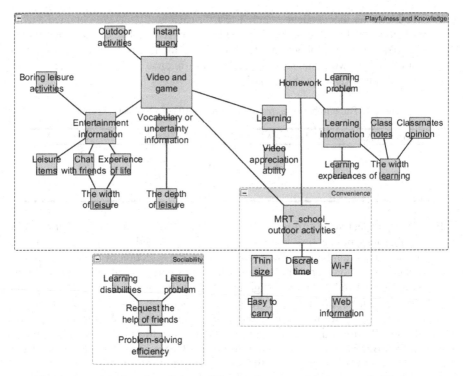

Fig. 4. A comparative about content difference between oA2 cluster and oB1 cluster

After observing 22 items, we obtain the following items with significant difference: convenience (2, 4, 5) 3 items, entertainment and information (7, 8, 11, 13, 14, 19, 20, 21, 22) 9items, sociability (17, 18) 2 items. Of them, entertainment and information account for 41%, which outnumbers the other items. Next, when we compare with figure 4, we may find out that convenience and sociability form individually their scattered sub-clusters. Through the interpretation, the experts perceive that the respondents' realization of using netbooks, between these two dimensions and the early adopters, has reached partial consensus. In other words, there is a certain gap difference connection established between early majority and early adopters. We need simply reinforce the information presented in the fig. 4. As to entertainment and information, they form a larger convergent cluster. Through the experts' interpretation, they think the early majority has not yet reached a consensus between these two dimensions and the early adopters. Furthermore, based on the size of terms squares shown in the fig. 4, we may realize that the difference between them remains significant. They need larger scale of construction in building the bowling alley. In the next section, we will explain how to fill the gap difference.

4.4 Alternative Design

In order to fill the gap difference between both ends of a bowling alley, two experts (one is a professor of business & management, and the other is a Ph.D. program student who major in business & management) joined together to attend a meeting for brainstorming. In the meeting, they compare the variances difference for the majority's acceptance of netbooks in oA2 and oB1. At the same time, we also find operational definition for these variances. Lack of actual application in the real time, it is hard to make freshmen get used to the use situation. Therefore, in order to make up the insufficiency of this information, for example, the description of being light and convenient, the experts not only use physic measurement to describe actual lightness, thinness, and sizes, shown in fig. 4 and fig. 2, but also adopt certain measurement to describe the characteristics of being light and thin of netbooks to enable their identification. For example, we use the thickness of a ten N.T. (new Taiwan dollars) coin which students tend to use while taking the train to school or attending classes to make it easier for them to refer to their daily life experience. In the entertainment and information aspects, as shown in fig. 4, it talks about audio-visual entertainment and information learning and the width of information. From fig. 2, we may acquire much experience about tools uses such as MSN, Wiki, movie, and word processor so on. These tools uses are related to the ways of uses in real life.

Employing these terms may lead students to getting involved in their own living context. About the last scenario of sociability, in fig. 4, it simple shows the contact with friends whereas in fig. 2, it shows MSN, Skype, webcam, and meeting. As the ways students communicate with their friends in real life, and by means of these two figures, the experts are made possible to quickly extend variance operational definition to daily life's operation. This helps to design a scheme for an everyday use context, which may contribute to making students get involved in the operational situation. This scheme can serve as a way to fill the gap difference between both ends of a bowling alley. Finally, using one item with two operational definition ratios, the experts are able to quickly make outlines for four stories.

4.5 Choice

Through the data analysis by TAM testing, we find that although there is no significant change about the number of students in each cluster. However, there is a significant increase in the level of students' acceptance. This explains that our remedial scheme has certain influence on every student. In addition, the purpose of this research is to see whether the target cluster (nB1) is able to move toward the innovative cluster (oA2). Therefore, we remain 3 clusters when conducting the division of the samples collected for the second time. We analyze the sub-cluster (nA2 and nB1) of the third cluster and the sub-cluster (oA2 and oB1) of the third cluster obtained from the data collected in the first-round sampling. The comparing result is shown in table 2.

Table 2. A comparative of new and old sub-clusters

Sig.	oA2 - oB1	oA2 - nB1	oB1 - nB1
-	8	16	8
*	0	1	2
**	4	2	5
***	10	3	7

From the testing result, we find that the significant difference is lesser between nB1 and oA2 than that between oB1 and oA2. This explains than nB1 is much closer to oA2 than oB1. At the same time, there is a significant difference between nB1and oB1, which means both of them have a significantly different point of view in the acceptance of innovative products. The results of these two experiments all show evidence that nB1 moving toward oA2. Moreover, oA2 belongs to the early adopters, which also proves that this research is surely able to define the target cluster of innovation diffusion. Additionally, through the difference of environmental variance analysis and the early adopters' creative operational variance, we find that the constructed remedial scheme is surely able to build a bowling alley over the chasm, which takes on a chance to cross the chasm.

5 Conclusions

If an enterprise, under the limitation of business resources, can identify earlier what product may cross the chasm, and make an arrangement and fill the position in the market, this may help bring great profit to the enterprise. So, crossing the chasm is not simply an important issue for innovation diffusion. It is also an important issue for enterprises to plan marketing strategies. However, before crossing the chasm, we must determine what the next target cluster of innovation diffusion is, or which target cluster is closest to the early adopters. Next, when doing the crossing, we need to fill the gap difference between both ends of the chasm. Then, the smooth diffusion can be expected. Therefore, this search uses TAM to test data. In other words, we utilize the variance of the questionnaire to test the response of the public. We also conduct the public's clustering and determine the target cluster for innovation diffusion. Although TAM variance is able to define the difference between both ends of the bowling alley; it lacks the definition for everyday operational variance. Therefore, this research uses HCCS to analyze the creative use value created by early adopters in the blogs. Through experts' brainstorming meeting to extract some appropriate everyday operational variance, plus writing up four everyday operational stories (difference remedial scheme) based on the difference between both ends of the bowling alley, which may reinforce the insufficiency of the TAM variance operation. Finally, we conduct once again the TAM test to both analyze and interpret the degree of acceptance by the target cluster. We find that the remedial scheme surely has an impact on the target cluster to construct the bowling alley. This is to prove that this research method can provide chance for enterprises to cross the chasm.

References

1. Anderson, W.T., Golden, L.L.: Lifestyle and Psychographics: A Critical Review and Recommendation. Advances in Consumer Research 11(1), 405–411 (1984)
2. Assael, H.: Consumer Behavior: A Strategic Approach (, 1st edn. Houghton Mifflin, South-Western College Pub. Boston (2003)
3. Bass, F.M.: A new product growth for model consumer durables. Management Science 15(5), 215–227 (1969)

4. Davis, F.D.: Perceived Usefulness, Perceived Ease of Use, and User Acceptance of Information Technology. MIS Quarterly 13, 319–340 (1989)
5. Egmond, C., Jonkers, R., Kok, G.: A Strategy and Protocol to Increase Diffusion of Energy Related Innovations into the Mainstream of Housing Associations. Energy Policy 34(18), 4042–4049 (2006)
6. Engel, J.F., Kegerreis, R.J., Blackwell, R.D.: Word-of-Mouth Communication by the Innovator. Journal of Marketing 33, 15–19 (1969)
7. Herr, P.M., Kardes, F.R., Kim, J.: Effects of Word-of-Mouth and Product-Attribute Information on Persuasion: An Accessibility-Diagnosticity Perspective. Journal of Consumer Research 17, 454–462 (1991)
8. Hippel, E.V.: Lead Users: A Source of Novel Product Concepts. Management Science 32(7), 791–805 (1986)
9. Hirschman, E.C.: Innovativeness, Novelty Seeking and Consumer Creativity. Journal of Consumer Research 7, 283–295 (1980)
10. Moore, G.A.: Inside the Tornado: Marketing Strategies from Silicon Valley's Cutting Edge. Harper Business, New York (1999)
11. Morrison, P.D.: Determinants of User Innovation and Innovation Sharing in a Local Market. Management Science 46(12), 1513–1527 (2000)
12. Simon, H.A.: A Behavioral Model of Rational Choice. Quarterly Journal of Economics 79, 99–118 (1955)
13. Ram, S., Jung, H.-S.: Innovativeness in Product Usage A Comparison of Early Adopters and Early Majority. Psychology and Marketing 11(1), 57–67 (1994)
14. Rogers, E.M., Shoemaker, F.F.: Communication of Innovations: A Cross-Cultural Approach, 2nd edn. Free Press, New York (1971)
15. Rogers, E.M.: Diffusion of Innovations. Free Press, New York (2003)

Visualization of the Technological Evolution of the DVD Business Ecosystem

Yan-Ru Li

Department of Information Management, Aletheia University, No.32, Zhenli St.,
Danshui Dist., New Taipei City 251, Taiwan (R.O.C.)
yrli@email.au.edu.tw

Abstract. Business ecosystem presents a new perspective for repositioning the research and development portfolio with external linkage. The rapid evolution of technology has changed the ways of our resources allocation and investment. Technological evolution provides forecasting capability for the potential dominant paradigm. However the analysis of the technological evolution often has to rely on a lot of manpower. This paper would like to provide a rapid visualization of the fierce R&D competition. Previous studies have focused on patent indicators and classification systems, ignoring the diversification and path in the technology development. Therefore, this study proposes a process to illustrate the visualization of optical storage. Finally, this study provides the visualization of technological convergence and diversification.

1 Introduction

Those people work in the tech world are familiar with Moore's Law, which presents that the numbers of processors that can be placed on an integrated circuit will double every two year. The rapid change also challenges the work hard scientists and engineers. The open innovation and business ecosystem perspectives push the companies to seek suitable repositioning beyond pure mergers & acquisitions and co-operations to gain further competitive advantages (Moore, 1993; Iansiti and Levien, 2004; Li, 2009). However, analyzing the evolution of technologies has to rely on a lot of manpower, many projects commissioned by the professional institutions. Patent map is one of the tools that commonly used in statistical methods to present statistical tables, for example, bar chart, pie chart, line map and radar map and so on. The main drawback of the general statistical methods is the only show a small number, the number of non-continuous data, comparison and ratio of basic change in trend. Some patent analysis software for further use of this self-organizing map (SOM) technique into visual graphic. In order to deal with the amount of data, scholars use data mining as a technique for analysis. Nowadays the tools still have difficulties to show a clear path for a technological ecosystem. Fortunately, the aid of computer helps us to understand the evolution process easier. In the paper, we analyze technology development by collecting all US patents filed by DVD 6C from an ecosystem perspective. Finally, we provide some explorative discussions for this case study.

R. Katarzyniak et al. (Eds.): Semantic Methods, SCI 381, pp. 231–237.
springerlink.com   © Springer-Verlag Berlin Heidelberg 2011

2 Literature Review

In this study, literature review will be divided into three parts: the section 2.1 is the business ecosystem; in section 2.2, we introducing the meaning and practice of the patent map, and explain why the current patent map not track the reasons of industrial technology introducing text mining and applications; in section 2.3, describes the technical specifications of optical storage industry, the evolution of the various stages of the process.

2.1 Business Ecosystem

Market and hierarchies have dominated our thinking about an economic organization (Moore, 2006). However, a business ecosystem is an emerging concept analogized from biology. Moore (1996) gives the description of business ecosystem as the following:"An economic community supported by a foundation of interacting organizations and individuals—the organisms of the business world. The economic community produces goods and services of value to customers, who are themselves members of the ecosystem. The member organisms also include suppliers, lead producers, competitors, and other stakeholders. Over time, they coevolve their capabilities and roles, and tend to align themselves with the directions set by one or more central companies. Those companies holding leadership roles may change over time, but the function of ecosystem leader is valued by the community because it enables members to move toward shared visions to align their investments, and to find mutually supportive roles."

The concept presents the companies that focus primarily on their internal capabilities have pursued strategies that not only aggressively further their own interests but also promote their overall ecosystem health—that is to say, networking capabilities enhance the creation and delivery of a company's own offerings (Li, 2009). Successful firms also have done this by creating a "platform"–services, tools, or technologies–that other members of the ecosystem can use to enhance their own performance (Moore, 1993; Iansiti and Levien, 2004). Chesbrough and Schwartz(2007) also emphasize the co-development strategy that has been important for a company like Intel, which provides core complementary microprocessor technology for Microsoft software.

2.2 Patent and Text Mining

Text mining(TM), also known as data mining, document mining, or knowledge discovery in text, in general, is removed from the unstructured text in the important information and useful knowledge, is a fledgling discipline in the field. TM can be used for technological analysis such as correlation analysis, classification, clustering, summarization, forecasting and so on. Text mining are useful for considerable quantities of data, while some authors regard that several years from now, techniques for applying these concepts will be routine (Yoon and Park, 2004). These algorithms are discussed well in Yoon and Park (2004), Kostoff et al. (2001) and Losiewicz et al. (2000). Nowadays, patent information is an important source for measuring human resource, technological capability and performance (Li et al., 2009; Coombs and Bierly, 2006; Moehrle et al. 2005; Lin and Chen, 2005; Yoon et al. 2002; Britzman et al. 2002).

2.3 Optical Storage Industry

The DVD standard has two main alliances: the DVD+ and DVD- series which were formed separately by 3C (Sony, Philips, and Pioneer) and 6C (Hitachi, Matsushita, Mitsubishi, Time Warner, Toshiba, and JVC). In the following years, IBM, Sanyo and Sharp joined the 6C alliance consecutively. In August 2005, IBM quit DVD development and IBM's essential patents were acquired by Mitsubishi. For this reason, there are eight firms in this alliance currently. All patents in this pool have been reviewed by experts who determine these patents are "necessarily infringed when implementing the DVD standard specifications or claiming technologies for which there is no realistic alternative in implementing the DVD Standard Specifications."

3 Methods

We divide the method into three steps in Fig. 1. The first step is to collect the patent data from the DVD patent pool. The second step is to transfer the patent documents into organized data such as the features matrix. The third step is to visualize the technological trajectory.

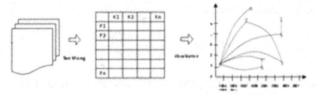

Fig. 1. Research process

3.1 Patent Data

The patent data is based on the DVD player category in the DVD 6C patent pool. This paper chooses 229 patents from U.S. Patent Trademark Office during 1987 to 2004.

3.2 The Vector Space

In this study, we choose the international classifications as example. The text mining is used to extract the three levels of the classifications from 3133 row data. And then we can reduce to 16 top three level classifications from 236 classifications in five levels. Each classification means a special technological domain. Finally, we transfer the classification in to vectors as the following.

$$\text{Company}_1 = (\text{IPC}_{11}, \text{IPC}_{21}, \text{IPC}_{31}, \text{IPC}_{41}, \text{IPC}_{51} \ldots\ldots\ldots\ldots\ldots, \text{IPC}_{i1})$$

$$\text{Company}_2 = (\text{IPC}_{12}, \text{IPC}_{22}, \text{IPC}_{32}, \text{IPC}_{42}, \text{IPC}_{52} \ldots\ldots\ldots\ldots\ldots, \text{IPC}_{i2})$$

$$\text{Company}_3 = (\text{IPC}_{13}, \text{IPC}_{23}, \text{IPC}_{33}, \text{IPC}_{43}, \text{IPC}_{53} \ldots\ldots\ldots\ldots\ldots, \text{IPC}_{i3}) \qquad (1)$$

$$\ldots\ldots$$

$$\text{Company}_j = (\text{IPC}_{1j}, \text{IPC}_{2j}, \text{IPC}_{3j}, \text{IPC}_{4j}, \text{IPC}_{5j} \ldots\ldots\ldots\ldots\ldots, \text{IPC}_{ij})$$

The original 229 patents are also coded into vectors as Patent (Y, IPC) as the following:

$$\text{Patent}_1 = (\text{Year}_1, \text{IPC}_{11}, \text{IPC}_{21}, \text{IPC}_{31}, \text{IPC}_{41}, \text{IPC}_{51} \dots\dots\dots\dots, \text{IPC}_{i1})$$

$$\text{Patent}_2 = (\text{Year}_2, \text{IPC}_{12}, \text{IPC}_{22}, \text{IPC}_{32}, \text{IPC}_{42}, \text{IPC}_{52} \dots\dots\dots\dots, \text{IPC}_{i2})$$

$$\text{Patent}_3 = (\text{Year}_3, \text{IPC}_{13}, \text{IPC}_{23}, \text{IPC}_{33}, \text{IPC}_{43}, \text{IPC}_{53} \dots\dots\dots\dots, \text{IPC}_{i3}) \tag{2}$$

$$\dots\dots$$

$$\text{Patent}_j = (\text{Year}_j, \text{IPC}_{1j}, \text{IPC}_{2j}, \text{IPC}_{3j}, \text{IPC}_{4j}, \text{IPC}_{5j} \dots\dots\dots\dots, \text{IPC}_{ij})$$

We develop a visualization algorithm on the basis of co-occurrence of the international classifications by time sequence. And then, we can draw the y-axis by the feature frequency and x-axis by the issued year. Again, the connection means the relationship between classifications.

The first step is to extract the first filing year, min(Patent(Y) for each IPC. The data construct the x-axis for the plot.

The second step is to accumulate the Patent (IPC) for each filing year that constructs the y-axis.

The third step is to calculate the IPC association. The IPC association equation is defined for a succeeding procedure:

$$assoc(IPC_i, IPC_j) = \frac{\left\| IPC_i \cap IPC_j \right\|}{\left\| IPC_i \cup IPC_j \right\|} \tag{3}$$

The fourth step gives a threshold value for the association to plot the linkage in the picture.

4   Results

4.1   The Basic Statistics

The statistic result is in the Table 1. We can see the G11B "Records converter based carrier and the relative motion between the implementation of information storage memory" that is the top one classification. The second is the H04N "Facsimile telegraph." Other classifications between frequency 30 and 50 are the G06F (General image data processing or production), G06K (Data identification; data representation; record carriers: recording carrier of the treatment), G09B(Educational or demonstration appliances) and H03M(General coding; general decoding or code conversion).

4.2   The Evolution of DVD Technologies

The Visualization provides a clearer roadmap of the technology for DVD player in Fig. 2. For example, we can understand the core technology is the G11B which provides the technological diversifications to all other classifications. And then, we can see the main technological focus in on the G01D, H04N, G06F and H03M before 1993. The engineers try to file the patents related to broadcast communication, electronic digital data processing and general coding.

Table 1. The Basic Statistics of Technological Classification

Year	B32B	C09K	F42B	G01D	G06F	G06K	G06T	G09B	G09G	G10H	G11B	G11C	H03M	H04H	H04L	H04N
1987			1								10					
1990					5											
1991			1								1		1			6
1993											17					
1994											9					12
1995								1			80			1		29
1996					2		2	3			67		5		2	64
1997	1										76					41
1998		2			2	2	3	4		6	193		9		1	196
1999					3	3	3	2	1	4	294		11		4	199
2000		2		1		15	9	6		4	224		9		4	196
2001	1				2	18		6			225	1	4		1	121
2002	1	2			19			16		8	203					218
2003					15	6		4		4	186					210
2004											7					4
Total	3	2	4	1	48	44	17	40	1	30	1592	1	39	1	12	1296

After 1993, there is a significant increase in the DVD player technologies. The diversifications are B32B, C09K, F42B, G06T, G09B, G09G, G10H, G11C, H04H and H04L. The technological relationship is not significant such as the G11B and other classifications. The diversifications grow their technological trajectories after 1993. For example, H04N connect to H04L, G06T and G09B. Finally, we can be easy to realize the core-tech, diversification and convergence periods for DVD technology.

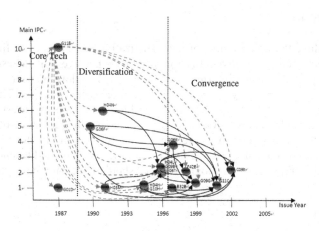

Fig. 2. Visualization of the technological evolution

4.3 The DVD Business Ecosystem

The structure in Fig. 3 illustrates the firms' strategy and the behaviors of the community of firms. the visualization provides clearer relationships among the DVD firms. Matsushita not only cooperates with Toshiba and Mitsubishi but also plays as a core role of the DVD player technology. These relationships among firms and the linkage to the technology co-evolve specific moves made by the web of DVD business ecosystem.

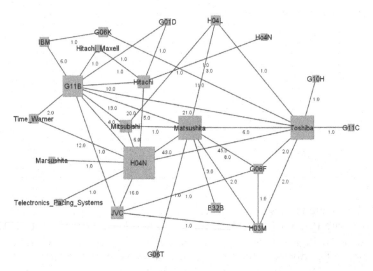

Fig. 3. Visualization of the business ecosystem

5 Conclusion

We can conclude the effectiveness of the visualization that helps us to find the technological convergence and diversification without expensive human resources. The visualization provides us the information of technological development in three periods, they are core technology, diversification and convergence period. The visualization of business ecosystem shows how each   technology affects and is affected by the others in the technological lifecycle. This paper is the preliminary study of the growth of the technology ecosystem; the more specific analysis should be done in the future about the technological paradigm, technological diversification and technological trajectory. The evolution of a technological regime should have paid more attention.

References

1. Breitzman, A., Thomas, P., Cheney, M.: Technological powerhouse or diluted competence: Techniques for assessing mergers via patent analysis. R&D Management 32(1), 1–10 (2002)
2. Chesbrough, H., Schwartz, K.: Innovating business models with co-development partnerships. Research Technology Management 50(1), 55–59 (2007)
3. Coombs, J.E., Bierly, P.E.: Measuring technological capability and performance. R&D Management 36(4), 421–438 (2006)
4. Iansiti, M., Levien, R.: Strategy as ecology. Harvard Business Review 82(9), 69–78 (2004)

5. Kostoff, R.N., Toothman, D.R., Eberhart, H.J., Humenik, J.A.: Text mining using database tomography and bibliometrics: A review. Technological Forecasting and Social Change 68(3), 223–253 (2001)
6. Li, Y.R.: The technological roadmap of Cisco's business ecosystem. Technovation 29(5), 379–386 (2009)
7. Li, Y.R., Hong, C.F., Wang, L.O.: Extracting the significant rare keywords for patent analysis. Expert Systems with Applications 36(3), 5200–5204 (2009)
8. Lin, B.W., Chen, J.S.: Corporate technology portfolios and R&D performance measures: a study of technology intensive firms. R&D Management 35(2), 157–170 (2005)
9. Losiewicz, P., Oard, D.W., Kostoff, R.N.: Textual data mining to support science and technology management. Journal of Intelligent Information Systems 15(2), 99–119 (2000)
10. Moehrle, M.G., Walter, L., Geritz, A., Müller, S.: Patent-based inventor profiles as a basis for human resource decisions in research and development. R&D Management 35(5), 513–524 (2005)
11. Moore, J.F.: Predators and prey:a new ecology of competition. Harvard Business Review 71(3), 75–86 (1993)
12. Moore, J.F.: The death of competition: Leadership & strategy in the age of business ecosystems. Harper Business, New York (1996)
13. Moore, J.F.: Business ecosystems and the view from the firm. The Antitrust Bulletin 51(1), 31–75 (2006)
14. Yoon, B., Yoon, C., Park, Y.: On the development and application of a self–organizing feature map–based patent map. R&D Management 32(4), 291–300 (2002)
15. Yoon, B., Park, Y.: A text-mining-based patent network: Analytical tool for high-technology trend. Journal of High Technology Management Research 15(1), 37–50 (2004)

Part IV
Models and Environments
for Human-Centered Computing

Discovering Students' Real Voice through Computer-Mediated Dialogue Journal Writing

Ai-Ling Wang[1] and Dawn Michele Ruhl[2]

[1] Tamkang University, Tamsui, New Taipei City, Taiwan
wanga@mail.tku.edu.tw
[2] Nagasaki University, Nagasaki, Nagasaki City, Japan
michele@nagasaki-u.ac.jp

Abstract. This study aims at exploring how students might reveal their real voice through computer-mediated dialogue journal writing. College English majors in Taiwan taking a composition class were assigned to write dialogue journals online to communicate with the teacher. They were allowed to write whatever topics that interested them. Findings of the study showed that students wrote differently in writing traditional essays and in writing online dialogue journals. Students did reveal their real voice in writing dialogue journals. Finally, some limitations and recommendations for further study were presented.

1 Introduction

This study initially concerns the development of college students' writing skills as they are facing a greater demand of academic writing and an increasing loss of their voice in their writing process. On the other hand, this study also concerns how the use of computers may have impacts on students' perception of writing. This introductory section will provide an overview of the nature of the research and how the research is developed. Based on the study at college level, the researcher finally recommends some alternatives of applying dialogue journal writing to students at secondary level.

In a traditional college writing class, writing teachers teach students different literary patterns, such as narrative, descriptive, process analysis, comparison and contrast, argument, and so on, in order to prepare students for a more advanced academic writing. In this instructional mode of teaching writing, students are assigned a topic to write about after the instruction of a particular literary pattern. Their essays are then graded and discussed in class. The downside of this instructional pattern has been that students are assigned a fixed topic and a literary pattern and are generally writing for the teacher. Writing under the circumstances, students generally lack a real purpose for writing. For example, they don't really write an argumentative essay in order to persuade people or to propose a course of action. Rather, they adjust their point of view to the teacher's expectations. The teacher evaluates students' essays based on the grammar, spelling, structure, and development presented in the entire essay. How the students involved in the entire writing process and how strong the students feel they are committed to the topic assigned to them are seldom taken into account in the evaluation process.

R. Katarzyniak et al. (Eds.): Semantic Methods, SCI 381, pp. 241–251.
springerlink.com   © Springer-Verlag Berlin Heidelberg 2011

The traditional pattern of writing instruction is an essential step to guide students towards academic writing. However, an approach that provides students with a real purpose of writing and a departure of their attitude toward writing from an academically-defined view to a purpose-oriented perspective may be crucial to prepare students for writing in their professional disciplines in the future. An online journal writing activity is proposed in this study to serve as an opportunity for students to write for their own purposes and, eventually, to gain awareness of purposes for writing and control over writing in general. But, why do we write online and why do we write dialogue journals?

The researcher is convinced that online dialogue journal writing should play an essential role in the writing class and is motivated to explore its impacts on students in their development of writing skills.

1.1 Why Should We Write Online

It seems inevitable that some technophobes may argue "that our increasing reliance on technology is a fatal flaw that will ultimately lead to the demise of society" (Kearsley 2000, p. 137). Others may feel the threat that technologies will eventually take the place of the teacher in the educational arena and that they will lose their power of prestige due to changed role. As a matter of fact, "at some point in the 21st century, online education will become the norm and traditional classroom instruction will be an anachromism." (Kearsley, 2000, p. 148) As Kearsley and Shneiderman (1998) have pointed out, current usage of computers in education have move away its emphasis on individualized instruction and interaction with an instructional program into an emphasis on human interaction in the context of group activities. Computers in education are regarded as communication tools rather than some form of media delivery devices. In this scenario, a teacher's role as facilitator is especially needed.

As pointed out by Warschauer (1998), "in the networked society,…much if not most real writing will take place on screen," and second language teachers will need to teach students effective online writing skills, including both the genres of electronic communication as well as the relationship of texts to other media. Gray (1988) also pointed out that technology "and telecommunications are rarely introduced to classrooms as isolated innovations. Rather, they are accompanied by, and — in fact — serve as a catalyst for other instructional changes." (p. 17) Among many electronic utilities available online, email is one of the most important collaborative tools and it usually serves as the communication backbone for all activities (Kearsley & Shneiderman 1998).

There are some benefits of using the computer as a vehicle of communication. First, communications through the computer is asynchronous. It helps bridge the gap of time and space between teachers and students. The teacher and the students may choose whatever time convenient for them to write or to respond. In a sense, writing online breaks down old barriers of time and space. Second, writing online makes it possible to generate, edit, store, transmit, access, retrieve and process texts at increased speeds and with increased efficiency and flexibility. (Wright, 2000) Third, writing online provides students with a sense of ownership of their own writing. In the mid-1980s, there was awareness of writing as a process, as opposed to writing as a product, among educators. Writing "teachers learned to guide students through a process for writing that encouraged an interactive set of activities focused on planning, drafting, revising, editing, and publishing." (p. 241) Fortunately, technology resources were available to teachers and students for all stages of the writing process. (Roblyer, 2003) By their writing products published online, students feel responsible for their own writing and gain a sense of ownership of their writing products. Although the invention of printing opened up and made available new opportunities for mass communication, it is the more powerful people who determine what counts as worthwhile knowledge that should be published for educational use. In the case of online writing, students can easily produce and disseminate their message. They have control over their own texts, not just reduced to being passive consumers of other people's knowledge. (Wright, 2000)

1.2 Why Should We Write Dialogue Journals

Peyton (1993) defines a dialogue journal as "a written conversation in which a student and teacher communicate regularly over a semester, school year, or course." He suggests that, in dialogue journal writing, the teacher is a participant in an ongoing, written conversation with the student, rather than an evaluator who corrects or comments on the student's writing. As Wright (2000) has pointed out, although the invention of writing offered distinct communication advantages over simple "word of mouth," it also have disadvantages, such as resulting in a loss of the interactive dimension, which is a significant feature of face-to-face encounters. Dialogue journal writing, with its being communicative in nature, essentially enhances interaction between teacher and student.

In sum, dialogue journal writing via email is one type of online learning. It features asynchronous communications, which may provide learners with a greater amount of freedom to allocate their time for learning, and a student-generated learning environment, which may encourage and motivate EFL students to use the target language in their daily life. This research study intends to explore how students perceive dialogue journal writing in comparison with traditional composition writing and how, if any, dialogue journal writing changes the nature of class interaction.

2 Methodology

This study was designed mainly in the qualitative format, although some quantitative components are also presented in this report. As Bogdan and Biklen (1992) have pointed out, the goal of qualitative researchers is to better understand human behavior and experience and "to grasp the process by which people construct meaning and to describe what those meanings are," (P 49) rather than to collect the fact "of human behavior, which when accumulated will provide verification and elaboration on a theory that will allow scientists to state cause and predict human behavior." (p 48) That is, qualitative researchers see themselves as interpreters of human behaviors or social phenomenon. They would, for example, make writing an object of study, rather than take it in the form of text in papers or books for granted. The researcher of the present study is convinced that the complexity of human behaviors, contexts, and meaning formation is not easy to be handled by statistical procedures. Thus, student's writing behavior and writing products and the electronic setting in this study were considered cognitively and socially complicated and were analyzed qualitatively.

On the other hand, a smaller part of data, which belong to the affective domain such as those collected from the questionnaire, were processed under the paradigm of quantitative analysis. Bogdan and Biklen (1992) have pointed out, there are studies with both qualitative and quantitative components, and it is quite often that descriptive statistics and qualitatively findings are presented together.

2.1 Theoretical Framework

There are two main elements involved in this study: writing online and communicative writing. These two elements are evolved from Kearsley and Shneiderman's (1998) engagement theory and Brookes & Grundy's (1990) theoretical framework of a new approach to teaching writing respectively.

Engagement Theory
Engagement theory is formulated basically in electronic and distance education environment. "The fundamental idea underlying engagement theory is that students must be meaningfully engaged in learning activities through interaction with others and worthwhile tasks." There are three elements involved in this theory: collaboration, project-oriented learning, and authentic focus. Collaboration emphasizes team efforts that involve communication, planning, management and social skills. Collaborative learning is a technique that may apply to any domain. For example, English majors may review each other's work.

Project-oriented learning makes learning a creative, purposeful activity. Because students have to define the nature of the project, they have a sense of control over their learning which is absent in the traditional classroom instructions. In the case of online dialogue writing, students can choose whatever topics they want to communicate with their teacher or fellow students or whatever questions they want to ask.

Authentic focus stresses the value of making a useful contribution while learning. According to Kearsley and Shneiderman, doing "authentic projects provides a higher level of satisfaction to students than working on artificial problems since they can see the outcomes/impact of their work on people and organizations." Students working on an online dialogue writing project, for example, may ask questions of their concern or interest, suggest ideas for practice in the classroom, request clarification that may facilitate their learning, and share important information or interesting experiences or findings in writing. In other words, all the writing activities are done with real purposes.

New Approach to Teaching Writing
Brookes and Grundy propose a new approach to teaching writing, which combines communicative practice, an integrated approach, and humanistic principles. They argue that writing teachers should teach writing communicatively. "It is central to the communicative approach that learners exchange meanings and express opinions that are their own." (p. 7) Learners will need to be able to get their messages across to other people and make themselves fully understood by others. In communicative practice, learners will also develop register awareness. They will encode perceived distance and any power relations and write appropriately and naturally.

By integrated approach, Brookes and Grundy mean that one has to recognize that in real world we are rarely exercising only one skill, say writing, at a time. For example, if we take a telephone message, we are actually listening, speaking, writing the message down and probably reading it back. The key point of an integrated approach does not mean to work on the same text in each of the four skills in turn. Rather, it stresses on transferring naturally between one skill and another.

Finally, Brookes and Grundy proposed some humanistic principles to their new approach to teaching writing. They argue that the "learner's freedom to express himself or herself is clearly a central humanistic principle." That is, teachers should promote freedom to express self, recognize the learner as resource, ensure learner freedom from authority, value self-expression as intelligent, recognize centrality of personal discovery, and respect individual learning styles.

2.2 Research Questions

In developing the present study, the researcher is particularly interested in investigating how the students perceive online dialogue journal writing and how they respond to the task. On the other hand, the researcher is also interested in examining the role the teacher plays in an online dialogue journal writing project. The following research questions are thus raised for investigation:

1. Does dialogue journal writing change the way students perceive writing in English?
2. What are the characteristics particularly apparent in dialogue journals, if any, and are not commonly detected in academic compositions.
3. How do the role of the teacher and the way the teacher and student interact change, if any, in an online dialogue journal writing project.
4. How do students perceive online dialogue journal writing in general?

2.3 Participants

Participants of this study were 27 college juniors in the English Department at Tamkang University who were enrolled in the English Composition III class offered by the researcher. Among them, there were 5 males and 22 females. All the participants took Composition I, & II before as a prerequisite for Composition III. Furthermore, each of the participants holds at least an e-mail account and has basic knowledge of e-mailing.

2.4 The Research Project

The entire study lasted one semester long. On the first day of the class when the study started, the researcher announced the writing assignment. This writing assignment was entitled "dialogue journal writing" and was defined as an add-on to the core course. That is, in addition to writing about the main topics discussed in the class, such as descriptive, narrative, comparison and contrast, argument, and so on, the students were assigned to write at least one journal a week. The assignment is somewhat derailed from the traditional track of dialogue journal writing activity, which considers limiting journal topics to personal reflection on course materials is the norm as suggested by Brinton , Holten , and Goodwin (1993).

The students were allowed to choose whatever topics they are interested in or feel comfortable to write about, which may include arguments, debates, or inquiries on academic issues and personal reflections on individual, family, or social phenomena, and e-mail what they have written to the teacher. They were also encouraged to write journals to each other and send a copy to the teacher. There was not limit to the length of the journal. The teacher would then respond to each of the messages she received from the students. Students were told that their messages sent to the teacher would not be graded and corrected, although a holistic assessment based on their involvement in or contribution to the project would be taken into account and would take a certain percentage of their final grade for the course.

2.5 Data Collection and Analysis

Data needed for the present study were collected from the following sources:

1. Students' journals sent to the teacher on the computer and the teacher's responses to these journals
2. Questionnaires completed by the students
3. The researcher's observations during the class time.
4. The researcher's fieldnotes taken within the entire period of research
5. Informal conversations between the researcher and the students

To analyze qualitative data collected in this study, the researcher followed the two steps of analyzing data collected for qualitative research proposed by Bogdan and Biklen (1992): developing coding categories and sorting data. Having developed coding categories, the researcher adopted the cut-up-and put-in-folders approach to sort data. All data or notes collected in this study were looked over, and important units of the data were marked and labeled with coding numbers. These units were then cut up and placed in folders labeled with one code. Cross-coded units were duplicated and put into relevant folders.

3 Findings

There were altogether 178 e-mail messages collected in the research period. Among them, there were 92 student-generated messages and 86 teacher's responses. Students generally generated messages in three directions: things that happened on the campus or related to their studies, things that related to their families, friends, or relatives, and personal thoughts or reflections on daily activities.

Students' messages related to things happened on the campus included student clubs or associations, examinations, grading, group work, part-time jobs, friendship, career planning, book fairs, academic papers, and questions related to their academic study. There were also messages in which the students shared their family with the teacher. These messages covered things concerning family disputes, financial problems, family ties, fraternity, generation gap, and pressure from parents. Data collected from dialogue journals also showed that students were willing to share their personal thoughts or reflections on what they are concerned about or interested in with the teacher. They reflected on TV programs, movies, their readings, their attitude toward life and personal relations.

Some patterns emerged as written conversations between the teacher and students went on. This section will present findings of the study in response to the research questions stated previously.

3.1 Power Relation between Teacher and Student Has Changed

This study showed that dialogue journal writing indeed changed teacher-student interaction and power relation between teacher and student. In his second message sent to the researcher, Eric wrote, "....Can I ask you a question? What is your English name?" The researcher wrote in reply:

> I used to have an English name when I was in college. However, I used my Chinese name, Ai-Ling Wang, when I was in the United States. American professors and classmates had no problem pronouncing my name. Most importantly, this name, Ai-Ling Wang, had accompanied me all the way through my journey toward a Ph.D. That's the reason why I love to use my Chinese name, Ai-Ling Wang.

After receiving this message, Eric always began his messages to the researcher with "Hi! Ai-Ling Wang" instead of "Dear teacher" as he did in his first message. On the other hand, Iris, who always began her message with "How are you today?" wrote "...As a matter of fact, I like this assignment, dialogue journal, very much although sometimes I didn't know what to write about....You know what? I felt like I have a foreign friend and I could write down everything I want to tell her. It's terrific."

3.2 Students Wrote Dialogue Journal for More Genuine Purposes

Findings of this study also showed that most of the time students wrote journals with a purpose in mind. They might write to complain, to suggest, to request, to justify, or to share their excitement, frustration, or appreciation. For example, Daphne was badly in need of a part-time job, and she wrote to the researcher and asked if the researcher could write a letter of recommendation for her. In another situation, Rosa explained why she didn't come to the class last week. It might be true that students did not feel at ease to explain their misconduct or to make excuses face to face with the teacher. However, they felt more comfortable to justify themselves in their journals.

More purposes of writing dialogue journals could also be found in students' journals. For example, Daphne complained about some of her group members in a group work project. She wrote:

> This morning I was awakened by a phone call. Our group need to hand in a paper in the Fiction class today, but one of our group members told me that two of them didn't sent the part they need to finish. What I can do was to help our group to complete the paper even though I have done mine. Actually, I am a little mad about that two members. They usually forget to do the paper .... If they forget to mail the homework to the person who is responsible for putting all the parts together, we need to finish that part for them. But it's very urgent to do it. For

example, we need to hand in the paper at 10:00 but I got the phone call at 8:30. We have only one hour to complete. Fortunately, one of the group member and I have done it minutes ago. The feeling is so terrible. What I know is Don't put their name on the cover according to Dr. Smith said.

Students also took advantage of dialogue journal writing to make constructive suggestions to the class. For example, Rosa wrote:

Today is the first time we go to T303 for the composition class. It's a very comfortable class. Actually, it's too comfortable that everyone is using computers. Besides, we can't hear very clearly because the class is big and your voice is not loud enough. I think it's better for us to change to another classroom with no distraction. It's better for us to learn. Thanks!

3.3 Dialogue Journal Writing Is More Problem-Solving Than Skill-Building

In this dialogue journal writing activity, students were aware that their writing was meaningful and purposeful and that their messages would be responded by the researcher. Therefore, there's no wonder that some students raised their personal problems in their journals and expected assistance. For example,

Participants were facing the issue of career planning at the time of this study. Some of them took the opportunity of journal writing to ask for the researcher's advice. Reason's voice seemed representative of his classmates'. He wrote:

What should I do in the last year of my college life? I haven't decided whether I should continue my study of English literature or transfer to another field, business administration, or even to study what I truly love, graphic design.

I felt a litter bit confused. I was wondering whether I was the only one who felt the pressure and uncertainty of the future. It turned out I was not the one. My classmates felt exactly the same way. Most of us didn't know what to do after we graduate.

In response to Reason's message, the researcher made the following suggestions:

It's not unusual that college seniors feel this way. It's true that it is not easy for you to compete with business majors if you want to change your field of study. However, it is not really an impossible mission for you to study MBA. Several years ago, a T.A. in our English Department passed the graduate school entrance exam to study psychology at Taiwan University. Can you imagine that he ranked number one on the exam? I think it is a matter of "willingness", not "ability." If you really don't like to teach in the future, then go for your MBA.

Some participants got their problems in their family and wrote journals, seeking for the researcher's assistance. Kelley's case was a typical one. She wrote:

I have been very upset recently because there was something terrible happened in my family. About one week ago, my dad held a knife, trying to kill my mom. My dad has been gambling for a long time and was deeply in debt. So my dad threatened my mom to give him money....I love my mom, who always take great care of me. I really hope that I can have a happy family....I felt very sad and scared.

Sensing the urgency of the help Kelley might need, the researcher responded:

....My suggestion was that you go to the Counseling Section in the Office of Student Affairs, which is located in Room xxx. Their extension number is xxxx, and you can talk to Ms. xxx. I understand very well that she will not be able to solve your problem. However, after talking to her, you will feel better and will have a better idea as to how to adjust yourself. I will always be here to listen to you, too.

4 Conclusion

The findings presented above show that this online dialogue journal exchange activity serves at least three functions: social function, pedagogical function, and affective function. This section will discuss the findings from the three different perspectives. First, when looked at from a social perspective, dialogue journal exchanges change the power relation between the teacher and students. Unlike those in a traditional classroom who control the process of the class and who evaluate students' language progress like a judge, the teacher who corresponds with students shares his or her own experiences, provides timely assistance, and respond directly and personally. In this type of writing environment, the teacher is actually leading the students to develop a learning community. In the community, the teacher serves as a facilitator and a problem-solver and students regard the teacher as a friend, a pen pal or a consultant rather than an instructor or an evaluator as shown in this study.

Pedagogically, the teacher and the more talented peers in a dialogue journal exchange activity serve as models of better language usage. On the other hand, students participating in the activity are writing in a genuine context and for a real purpose. For language teachers, dialogue journals can not only add variety to their writing instruction, but also help enhance their students' language acquisition as students participating in written communication experiment with what they have learned about vocabulary or grammar in a real communicative situation.

It seems to many language teachers that individualized instruction is an ideal rather than an achievable goal. However, Peyton (1993) have pointed out that one of the benefits for dialogue journal writing has been that, by keeping reading students' journal entries, the teacher may continually receive information about his or her students' levels of English proficiency and their language progress. Therefore, dialogue journal writing activities may "lead to individualized instruction for each student."

Roblyer (2003) has raised the issue in foreign language instruction: direct versus contextual language instruction. Direct language instruction focuses on teaching discrete language skills (grammar, vocabulary, translation), whereas contextual language instruction provides students with a more contextualized approach to learning the language. That is, in contextual language instruction, language is introduced and taught in context and real situations are used that encompass all aspects of a conversational exchange, including the physical setting, the purpose of exchange, the roles of the participants, the socially accepted norm. Online dialogue journal writing provides students with opportunities to experiment on what they have learned in the class, such as appropriate ways to address to the reader, to request, or to apology.

In the affective domain, there are also some advantages accrues to the writing instruction from dialogue journal writing activities. First, as Brinton, Holten, and Goodwin (1993) has pointed out, holding back criticism of grammar, development, and, and organization encourages students' ideas and introspection and helps them gain confidence in themselves as writers. A quality journal response may put student at ease and lead to richer journal exchanges. Ungraded, uncorrected journals can provide a non-threatening way for students to express themselves in written English (Spack & Sadow, 1983). Second, students' motivation to learn English and to write in English increased along with their writing dialogue journals. Through the ongoing communication in writing between the teacher and the students, the students got encouragement, reinforcement, solutions, and suggestions from the teacher. Lile (2002) has pointed out, "A teacher's positive energy could lead to the students becoming more motivated.... Positive approval and praise for student efforts is very effective, even if the student is wrong. Let the students know that you're glad they tried and being wrong isn't such a big problem."

In sum, college students might benefit from dialogue journal writing from different perfective. Writing teachers may make some adjustments to accommodate students' level of English proficiency, cognitive development, and context of acquisition.

References

1. Bogdan, R.C., Biklen, S.K.: Qualitative research for education: An introduction to theory and methods. Allyn and Bacon, Needham Heights (1992)
2. Brinton, M., Holten, C.A., Goodwin, J.M.: Responding to dialogue journals in teacher preparation: What's effective? TESOL Journal 2(4), 15–19 (1993)
3. Brookes, A., Grundy, P.: Writing for study purposes: A teacher's guide to developing individual writing skills. Cambridge, New York (1990)
4. Cole, R., Raffier, L.M., Rogan, P., Schleicher, L.: Interactive group journals: Learning as a dialogue among learners. TESOL Quarterly 32(3), 556–568 (1998)
5. Dolly, M.R., Adult, E.S.L.: students' management of dialogue journal conversation. TESOL Quarterly 24(2), 317–321 (1990)
6. Gray, T.: Online environments for teacher professional development: A pilot study. Unpublished dissertation, Pepperdine University, Malibu, California (1998)
7. Kearsley, G.: Online education: Learning and teaching in cyberspace. Wadsworth, Belmont (2000)
8. Kearsley, G., Shneiderman, B.: Engagement theory. Educational Technology 38(3) (1998), http://home.sprynet.com/~gkearsley/engage.htm
9. Lile, W.T.: Motivation in the ESL classroom. The Internet TESL Journal 53(1) (2002), http://iteslj.org/Techniques/Lile-Motivation.html
10. Nagel, P.S.: E-mail in the virtual ESL/EFL classroom. The Internet TESL Journal 7 (1999), http://iteslj.org/Articles/Nagel-Email.html
11. Peyton, J.K.: Dialogue journals: Interactive writing to develop language and literacy. ERIC Digests (ERIC Document Reproduction Service No. 354 789) (1993)
12. Roblyer, M.D.: Integrating educational technology into teaching, 3rd edn. Merrill Prentice Hall, Columbus (2003)
13. Santoro, G.M.: The Internet: An overview. Communication Education 43(2), 73–86 (1994)

14. Spack, R., Sadow, C.: Student-teacher working journals in ESL freshman composition. TESOL Quarterly 17(4), 575–593 (1983)
15. Warschauer, M., Healey, D.: Computers and language learning: An overview. Language Teaching 31, 57–71 (1998)
16. Wright, C. (ed.): Issues in education & technology: Policy guidelines and strategies. Commonwealth Secreatariat, London (2000)

The $\mathcal{ALCN}$ Description Logic Concept Satisfiability as a SAT Problem

Adam Meissner

Institute of Control and Information Engineering,
Poznań University of Technology, pl. M. Skłodowskiej-Curie 5,
60-965 Poznań, Poland
Adam.Meissner@put.poznan.pl

Abstract. In this paper we propose a procedure for encoding the $\mathcal{ALCN}$ Description Logic concept satisfiability into SAT problem. In particular, a concept description is translated into the set of propositions, which is satisfiable iff the description is satisfiable, as well. The satisfiability of the resulting propositions is tested by SAT solving tools based on the DPLL procedure and widely known for their efficiency. We describe and analyse the results of preliminary experiments aimed at evaluating the performance of the presented approach.

Keywords: $\mathcal{ALCN}$ description logic, SAT encoding, DPLL procedure.

1 Introduction

The term "Description Logics" (DLs) [1] refers to the varied group of formalisms mainly used for representing and processing an ontological knowledge. A particular interest is given to a class of DLs with the language $\mathcal{ALC}$ as the core ($\mathcal{ALC}$ DLs). Logics from this class have a reasonable expressivity and, in many cases, are decidable. Hence, they have been successfully applied in various domains, such as software engineering, object databases, control in manufacturing, action planning in robotics, medical expert systems, etc. Furthermore, an extension of $\mathcal{ALC}$, namely the language $\mathcal{ALCN}$ stands behind the OWL language [11], which is widely used to represent the semantics of documents available in the WWW network.

One of the basic inference problems in DLs is checking for concept satisfiability. The problem is PSPACE-complete for both $\mathcal{ALC}$ and $\mathcal{ALCN}$, which means it can not be solved in the polynomial time (assuming that $P \neq \text{PSPACE}$). For this reason, various methods have been developed in order to make the concept satisfiability checking tractable. According to the classification given in [9], all these techniques can be divided into two general groups. The first one represents the "classic" tableau-based approach [1] originating from the semantic tableau calculus. The latter class comprises all methods, which translate the DL formulas into the first order logic (FOL) representation. The resulting formulas are processed by FOL-specific reasoning algorithms based on such approaches as constraint satisfaction problem solving, binary decision diagrams or resolution

R. Katarzyniak et al. (Eds.): Semantic Methods, SCI 381, pp. 253–263.
springerlink.com   © Springer-Verlag Berlin Heidelberg 2011

principle. Some of these methods consider the transformation of DL formulas into propositions.

In other words, the problem of checking for DL concept satisfiability is reduced to the boolean satisfiability problem (in abbreviation: *SAT problem*). This approach was proposed by Giunchiglia and Sebastiani [4], who actually combined tableau and translation methods. Their proposition resulted in various extensions (e.g. [5,2]). In particular, the purely translational procedure, namely the *SAT encoding*, was defined by Sebastiani and Vescovi [8,9] for the language $\mathcal{ALC}$. The main idea standing behind this method is to take advantage on the computational power of contemporary SAT solvers.

Admittedly, the encoding is exponential in the worst case, since SAT is an NP-complete problem (assuming that NP $\neq$ PSPACE). However, it turns out that the overall performance of the reasoning tools based on this approach is comparable to the one of the current state-of-the-art DL reasoners. Furthermore, in some cases it is even better. This effect follows from the significant progress in SAT solving techniques and systems, which has been made for the last twenty years. Modern SAT-solvers [7], although still using the original DPLL (*Davis-Putnam-Logemann-Loveland*) procedure [3] as the main algorithm, thanks to sophisticated optimizations can handle sets of formulas with the number of propositional variables and clauses in the order of 10^7. As the result, many difficult practical problems may be solved by transforming them into SAT (e.g. formal verification of microprocessors, cryptanalysis or genome mapping). In this paper we extend the results presented in [9], namely we define the SAT encoding procedure for the language $\mathcal{ALCN}$. We also summarize preliminary experiments intended for estimating the performance of the proposed reasoning method.

The rest of the paper is organised as follows. Section 2 provides the reader with principles of the $\mathcal{ALCN}$ DL. In Section 3 we propose the procedure by which the $\mathcal{ALCN}$ DL concept satisfiability can be encoded and solved as the SAT problem. In Section 4 we discuss the preliminary evaluation of the proposed method. Section 5 concludes with some final remarks.

2   Principles of the $\mathcal{ALCN}$ DL

We start form outlining the syntax and the semantics of the language $\mathcal{ALCN}$. Then, we briefly present one of basic inference problems for $\mathcal{ALCN}$ DL, namely testing for concept satisfiability. Finally, we introduce some elementary concepts and terms, which are used in the description of the SAT encoding procedure given in the next section.

The $\mathcal{ALCN}$ language contains two kinds of primary expressions, that is to say *atomic descriptions* (or synonymously, *names*) of *concepts* and atomic descriptions of *roles*. A concept is a set of individuals called *instances* of this concept. A role is a binary relation holding between individuals. Any element of a role is called an *instance* of this role. Concepts, besides names, can also be represented by complex descriptions, which are built from simpler descriptions and special symbols called *concept constructors*. We use the letter A to denote a concept name and letters C or D as symbols of any concept descriptions; the letter

R stands for a role description. All these symbols can possibly be subscripted. The set of $\mathcal{ALCN}$ DL concept constructors comprises seven elements, namely *negation* ($\neg C$), *intersection* ($C \sqcap D$), *union* ($C \sqcup D$), *existential quantification* ($\exists R.C$), *value restrictions* ($\forall R.C$), *at-least restrictions* ($\geq n\ R$) and *at-most restrictions* ($\leq n\ R$); expressions written in parentheses are schemes of relevant concept descriptions. If it does not lead to misunderstanding, in the sequel we often use constructor names to call the descriptions created with them. For example, we say "a union" instead of "a concept description with a union as the main constructor". Expressions of the form $C(x)$ and $R(x,y)$ are called *concept assertions* and *role assertions*, respectively. The individual y is called a *filler* of the role R for x. Moreover, when an assertion is built from an atomic concept, it is called an *atomic assertion*.

The semantics of concept and role descriptions is defined by means of an interpretation $\mathcal{I}$, which consists of the interpretation domain $\Delta^{\mathcal{I}}$ and the interpretation function $\cdot^{\mathcal{I}}$. The interpretation function assigns a subset of $\Delta^{\mathcal{I}}$ to every concept name and a subset of $\Delta^{\mathcal{I}} \times \Delta^{\mathcal{I}}$ to every role description. The semantics of complex concepts is given below. The symbol $|\cdot|$ denotes the cardinality of a set and $n \geq 0$ is a natural number.

$$(\neg C)^{\mathcal{I}} = \Delta^{\mathcal{I}} \setminus C^{\mathcal{I}}$$
$$(C \sqcap D)^{\mathcal{I}} = D^{\mathcal{I}} \cap C^{\mathcal{I}}$$
$$(C \sqcup D)^{\mathcal{I}} = D^{\mathcal{I}} \cup C^{\mathcal{I}}$$
$$(\exists R.C)^{\mathcal{I}} = \left\{ x \in \Delta^{\mathcal{I}} \mid (\exists y)\, \langle x,y \rangle \in R^{\mathcal{I}} \wedge y \in C^{\mathcal{I}} \right\}$$
$$(\forall R.C)^{\mathcal{I}} = \left\{ x \in \Delta^{\mathcal{I}} \mid (\forall y)\, \langle x,y \rangle \in R^{\mathcal{I}} \rightarrow y \in C^{\mathcal{I}} \right\}$$
$$(\geq n\ R)^{\mathcal{I}} = \left\{ x \in \Delta^{\mathcal{I}} \mid |\{y \mid \langle x,y \rangle \in R^{\mathcal{I}}\}| \geq n \right\}$$
$$(\leq n\ R)^{\mathcal{I}} = \left\{ x \in \Delta^{\mathcal{I}} \mid |\{y \mid \langle x,y \rangle \in R^{\mathcal{I}}\}| \leq n \right\}$$

Furthermore, there are two special concept descriptions, that is to say $\top$ (*top*) and $\bot$ (*bottom*). The first one denotes the most general concept, that is $\top^{\mathcal{I}} = \Delta^{\mathcal{I}}$ while the second represents the least general concept, i.e. $\bot^{\mathcal{I}} = \emptyset$. It should be noticed that $(\geq 0\ R) = \top$ and $(\leq 0\ R) = (\forall R.\bot)$ for every R. In the sequel, we assume that number restrictions with 0 are always replaced by their equivalents.

We say that the assertion $C(x)$ (or the assertion $R(x,y)$) *holds*, if $x \in C^{\mathcal{I}}$ (or $< x,y > \in R^{\mathcal{I}}$), respectively. We also say that the interpretation $\mathcal{I}$ *satisfies* the description C if it assigns a nonempty set to it. Such an interpretation is called a *model* of the concept C. The concept is *satisfiable* if there exists a model of it, otherwise it is *unsatisfiable*. Moreover, the concept D *subsumes* the concept C ($C \sqsubseteq D$) if $C^{\mathcal{I}} \subseteq D^{\mathcal{I}}$ for every model $\mathcal{I}$ of C and D. The concept subsumption checking can be transformed to concept unsatisfiability, namely $C \sqsubseteq D$ holds iff the concept $C \sqcap \neg D$ is unsatisfiable. The correctness of such reduction follows from elementary properties of sets.

As has been stated before, every $\mathcal{ALC}$ concept description C can be *SAT encoded*, namely transformed to a logically equivalent set of FOL propositions S. The set S is satisfiable iff C is satisfiable [9]. Let an *atomic proposition* be any expression of the form $p(t_1, \ldots, t_n)$, where p is an n-ary first order predicate symbol and $t_1, \ldots, t_n$ are terms containing no individual variables. In particular, an atomic proposition can be a *propositional variable* (in short: a variable), i.e. a nullary predicate. A *proposition* (symbolized by ϕ or ψ) is an expression in one of the following forms: an atomic proposition, a *negation* ($\sim \phi$), a *disjunction* ($\phi \vee \psi$), a *conjunction* ($\phi \wedge \psi$), an *implication* ($\phi \rightarrow \psi$) or an *equivalence* ($\phi \leftrightarrow \psi$). A *literal* is an atomic proposition or the negation of an atomic proposition. Every symbol denoting a proposition can possibly be subscripted. We assume a typical FOL semantics for every connective occurring in a proposition. A *truth assignment* is a function mapping every proposition to the value *true* or *false*. We say, the proposition is *satisfiable* if there exists a truth assignment, which assigns the value *true* to it. In turn, a set of propositions is satisfiable if it consists of satisfiable elements only.

3   SAT Encoding Procedure

In this section we propose a SAT encoding procedure for the language $\mathcal{ALCN}$. Let us assume that the given input concept description C is in the *negation normal form* (NNF), namely the negation symbol occurs only in front of concept names. The input concept C is transformed to the set of propositions S, which is satisfiable iff the concept C is satisfiable, as well. Due to the lack of space we do not prove this fact formally. Instead, we describe basic intuitions justifying the correctness of our approach. The procedure is based on the algorithm defined in [9] for the language $\mathcal{ALC}$.

We extend this algorithm by adding steps that handle number restrictions. For this purpose we impose equality or inequality constraints on individuals being fillers of respective roles. The similar idea is applied in the classical, tableau-based reasoning approach [1]. In particular, two individuals with distinct names are regarded as distinct, unless they are explicitly declared as equal. This fact is expressed by a unique binary predicate symbol =. Its definition comprises transitivity axioms only, since the reflexivity is obvious and the axioms of symmetry can be omitted due to the assumption that the set of all individuals is ordered and $i < j$ for every proposition $y_i = y_j$ built from the predicate =. Therefore, the expression $y_i = y_j$ should be rather interpreted as a *substitution*, stating that every occurrence of the individual y_i can be replaced by the individual y_j. In consequence, if the concept assertion $D(y_i)$ (or the role assertion $R(x, y_i)$) holds and the substitution $y_i = y_j$ is true then the assertion $D(y_j)$ (and the assertion $R(x, y_j)$) must hold, as well. Moreover, we consider an unary predicate symbol $\pi_{R,x}$, which occurs in expressions of the form $\pi_{R,x}(y)$ declaring the individual y as an "explicit" filler of the role R for x. This means that the assertion $R(x, y)$ must hold and the individual y can not be substituted by any other filler of the role R.

The general idea of SAT encoding is inspired by the *propositional approximation of the modal formula* [2], since the $\mathcal{ALC}$ DL is isomorphic to the propositional modal logic $K_{(m)}$. This idea, applied to DLs, can be illustrated by the following example. Let us assume the input concept description $A_1 \sqcap (A_2 \sqcup \neg A_1)$. The description is satisfiable if there exists an arbitrary individual x such that

$(A_1 \sqcap (A_2 \sqcup \neg A_1))(x)$ holds. Furthermore, let us assume that for every atomic concept assertion $A(x)$ a propositional variable $V_{A(x)}$ is created such that $V_{A(x)}$ is true iff the assertion $A(x)$ holds. Thus, the input assertion can be converted to the proposition $\phi = V_{A_1(x)} \wedge (V_{A_2(x)} \vee \sim V_{A_1(x)})$, which is true iff the input assertion holds, as well. If the input concept contains the other constructors, namely existential quantifications, value restrictions or number restrictions, then they are also represented as propositional variables. However, for every such a variable a number of implications is created. These formulas reflect the definitions (i.e. the semantics) of respective constructors. For example, the implication corresponding to the variable $V_{(\exists R.A)(x)}$ has the form $V_{(\exists R.A)(x)} \rightarrow V_{R(x,y)} \wedge V_{A(y)}$, where y is a new, unique individual and the variable $V_{R(x,y)}$ represents the role assertion $R(x,y)$, similarly as in case of concept assertions.

The SAT encoding procedure takes five steps, which are described below. At first, the assertion $C(x)$ is created for the input concept C and converted into the *top-level proposition* $\phi_{C(x)}$ by the following transformation rules. The proposition $\phi_{C(x)}$ is added to the set S, namely $S = \{\phi_{C(x)}\}$.

$(\top(x)$ or $\neg\bot(x)) \Rightarrow true$

$(\bot(x)$ or $\neg\top(x)) \Rightarrow false$

$A(x) \Rightarrow V_{A(x)}$

$\neg A(x) \Rightarrow \sim V_{A(x)}$

$(D_1 \sqcap D_2)(x) \Rightarrow D_1'(x) \wedge D_2'(x)$

$(D_1 \sqcup D_2)(x) \Rightarrow D_1'(x) \vee D_2'(x)$

$(\exists R.D)(x) \Rightarrow V_{(\exists R.D)(x)}$

$(\forall R.D)(x) \Rightarrow V_{(\forall R.D)(x)}$

$(\geq n\, R)(x) \Rightarrow V_{(\geq n\, R)(x)}$

$(\leq n\, R)(x) \Rightarrow V_{(\leq n\, R)(x)}$

Symbols *true* and *false* denote the respective logical constants. A primed symbol $D'(x)$ stands for the proposition being a result of the exhaustive application of the transformation rules to the assertion $D(x)$. This step corresponds to the respective one in the $\mathcal{ALC}$ SAT encoding procedure, however we extend it by rules, which handle number restrictions. In particular, one of them replaces every concept assertion of the form $(\geq n\, R)(x)$ by the propositional variable $V_{(\geq n\, R)(x)}$. The other rule performs the analogous operation on at-most restrictions, namely it replaces an assertion $(\leq n\, R)(x)$ by the variable $V_{(\leq n\, R)(x)}$.

The second step concerns existential quantifications and is identical as for the $\mathcal{ALC}$ SAT encoding. In particular, the implication

$$V_{(\exists R.D)(x)} \rightarrow V_{R(x,y)} \wedge D'(y) \tag{1}$$

is added to the set S for every variable $V_{(\exists R.D)(x)}$ occurring in S. It should be noted, that every implication (1) introduces a new individual y, which can possibly be a filler of the role R for x (it happens when the variable $V_{(\exists R.D)(x)}$ is true). We mark the set of such individuals as $\Upsilon(R,x)$.

In the next two steps, variables representing number restrictions are processed. For the sake of brevity, we use the expression $\bigotimes \Phi$ as an abbreviation for $\phi_1 \otimes$

$\dots \otimes \phi_n$, where $\otimes$ stands for the symbol $\wedge$ or $\vee$ and $\Phi = \{\phi_1, \dots, \phi_n\}$ is a set of propositions for $n > 1$. If Φ is one-element set (i.e. $\Phi = \{\phi\}$) then $\bigotimes \Phi = \phi$, also $\bigotimes \emptyset$ is equal to the logical constant *true*. The symbol $\sim \Phi$, in turn, is an abbreviation for $\{\sim \phi_1, \dots \sim \phi_n\}$. Also, we write $y_1 \neq y_2$ instead of $\sim (y_1 = y_2)$. Moreover, we often say "the assertion occurring in the set S", instead of "the assertion corresponding to the variable, which occurs in S".

The third step consists in processing variables representing at-least restrictions. Furthermore, the following actions are performed on every propositional variable $V_{(\geq n R)(x)}$ occurring in S. At first, the cardinality of the set $\Upsilon(R, x)$ is compared to m, which is the maximal value of n over all assertions $(\geq n R)(x)$ in S for the given role R and the given individual x. If $|\Upsilon(R, x)| < m$ then the set $\Upsilon(R, x)$ has to be completed up to m by new, unique individuals. Then, the set S is extended by implications of the form

$$V_{(\geq n R)(x)} \rightarrow \bigvee \{\bigwedge \theta | \theta \in \mathcal{P}(\Pi(\Upsilon(R, x)), n)\} \tag{2}$$

Roughly speaking, the formula (2) postulates that if the variable $V_{(\geq n R)(x)}$ is true, then there must exist at least n distinct fillers of the role R for the individual x. The symbol $\mathcal{P}(\Pi(\Upsilon(R, x)), n)$ denotes the set encompassing all n-element subsets of the set $\Pi(\Upsilon(R, x)) = \{\pi_{R,x}(y) | y \in \Upsilon(R, x)\}$. The predicate $\pi_{R,x}$ has been intuitively described in the first paragraph of this section. Its definition is given in the next step, since it becomes valid when at-most restrictions are handled. Otherwise it is unnecessary, as individuals are considered to be pairwise distinct by default. Also, it should be underlined that at-least restrictions have to be processed strictly after existential quantifications due to the possible completion of the set $\Upsilon(R, x)$, which depends on the number of role assertions added to the set S in the second step of the transformation. Below, we present an example of the formula (2) constructed by the assumption that the variable $V_{(\geq 2 R)(x)}$ is present in the set S and $\Upsilon(R, x) = \{y_1, y_2, y_3\}$.

$$V_{(\geq 2 R)(x)} \rightarrow \pi_{R,x}(y_1) \wedge \pi_{R,x}(y_2) \vee \pi_{R,x}(y_1) \wedge \pi_{R,x}(y_3) \vee \pi_{R,x}(y_2) \wedge \pi_{R,x}(y_3)$$

At-most restrictions are handled in the step four. The actions described below are taken for every concept assertion of the form $(\leq n R)(x)$ occurring in the set S. Strictly speaking, the set is extended by formulas listed below. At first the implication

$$V_{(\leq n R)(x)} \rightarrow \bigvee \{\bigwedge \sim \theta | \theta \in \mathcal{P}(\Pi(\Upsilon(R, x)), m - n)\}\} \tag{3}$$

is added. The intuitions about the meaning of this formula are as follows. The set $\Upsilon(R, x)$ encompasses all individuals, which can be fillers of the role R for the individual x (in other words, these are all possible fillers). Since $|\Upsilon(R, x)| = m$, then at least $m - n$ elements of this set has to be declared as "non-fillers" of the considered role if the variable $V_{(\leq n R)(x)}$ is true. This is expressed by means of the predicate $\pi_{R,x}$, to wit the formula $\sim \pi_{R,x}(y)$ states that the individual y is not an "explicit" filler of R for x. The definition (4) of this predicate has to be added to the set S for every proposition $\pi_{R,x}(y_i)$ occurring in S.

$$\pi_{R,x}(y_i) \leftrightarrow V_{R(x, y_i)} \wedge \bigwedge \{y_i \neq y_j | i + 1 \leq j \leq m \wedge y_j \in \Upsilon(R, x)\} \tag{4}$$

It has a form of equivalence since the predicate $\pi_{R,x}$ generally appears both in positive and negative literals.

The next three types of formulas define properties of the predicate $=$, which are informally described in the first paragraph of this section. In particular, implications (5) express transitivity axioms, which have to be added to the set S for every set $\Upsilon(R,x)$.

$$y_i = y_j \wedge y_j = y_k \rightarrow y_i = y_k \text{ for } 1 \leq i < j < k \leq m \wedge \{y_i, y_j, y_k\} \subseteq \Upsilon(R,x) \quad (5)$$

Furthermore, implications of the form (6) are created for every role R and every pair of individuals $< x, y_i >$, such that $y_i \in \Upsilon(R,x)$ for $1 \leq i \leq m$

$$\bigvee \{V_{R(x,y_i)} \wedge y_i = y_j | i+1 \leq j \leq m \wedge x_j \in \Upsilon(R,x)\} \rightarrow V_{R(x,y_j)} \quad (6)$$

A formula (6) generally states that if the individual y_i is a filler of the role R for the individual x and the substitution $y_i = y_j$ holds, then also y_j is a filer of R for x. Analogously, the same principle is expressed for concept assertions by implications (7). In particular, if the assertion $D(y_i)$ appears in the set S and $y_i \in \Upsilon(R,x)$ for some role R and some individual x, then the following formulas are added to the set S for $i \leq j \leq m$

$$V_{D(y_j)} \wedge y_j = y_k \rightarrow D'(y_k) \qquad \text{for } j+1 \leq k \leq m \wedge y_k \in \Upsilon(R,x) \quad (7)$$

One should observe that every individual y occurring in the set S belongs to at most one set $\Upsilon(R,x)$. Also, it should be noted that all formulas constructed in the considered step depend on the set $\Upsilon(R,x)$ and therefore at-most restrictions have to be processed strictly after existential quantifications and at-least restrictions. Let us illustrate the transformations performed in the step four by the example. For this purpose, we assume that the concept assertions $(\leq 2\,R)(x)$ and $A(y_2)$ are present in the set S and $\Upsilon(R,x) = \{y_1, y_2, y_3, y_4\}$. Then, the following formulas are constructed.

$$
\begin{aligned}
V_{(\leq 2\,R)(x)} \rightarrow &\sim \pi_{R,x}(y_1) \wedge \sim \pi_{R,x}(y_2) \vee \sim \pi_{R,x}(y_1) \wedge \sim \pi_{R,x}(y_3) \vee \\
&\sim \pi_{R,x}(y_1) \wedge \sim \pi_{R,x}(y_4) \vee \sim \pi_{R,x}(y_2) \wedge \sim \pi_{R,x}(y_3) \vee \\
&\sim \pi_{R,x}(y_2) \wedge \sim \pi_{R,x}(y_4) \vee \sim \pi_{R,x}(y_3) \wedge \sim \pi_{R,x}(y_4)
\end{aligned}
$$

$$\pi_{R,x}(y_1) \leftrightarrow V_{R(x,y_1)} \wedge y_1 \neq y_2 \wedge y_1 \neq y_3 \wedge y_1 \neq y_4$$
$$\pi_{R,x}(y_2) \leftrightarrow V_{R(x,y_2)} \wedge y_2 \neq y_3 \wedge y_3 \neq y_4$$
$$\pi_{R,x}(y_3) \leftrightarrow V_{R(x,y_3)} \wedge y_3 \neq y_4$$
$$y_1 = y_2 \wedge y_2 = y_3 \rightarrow y_1 = y_3$$
$$y_1 = y_2 \wedge y_2 = y_4 \rightarrow y_1 = y_4$$
$$y_1 = y_3 \wedge y_3 = y_4 \rightarrow y_1 = y_4$$
$$y_2 = y_3 \wedge y_3 = y_4 \rightarrow y_2 = y_4$$
$$V_{R(x,y_1)} \wedge y_1 = y_4 \vee V_{R(x,y_2)} \wedge y_2 = y_4 \vee V_{R(x,y_3)} \wedge y_3 = y_4 \rightarrow V_{R(x,y_4)}$$
$$V_{R(x,y_1)} \wedge y_1 = y_3 \vee V_{R(x,y_2)} \wedge y_2 = y_3 \rightarrow V_{R(x,y_3)}$$
$$V_{R(x,y_1)} \wedge y_1 = y_2 \rightarrow V_{R(x,y_2)}$$

$$V_{A(y_2)} \wedge y_2 = y_3 \rightarrow V_{A(y_3)}$$
$$V_{A(y_2)} \wedge y_2 = y_4 \rightarrow V_{A(y_4)}$$
$$V_{A(y_3)} \wedge y_3 = y_4 \rightarrow V_{A(y_4)}$$

The last, fifth step, consists in processing value restrictions and it is identical as for the $\mathcal{ALC}$ SAT encoding procedure. More precisely, the implication

$$V_{(\forall R.D)(x)} \wedge V_{R(x,y)} \rightarrow D'(y) \tag{8}$$

is added to the set S for every variable $V_{(\forall R.D)(x)}$ and for every variable $V_{R(x,y)}$, where $y \in \Upsilon(R, x)$. Similarly as the previous step, also this step has to be performed after the set $\Upsilon(R, x)$ has been constructed. The transformation procedure is exhaustively applied to all new propositions added to S until no more steps are to be taken. Then, the set S is the SAT encoding of the input concept C.

The whole process of constructing the set S can be demonstrated by the following example. Let us assume that we want to check if the subsumption $\exists H.G \sqcap \exists H.W \sqcap (\leq 1\,H) \sqsubseteq \exists H.(G \sqcap W)$ holds. Intuitively, the role H can be interpreted as the relation *has a son* and the concept names G and W denote individuals who are good and wise, respectively. Thus, the description on the right-hand side of the symbol $\sqsubseteq$ represents all individuals with at least one son, who is good and wise. The left-hand side description, in turn, corresponds to all individuals having at least one son who is good and at least one son who is wise and one son at most. It should be observed that the given subsumption holds due to the presence of the concept $(\leq 1\,H)$ in the left-hand side description. The son must be both wise and good since he is the only one. In consequence, the transformation of the problem to concept unsatisfiability and then to NNF leads to the description $(\exists H.G) \sqcap (\exists H.W) \sqcap (\leq 1\,H) \sqcap \forall H.(\neg G \sqcup \neg W)$, which is unsatisfiable. Therefore, the set of propositions S being a SAT encoding of the considered description is unsatisfiable, as well. And indeed, one can observe that there exists no truth assignment satisfying all the propositions from the set S, which are listed below.

$$V_{\exists H.G} \wedge V_{\exists H.W} \wedge V_{(\leq 1\,H)} \wedge V_{\forall H.(\neg G \sqcup \neg W)}$$
$$V_{\exists H.G} \rightarrow V_{H(x,y_1)} \wedge V_{G(y_1)}$$
$$V_{\exists H.W} \rightarrow V_{H(x,y_2)} \wedge V_{W(y_2)}$$
$$V_{(\leq 1\,H)} \rightarrow \sim \pi_{H,x}(y_1) \vee \sim \pi_{H,x}(y_1)$$
$$\pi_{H,x}(y_1) \leftrightarrow y_1 \neq y_2 \wedge V_{H(x,y_1)}$$
$$\pi_{H,x}(y_2) \leftrightarrow V_{H(x,y_2)}$$
$$V_{H(x,y_1)} \wedge y_1 = y_2 \rightarrow V_{H(x,y_2)}$$
$$V_{W(y_1)} \wedge y_1 = y_2 \rightarrow V_{W(y_2)}$$
$$V_{\forall H.(\neg G \sqcup \neg W)} \wedge V_{H(x,y_1)} \rightarrow \sim V_{G(y_1)} \vee \sim V_{W(y_1)}$$
$$V_{\forall H.(\neg G \sqcup \neg W)} \wedge V_{H(x,y_2)} \rightarrow \sim V_{G(y_2)} \vee \sim V_{W(y_2)}$$

4   Experimental Results

We have preliminary evaluated the performance of the presented method on the exemplary testing problem, which is a simple generalization of the example given in the previous section. Furthermore, the input concept has the form $(\exists R.A_1) \sqcap \ldots \sqcap (\exists R.A_n) \sqcap (\leq 1\,R) \sqcap \forall H.(\neg A_1 \sqcup \ldots \sqcup \neg A_n)$. The concept is unsatisfiable, however, in the experiments we have also considered its satisfiable variant with the description $\neg A_n$ replaced by $\neg B$ where $A_n \neq B$.

Checking for the input concept satisfiability consists of two general phases. At first, the input description is SAT encoded, namely it is transformed to the set of propositions. Then, the set is converted to the DIMACS format [6], which is a particular representation of the *conjunctive normal form* (CNF). A proposition in CNF is a conjunction of (or a set of) clauses i.e. disjunctions of literals. DIMACS is a standard format of the input data for SAT solvers. It should be noted, that the conversion of a formula to CNF may lead to an exponential number of clauses. In the second phase, the satisfiability of the resulting set is tested by a SAT solver.

The experiments have been performed on the computer equipped with the processor Pentium P4D 3.4 GHz, 1 GB RAM and powered by Linux 2.6.32. The SAT encoding procedure, with an additional conversion to DIMACS format, has been implemented in SWI Prolog 5.8.0. In the second phase the solver RSAT 2.01 has been used, which is a newer version of the tool suggested in [9]. The obtained results are collected in Table 1. We have taken into consideration the time of converting the input concept to DIMACS format (*SAT Enc.*), the number of generated clauses (*Clauses*), the number of propositional variables (*Vars*) and the SAT solver running time (*SAT Sol.*), all times are given in seconds. For

Table 1. Results for satisfiable and unsatisfiable concept descriptions

n	\multicolumn{4}{Satisfiable}				Unsatisfiable			
	SAT Enc.	Clauses	Vars	SAT Sol.	SAT Enc.	Clauses	Vars	SAT Sol.
14	1.47	17092	332	0.18	1.49	17092	331	0.18
15	3.18	33614	378	0.70	3.20	33614	377	0.70
16	7.20	66537	427	2.72	7.32	66537	426	2.70
17	16.58	132246	479	10.71	16.66	132246	478	10.70
18	36.78	263510	534	43.11	36.85	263510	533	42.94

$n < 14$ times have been too small to measure while the input data with $n > 18$ have caused a hang up in the Prolog runtime system. As one can observe, the results for satisfiable and unsatisfiable variant of the input concept are nearby identical, to wit they differ less than 2% for every parameter, respectively. The number of variables generated at the first phase is a linear function of n, while the computational time is proportional to the number of constructed clauses and is roughly $O(2^n)$. It should be remarked that the function relating the exact number of clauses to n, for the given example, is rather not hard to find by theoretical analysis. The computational time of the second phase is about $O(4^n)$.

The presented example shows that handling number restrictions leads to the exponential number of computations for a certain class of problems. In particular, the number of operations depends, on the one hand, on the cardinality of the set of fillers for the given role and the given individual and, on the other hand, on the number occurring in the restriction. This limits the application of the method. However, it still can be useful, since in many practical cases (e.g. TONES ontologies expressed in $\mathcal{ALCN}$ [10]), both these numbers are relatively small. Also, various optimizations can be applied during the SAT encoding, which significantly reduce the number of computations. For instance $(\geq n_1\ R) \sqcap (\leq n_2\ R) \Rightarrow \bot$ if $n_1 > n_2$. We plan to take advantage on these properties in future.

5   Final Remarks

We present a method of checking the satisfiability of concepts in the $\mathcal{ALCN}$ description logic. For this purpose, the input concept description is SAT encoded, that is to say, transformed to the set of propositions, which is satisfiable iff the the input description is satisfiable, as well. The satisfiability of the resulting set is tested, in turn, by a SAT solver. The SAT encoding procedure extends the Sebastiani's and Vescovi's algorithm [9] by steps that handle number restrictions. Preliminary tests showed that the presented approach is essentially exponential for a certain class of problems. However, the size of the output set of propositions as well as the SAT solving time can be still acceptable in many practical cases. Also, both of these parameters can be significantly reduced by various optimizations. Thus, the practical value of the presented method should be established on realistic data. These tests are planned for future. The work was supported by the Poznań University of Technology under Grant DS 45-083/11.

References

1. Baader, F., McGuinness, D.L., Nardi, D., Patel-Schneider, P.F. (eds.): The Description Logic Handbook: Theory, implementation, and applications. Cambridge University Press, Cambridge (2003)
2. Brand, S., Gennari, R., de Rijke, M.: Constraint Methods for Modal Satisfiability. In: Apt, K.R., Fages, F., Rossi, F., Szeredi, P., Váncza, J. (eds.) CSCLP 2003. LNCS (LNAI), vol. 3010, pp. 66–86. Springer, Heidelberg (2004)
3. Davis, M., Logemann, G., Loveland, D.: A Machine Program for Theorem-Proving. Commun. ACM 5(7), 394–397 (1962)
4. Giunchiglia, F., Sebastiani, R.: Building Decision Procedures for Modal Logics from Propositional Decision Procedures: The Case Study of Modal K(m). Inform. Comput. 162(1/2), 158–178 (2000)
5. Giunchiglia, E., Tacchella, A., Giunchiglia, F.: SAT-Based Decision Procedures for Classical Modal Logics. J. Autom. Reasoning 28(2), 143–171 (2002)
6. Hoos, H.H., Stützle, T.: SATLIB: An Online Resource for Research on SAT. In: Gent, I.P., et al. (eds.) SAT 2000, pp. 283–292. IOS Press, Amsterdam (2000)

7. Nadel, A.: Understanding and Improving a Modern SAT Solver. PhD thesis, Tel Aviv University (2009)
8. Sebastiani, R., Vescovi, M.: Encoding the Satisfiability of Modal and Description Logics into SAT: The Case Study of K(m)/$\mathcal{ALC}$. In: Biere, A., Gomes, C.P. (eds.) SAT 2006. LNCS, vol. 4121, pp. 130–135. Springer, Heidelberg (2006)
9. Sebastiani, R., Vescovi, M.: Automated Reasoning in Modal and Description Logics via SAT Encoding: the Case Study of K_m/$\mathcal{ALC}$-Satisfiability. J. Artif. Intell. Res. 35, 343–389 (2009)
10. TONES Ontology Repository, http://owl.cs.manchester.ac.uk/repository/
11. World Wide Web Consortium, Semantic Web, http://www.w3.org/2001/sw/

Embedding the HEART Rule Engine
into a Semantic Wiki*

Grzegorz Jacek Nalepa and Szymon Bobek

AGH University of Science and Technology,
Al. A. Mickiewicza 30, 30-059 Krakow, Poland
{gjn,szymon.bobek}@agh.edu.pl

Abstract. Semantic wikis provide a flexible solution for collaborative knowledge engineering. The reasoning capabilities of most of the semantic wiki systems are limited to lightweight reasoning, e.g. with the use of RDF. Number of systems try to integrate strong reasoning with rules. Since most of the regular rule engines are hard to integrate, custom solutions are needed. In the paper a practical approach for embedding a flexible rule engine called HeaRT into a semantic wiki is given. HeaRT supports several inference modes, as well as rule base structuring, which makes it particularly suitable for a wiki.

1 Introduction

The Internet can be considered a biggest knowledge repository in the world. The amount of data available online is so huge, that searching it becomes a problem. The difficulties in finding relevant data are caused not by the lack of information, but by the great amount of unstructured data.

Due to this problems, systems that introduce structure and semantics to knowledge bases were developed. Semantic wikis [2] are one of the most popular Semantic Web systems that provide reasoning in such knowledge bases. Semantic wikis provide a simple formalism for semantically annotating its content, semantic search, and other manipulation of knowledge not available in regular wikis. Currently, many implementations of semantic wikis use mostly lightweight reasoning solutions based on RDF. Some more advanced systems include limited OWL support. However, there are limitations of these systems to perform rule-based reasoning required by more demanding knowledge engineering applications.

The original contribution of this paper is the proposal of combining a semantic wiki with a rule-based inference engine. The PlWiki [4] system is used, with the embedded HeaRT [8] rule engine. This is a flexible solution, that uses expressive rule language (XTT2 [9]). A rule-based recommender system case is considered.

The paper is organized as follows: In Sec. 2 important implementations of semantic wikis are described. The Sec. 3 contains problem definition and motivation for the solution presented in the following sections. In Sec. 4 the HEART

---

* The paper is supported by the BIMLOQ Project funded from 2010–2012 resources for science as a research project.

R. Katarzyniak et al. (Eds.): Semantic Methods, SCI 381, pp. 265–275.
springerlink.com

inference engine is described. The Sec. 6 contains description of combining rule-based reasoning engine with a semantic wiki system. An use case example of this solution is shown in Sec. 7. Summary and future work are presented in Sec. 8.

2   State-of-the-Art in Semantic Wikis

In recent years there have been a number of system proposals that aim at introducing reasoning within knowledge bases. One of the most popular systems are semantic wikis that combine collaborative knowledge acquisition with semantic annotations. They allow for flexible knowledge processing and simple reasoning. In this section popular semantic wiki systems are described. The emphasis of the description is on the knowledge representation and reasoning.

IkeWiki[10] was one of the first semantic wiki allowing semantic annotation with RDF and OWL support and OWL-RDFS reasoning. Research on the IkeWiki has been continued by the Kiwi [11] project, where an effort was put mainly into improving collaborative knowledge management, with not so much emphasis on reasoning.

Semantic MediaWiki [2] is one of the most popular semantic wiki. It is built on the top of the MediaWiki adding possibilities to embed ontological annotation within the text of wiki articles, that are translated into OWL and can be queried by users with a wiki-like syntax.

One of the most interesting approaches to semantic wiki systems is represented by the AceWiki[3]. The system uses ACE (*Attempto Controlled English*) language to represent knowledge. The ACE language is processed by the AceWiki engine and can be mapped bidirectionally to OWL language. AceWiki also provides an AceRules module that allows for forward chaining inference in rule-based knowledge bases. Inference in such knowledge bases is based on first-order logic.

KnowWE[1] is a semantic wiki system that was build as an extension to JSP wiki system. It allows for heterogeneous semantic knowledge representation with use of decision trees, rules and set-covering models. Although the rules representation is more powerful than in AceWiki, the inference strategies and capabilities to manage structured rule bases are still simple.

PlWiki [4] allows for adding ontological annotation within articles and provides reasoning capabilities using powerful Prolog resolution algorithm. Embedding Prolog code within wiki articles enables expressing knowledge in a rule-based way using Horn clauses. The Prolog source code can be added explicitly or by special *wiki-like* markups that are translated to Prolog terms. The PlWiki wiki system has been developed as a plugin to DokuWiki. In fact, PlWiki is a prototype of the general concept of Loki, a wiki that provides expressive knowledge representation with rules [6,7].

3   Motivation

Rules are one of the most popular and powerful knowledge representations. However, practical reasoning with rules is not a trivial task. Thus, most of the semantic wikis do not support them for inference tasks. In fact, reasoning task in semantic wiki systems can be understood in two ways: 1) *lightweight reasoning* – including mostly classification problems, and 2) *strong reasoning* – a reasoning that involves rules and facts processing.

To improve the reasoning capabilities of semantic wikis, there is a need for providing strong reasoning. Existing classic rule engines (e.g. Drools, Jess, CLIPS) are in most cases too heavy and hard to integrate for this task. Therefore, semantic wikis providing inference engines use mostly lightweight reasoning. Recently some of the wiki systems have been integrating ontology reasoners. In fact, with the new OWL 2 RL Profile it is possible to use simple rule semantics with OWL. However, that current support for this feature in DL reasoners as well as semantic wikis using them, is very limited.

These limitations give motivation to propose a new system, combining a semantic wiki flexibility in collaborative knowledge engineering with the power of rule-based representation, allowing for advanced inference. Due to the fact that embedding a rule engine such as CLIPS, or Drools within a wiki system is difficult, a new solution should be provided. In the next section integration of the PlWiki system with the HeaRT rule engine is proposed.

4   HeaRT Inference Engine

HeKatE RunTime (HeaRT) [5] is a lightweight embeddable rule inference engine built as a part of the HeKatE project[1] [9]. The distinctive features of the HeaRT engine are the following:

- support for an expressive rule language (XTT2) that has a complete formal definition in the ALSV(FD) logic [9],
- the use of modularized rule bases: rules working in the same context are clustered into decision units forming an inference network with advanced inference strategies including forward and backward chaining are provided.
- rule base verification mechanisms that allows for checking for logical completeness, and redundancy in the rule base, and
- lightweight and embeddable implementation using a fast Prolog compiler.

Knowledge in the XTT2 representation uses extended decision tables. The tables are connected between and create an inference network. An example of the decision tables network is presented in Fig. 1.

Visual representation of the rule base is human-readable, but difficult to parse and process. To address this issue the HMR language is used. It is a simple and readable text representation suitable for automated processing. The textual representation is automatically generated from the visual one, by a dedicated design tools [9]. HMR is the native rule language for the HeaRT rule engine.

The HMR language includes definitions of attributes and their types, rule and tables specification. An example excerpt of HMR and its interpretation is given below. This is a part of a basic movie recommendation system presented in

---

[1] See http://hekate.ia.agh.edu.pl.

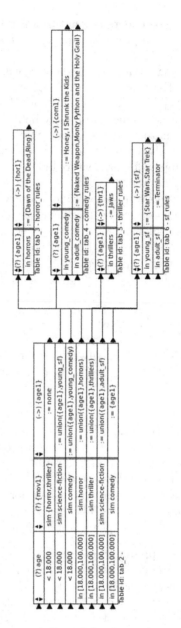

Fig. 1. XTT2 visual representation of simple movie recommendation system

Fig. 1. Using several attributes it tries to recommend a movie to a wiki user. The example includes two rules that depending on an age of the user and preferable movie genre, decides what types of movie are allowed for the person.

```
xattr [name: movie_types, type: genres, class: general].
xattr [name: age_filter, type:age_selection, class:general].

xrule filter/1:
  [age lt 18, movie_types sim [horror, thriller]]  ==>
  [age_filter set []].
xrule filter/2:
  [age lt 18, movie_types sim [science-fiction]]   ==>
  [age_filter set union(age_filter,[young_sf])]:sf_rules.
```

The meaning of rules from given examples is as follows: If the age of the user is less than 18, and s/he selected that s/he would like to watch a horror, or thriller, then set the filter to empty set, and stop. If the age is less than 18 and s/he selected science-fiction movie, then set the filter to science-fiction movies for young and go to rules responsible for making a suggestion on the movie title.

Rules in the form presented above, are directly processed by HeaRT. The engine supports several inference modes. *DDI* (a data-driven forward inference) identifies start tables, and puts all tables that are linked to the initial ones in the table network into a FIFO queue. When there are no more tables to be added to the queue, the algorithm fires selected tables in the order they are popped from the queue. *GDI* (goal-driven backward chaining) works backwards with respect to selecting the tables necessary for a specific task, and then fires the tables forward to achieve the goal. One or more output tables are identified as the ones that can generate the desired goal values and are put into a LIFO queue. As a consequence, only the tables that lead to the desired solution are fired, and no rules are fired without purpose. Moreover, a *TDI* (token-driven inference) suitable for complex inference networks is provided. This approach is based on monitoring the partial inference order defined by the design pattern structure with tokens assigned to tables. A table can be fired only when there is a token at each input. Intuitively, a token at the input is a flag, signalling that the necessary data generated by the preceding table is ready for use.

The engine is a stand alone application that can be easily integrated with a design environment, or a runtime framework. In fact, it can also be easily embedded into another application. The current implantation uses a fast and portable Prolog interpreter and compiler (SWI-Prolog[2]) which makes it easy to deploy.

5   Introduction to PlWiki

The main objectives of PlWiki are to enhance both representation and inference features, allow for a complete rule framework in the wiki. A decision has been made to build the new wiki with use of the Prolog language. This allows to provide rich knowledge representation, including rules, as well as allow for an efficient and flexible reasoning in the wiki, and expressive power equivalent to

---

[2] See http://swi-prolog.org.

Horn clauses. Thus it is possible the represent the domain specific knowledge and reasoning procedures with the same generic representation. The system provides a semantic layer, as well as the Semantic Media Wiki compatibility features.

PlWiki was built on the top of the DokuWiki. It is a simple, popular and fast wiki engine mainly for creating documentation, and storing information on-line. DokuWiki pages are stored on the server as text files and later parsed by the wiki engine which renders XHTML. DokuWiki architecture is extensible with extensions called plugins.

The current version of PlWiki implements the Syntax and Renderer plugin functionality. Text-based wikipages are fed to a lexical analyzer (Lexer) which identifies the special wiki markup. In PlWiki the standard DokuWiki markup is extended by a special `<pl>`...`</pl>` markup that contains Prolog clauses. The stream of tokens is then passed to the Helper plugin that transforms it to special renderer instructions that are parsed by the Parser. The final stage is the Renderer, responsible for creating a client-visible output (e.g. XHTML). In this stage the second part of the PlWiki plugin is used for running the Prolog interpreter and invoke HeaRT.

Below basic use examples of the generic Prolog representation are given:

```
<pl>movie_title('Terminator','StarWars').
    movie_type(science-fiction).</pl>
```

This simple statement adds two facts to the knowledge base. The plugin invocation is performed using the predefined syntax. To actually specify the goal (query) for the interpreter the following syntax is used:
```
<pl goal="movie_title(X),write(X),nl,fail"></pl>
```
It is possible to specify a given scope of the query (with wiki namespaces):

```
<pl goal="movie_title(X),movie_type(X),fail"
    scope="prolog:examples"></pl>
```

A bidirectional interface, allowing to query the wiki contents from the Prolog code is also available, e.g.:
```
<pl goal="consult('lib/plugins/prolog/plwiki.pl'),
wikiconsult('plwiki/pluginapi'),list."></pl>.
```
There are several options how to analyze the wiki knowledge base (that is Prolog files built and extracted from wiki pages). A basic approach is to combine all clauses. More advanced uses allow to select pages (e.g. given namespace) that are to be analyzed. On top of the basic Prolog syntax, semantic enhancements are possible. These are mapped to Prolog clauses for processing.

6   Embedding the HeaRT Engine in the Wiki

In this Section integration of powerful rule-based inference engine (HeaRT) with wiki engine (PlWiki) is proposed. HeaRT inference engine is written in Prolog, so it can be run using PlWiki. HMR language that is used to represent rule-based knowledge in HeaRT is also interpreted directly by Prolog and can be embedded on wiki pages as well. HMR language supports knowledge modularization which is inevitable in wiki systems, where information is spread over many pages or namespaces. HeaRT inference engine takes advantages of this modularization providing advanced inference strategies (see Sec. 4).

In Fig. 2 the architecture of PlWiki with HeaRT is presented. It is divided into two modules: the first module is responsible for rendering wiki pages, and extracting the HMR code, the second module is embedded within PlWiki engine and it is responsible for performing inference based on the HMR model passed to it by the PlWiki engine.

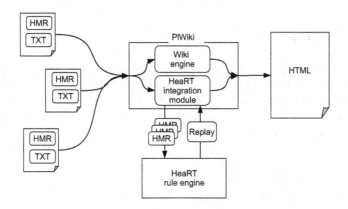

Fig. 2. Architecture of PlWiki and HeaRT system

The process of rendering a wiki page in PlWiki with HeaRT looks as follows:

1. Wiki engine parses the wiki page and extracts HMR code and reasoning queries (goals) for HeaRT.
2. Depending on a scope defined in the goal, PlWiki merges the HMR code from wiki pages in a given scope and passes it to HeaRT.
3. HeaRT performs the reasoning process and returns results to PlWiki engine.
4. PlWiki renders complete wiki page with previously parsed regular text and an answer to a given query (goal) produced by HeaRT.

HeaRT inference engine was added to the PlWiki as a part of a plugin responsible for parsing Prolog. HMR language is embedded on wiki pages with the `<pl></pl>` tag. To run reasoning a `<pl scope="" goal="">` tag is used. If the goal is valid HeaRT command for running inference process, then the reasoning is performed by the engine, result calculated and rendered on a wiki page.

To run inference in HeaRT, the *gox* command is used that takes three parameters: values of input attributes, rules to be process, and reasoning mode. Values of input attributes are passed as a name of the state element of HMR language. The state element stores values of attributes values. Rules that have the same attributes in conditional part and the same attributes in decision part are grouped in one table. Therefore, to pass to HeaRT rules that has to be processed, in fact names of the tables that contain them should be passed. An example of running invoking an inference in PlWiki is:

```
<pl scope="*" goal="gox(init,[result_table],gdi">
```

The meaning of the example is: run the Goal-Driven inference treating *result_ table* as a goal table and taking values of input attributes from the state called *init*. The *scope* parameter in *<pl>* tag is optional and it specifies a namespace from which types, attributes, tables and rules should be taken as an input for reasoning process. If not specified, the entire knowledge in wiki is processed. The *goal* parameter is mandatory, and it has to be a valid HeaRT command.

7   Use Example

The simple movie recommendation system presented in this section tries to recommend a movie set for a user of a given age and some film genre preferences. It uses separate namespaces for each user of the system and additional namespace for movie list. The XTT2 diagram containing rules and reasoning flow in the system is presented on Fig. 1.

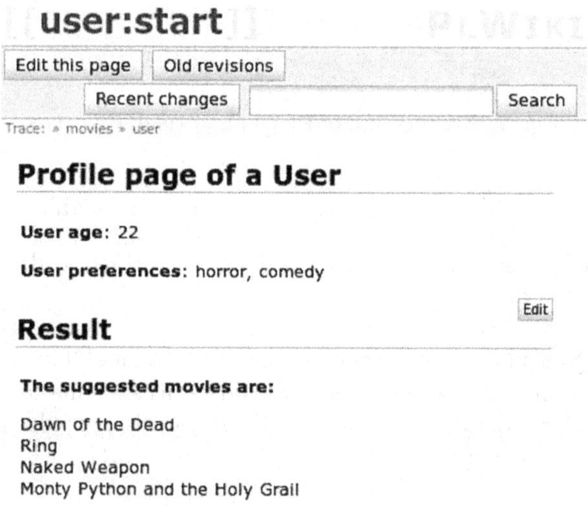

Fig. 3. Reasoning results on user page

In first step the system decides for which subsets of movies the user is allowed based on his age. For instance user that age is below 18, is not allow to watch horrors and thrillers, nor other movies that age limit is higher than 18. Secondly, based on this age filter, the system search for movies that best fit user preferences specified in his profile. At the end, the system response containing the list of recommended movies is printed on the PlWiki page. To initialize attribute values representing user age, and preferences, the *xstat* element from HMR language is used (See Fig. 4). In Fig. 3 a sample output from the system is presented.

Rules that correspond to the age filter are located in *movies* namespace. Rules responsible for matching movies to user preferences and the age filter are located on separate pages, finally the user profile is located in the user personal namespace. An example of rules, written in HMR language, that matches movies to user prefferences are shown below:

```
xrule filter/1:
   [age lt 18, movie_types sim [horror, thriller]]
   ==> [age_filter set [none]].
xrule filter/2:
   [age lt 18, movie_types sim [science-fiction]]
   ==> [age_filter set union(age_filter,[young_sf])]
   :sf_rules.
xrule filter/3:
   [age lt 18, movie_types sim [comedy]]
   ==> [age_filter set union(age_filter,[young_comedy])]
   :comedy_rules.
xrule filter/4:
   [age in [18 to 100], movie_types sim [comedy]]
   ==> [age_filter set union(age_filter,[adoult_comedy])]
   :comedy_rules.
xrule filter/5:
   [age in [18 to 100], movie_types sim [horror]]
   ==> [age_filter set union(age_filter,[horrors])]
   :horror_rules.
xrule filter/6:
   [age in [18 to 100], movie_types sim [thriller]]
   ==> [age_filter set union(age_filter,[thrillers])]
   :thriller_rules.
xrule filter/7:
   [age in [18 to 100], movie_types sim [science-fiction]]
   ==> [age_filter set union(age_filter,[adoult_sf])]
   :sf_rules.
```

To merge all this information, a scope has to be given when goal for the inference is specified. Scope accepts POSIX Regular Expressions, so to collect the HMR model located in several namespaces, the construction presented in Fig. 4 is used.

It is also possible to browse through the database of movies, and get information whether selected movie is recommended for the user or not. What is important, this functionality can be achived without modyfying a rule base. Running different inference mode (in this case Goal-Driven Inference, which is an implementation of backward chaining in HeaRT) will test if selected movie is one would be recommended for the user. Only one line of code is reuired to enable this functionality.

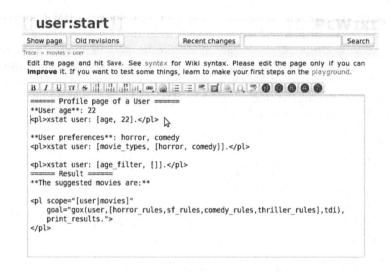

Fig. 4. Goal query on user profile page

An example below shows how to make a rule engine answer question whether there are comedies movied recommendation tor the user:

```
<pl scope="[user|movies]"
    goal="gox(user,[comedy_rules],gdi), print_results."></pl>
```

It is worth noting that this modularization of HMR model gives an opportunity to extend the system easily for other product recommendation (e.g. books). It would require from the user (or system developer) to change the scope parameter in the goal tag from *movies* to *books* and the system would use user profile data to make a suggestion on which books best match the user preferences.

8   Summary and Future Work

In the paper a system that combines a semantic wiki with the power of rule-based inference engines is presented. It allows for complex rule representation, knowledge base structuring, and advanced inference strategies. Presented solution is a prototype, implementing the following functionality: creating wiki articles with embedded Prolog code, and rule-based knowledge, organizing rules into modules and design inference flow within the knowledge using HMR language, querying system that performs reasoning on the knowledge and produces answers that are later rendered on the article page, and performing partial logical verification of the knowledge.

The PlWiki with HeaRT prototype is oriented towards practical applications in community sites, where individual users work together in a collaborative manner. Rule-based decision modules in a wiki can also be useful in e-commerce applications.

The functionality does not include any user interface that helps in creating knowledge, and querying the system. The complete knowledge of HMR language syntax is required. It is inconvenient for the random user even though the language is relatively human readable and intuitive. What is more current solution requires from user to manually organize rules into contexts (tables). In future it should be done automatically, so that the user would have to concentrate only on writing a rules.

References

1. Baumeister, J., Reutelshoefer, J., Puppe, F.: KnowWE: community-based knowledge capture with knowledge wikis. In: K-CAP '07: Proceedings of the 4th International Conference on Knowledge Capture, pp. 189–190. ACM, New York (2007), doi: http://doi.acm.org/10.1145/1298406.1298448
2. Krötzsch, M., Vrandecic, D., Völkel, M., Haller, H., Studer, R.: Semantic wikipedia. Web Semantics 5, 251–261 (2007)
3. Kuhn, T.: AceWiki: A Natural and Expressive Semantic Wiki. In: Proceedings of Semantic Web User Interaction at CHI 2008: Exploring HCI Challenges. CEUR Workshop Proceedings (2008)
4. Nalepa, G.J.: PlWiki – A generic semantic wiki architecture. In: Nguyen, N.T., Kowalczyk, R., Chen, S.-M. (eds.) ICCCI 2009. LNCS, vol. 5796, pp. 345–356. Springer, Heidelberg (2009)
5. Nalepa, G.J.: Architecture of the heaRT hybrid rule engine. In: Rutkowski, L., Scherer, R., Tadeusiewicz, R., Zadeh, L.A., Zurada, J.M. (eds.) ICAISC 2010. LNCS, vol. 6114, pp. 598–605. Springer, Heidelberg (2010)
6. Nalepa, G.J.: Collective knowledge engineering with semantic wikis. Journal of Universal Computer Science 16(7), 1006–1023 (2010)
7. Nalepa, G.J.: Loki – semantic wiki with logical knowledge representation. In: Nguyen, N.T. (ed.) TCCI III 2011. LNCS, vol. 6560, pp. 96–114. Springer, Heidelberg (2011)
8. Nalepa, G.J., Bobek, S., Gawędzki, M., Ligęza, A.: HeaRT Hybrid XTT2 rule engine design and implementation. Tech. Rep. CSLTR 4/2009, AGH University of Science and Technology (2009)
9. Nalepa, G.J., Ligęza, A.: HeKatE methodology, hybrid engineering of intelligent systems. International Journal of Applied Mathematics and Computer Science 20(1), 35–53 (2010)
10. Schaffert, S.: Ikewiki: A semantic wiki for collaborative knowledge management. In: WETICE 2006: Proceedings of the 15th IEEE International Workshops on Enabling Technologies: Infrastructure for Collaborative Enterprises, pp. 388–396. IEEE Computer Society, Los Alamitos (2006), doi: http://dx.doi.org/10.1109/WETICE.2006.46
11. Schaffert, S., Eder, J., Grünwald, S., Kurz, T., Radulescu, M.: KiWi – A platform for semantic social software (Demonstration). In: Aroyo, L., Traverso, P., Ciravegna, F., Cimiano, P., Heath, T., Hyvönen, E., Mizoguchi, R., Oren, E., Sabou, M., Simperl, E. (eds.) ESWC 2009. LNCS, vol. 5554, pp. 888–892. Springer, Heidelberg (2009)

The Acceptance Model of e-Book for On-Line Learning Environment

Wei-Chen Tsai and Yan-Ru Li

Department of Information Management, Aletheia University, No.32, Zhenli St., Danshui Dist., New Taipei City 251, Taiwan (R.O.C.)
{au5724,au5084}@mail.au.edu.tw

Abstract. For recent years, we have seen technology represent an increasing percentage of the core competencies of many industries. The challenge of product innovation is the possibility of effectively taking advantage of digitalization and the internet to create a new reading experience, rather than simply imitating the experience of reading a real book. Though the textbook industry has been resistant to change, it is undeniable that an electronic book (e-book) with a plausible and viable electronic distribution scheme is better than a physical book for several applications.

This paper describes a study on e-book among on-line readers. The main purpose of this study is to conceptual the on-line customers acceptance model of three pre-factors as digital structure, design, and content for intention to e-book acceptance. The acceptance model is based on on-line customers' perceptions on perceived usefulness of the on-line readers. To examine our hypotheses, we collected data from 336 respondents who focused their responses from using experience of e-book. Results demonstrate that the complementary effects of these infrastructural related factors significantly impact both perceptions of and commitments to e-book acceptance.

Keywords: e-book, digital content, technology acceptance model, partial least squares.

1 Introduction

E-learning environments in which on-line customers access digital content have changed significantly in recent years. The most important trend in this area is the inclusion of indexing information for digital content in popular search engines, such as the museum or college web platform. The emerging e-learning environment is also being shaped by new software ideas and increasing new product or application of those new technologies to the scholarly information world (Koufaris 2002; Langston 2003).

It's important to understand what a dramatic change this new environment represents for on-line customers (Koufaris 2002). Formerly, users would usually begin their search for digital articles or books on a personal computer in an abstracting and indexing database. After finding interesting article citations, on-line

R. Katarzyniak et al. (Eds.): Semantic Methods, SCI 381, pp. 277–287.
springerlink.com   © Springer-Verlag Berlin Heidelberg 2011

customers would then follow links from the database or digital website to full text residing on a publisher's server. But now we can download interesting articles or edited into digital files for consumers to read by e-books.

2 Literature and Hypothesis

2.1 On-Line Learning Environment

The Internet had an overwhelming impact on a number of industries and the growth in Internet usage has created much interest in online learning (Fong and Hui 2002). Tian (2001) argued that students can access resources globally through the Internet to assist them in their learning and that it has become an attractive alternative to traditional modes of communication.

Piccoli et al. (2001) refer to the leaner, facilitator, and e-learning technologies factors as antecedents of online learning effectiveness. E-Learning has increased the learner use of technology and has shifted control and responsibility upon the learners. Taylor (2002) agreed that facilitators are faced with steep learning curves and that the quality of their online lessons depends heavily on their knowledge of the required technologies and software. Dillon (2001) argued that the delivery of on-line instruction is heavily dependent on electronic technologies product. Piccoli et al. (2001) stressed that technology quality, tool reliability and accessibility is important characteristics within the online learning environment.

The e-book is regarded as a new kind of digital reading product or learning tool for young people or students (Roskos et al. 2009). Increasing in popularity and availability, e-books show promise as a new kind of interesting e-storybook reading for customers, piquing not only their interest in books, but also growing their knowledge and imagination. As above discussion, the influence factors of e-book acceptance behavior may be gathered from the consumers with using experience of e-books and the value of using e-books to learn in on-line environment.

2.2 The Acceptance Model and Influence Characteristics of e-Book

One of the earliest e-book studies (Hughes and Buchanan 2001) reviews cumulative collection use data over a three-month period from the initial launch of Questia (The Online Library of Books and Journals) in late January 2001 through to April 2001. The total collection consists of 35,000 full-text monographs in the social sciences and humanities. Business also attracts a higher level of e-book use than other subjects. Dillon (2001) examines usage reports for electronic books and finds that books on economics and business received higher usage than e-books in most other subject areas. The study also finds that these results were consistent across the various collections. Langston's (2003) study comes up with similar findings. Otherwise, Kim et al. (2010) discuss the digital content by three categories: design characteristics, scenario characteristics and structure (unity and conciseness). Based on above related studies about e-book and digital factors, we tend to build up a model of the customers acceptance process for e-book.

The main emphasis in the literatures is on the discussion of the comprehensive effect of e-book acceptance on customer behavior is too large a project to tackle in the context of a single article, this study try to provide a parsimonious but meaningful conceptual framework to address the technical aspects of the impact of e-book adoption with e-content for on-line learning environment across customers' using experience of the e-book (Figure 1).

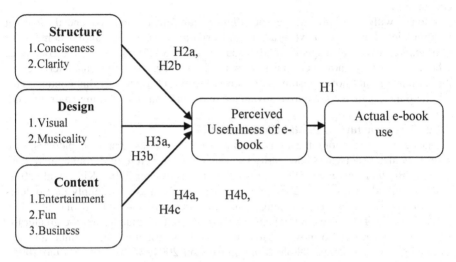

Fig. 1. Model of Characteristics and acceptance of e-book

2.2.1 Perceived Usefulness of e-Book

The construct of perceived usefulness is pivotal in the popular technology acceptance model (TAM) (Davis 1989), which describes relationships between the perceptions of an information technology (IT) and its acceptance by the users. Within the TAM, perceived usefulness refers to "the degree to which a person believes that using a particular system would enhance his or her job performance" (Davis 1989). It represents customers' expectations of the potential benefits from using an IT. Numerous studies have repeatedly confirmed the important role perceived usefulness plays in promoting the acceptance of an IT by generating a positive attitude toward using the technology, enhancing users' intentions to use the technology, and subsequently increasing the actual usage of the technology (Venkatesh et al. 2003).

Customers' value expectation is related to terms such as price and utility (Sullivan et al. 2006). Kim, Oh and Shin (2010) define value expectation as an interactive preference that can only be determined by taking into account the comparative and situational judgment of users: the way they compare it to other items of value and the situation in which they find themselves (Radhakrishnanet al. 2008; Waller et al. 2002). Working from this definition, the value expectation of digital content should be closely associated with the quality of the characteristics we discuss above (Radhakrishnan et al. 2008). Another way of looking at it is to say that the value expectation of digital content refers to how well the digital content satisfies the various needs or desires of users (Tsai et al. 2008). If the various elements of digital content catch customers' attention, they assign a great deal of value to that content (Radhakrishnan et al. 2008). Therefore, as the high level of quality and useful user interface of digital content, on-line customers will use them more frequently (Sullivan et al. 2006; Tsai et al. 2008). To the extent that consumers use acceptance decision making, we propose:

Hypothesis 1:The perceived usefulness is associated with the actual e-book use by consumers.

Along with the prevalence of digital equipment and advanced internet technologies, businesses have gradually incorporated the use of digital content for both marketing and as a specific end product (Tsai et al. 2008). To use the content of e-book effectively, however, businesses need to understand the characteristics of digital content and how consumers respond to different aspects of this content. This will help them gain a competitive advantage in the new on-line learning environment.

2.2.2 The Structure of Digital Characteristics

Structure means the organization of components such as graphics, pictures, and sound: we use conciseness and clarity to evaluate the structure of digital content with e-book (Benlian and Hess 2007; Jonsson and Gustavsson 2008; Kim et al. 2010; Sellitto et al.2007; Xie et al. 2005). Conciseness is called as how simply the contents are structured to deliver the meaning (Jonsson and Gustavsson 2008; Kim et al. 2010; Ni et al. 2007). The more comprehensible and concise the contents are, the greater the value. Clarity refers to how clearly the graphics, pictures and sound and are assembled into a coherent whole (Finn and Inman 2004; Welty and Guarino 2001). When the elements of digital content are properly structured, the value of content is increased (Kim et al. 2010). These arguments lead to the following hypotheses:

Hypothesis 2a: The conciseness of structure is associated with perceived usefulness of e-book by consumers.

Hypothesis 2b: The clarity of structure is associated with perceived usefulness of e-book by consumers.

2.2.3 The Design of Digital Characteristics

Design can be described in terms of the visual appeal of graphics, pictures and the musicality of sound (Calcaterra and Bennett 2003; Isler and Karnad 2008; Kim et al. 2010; Seppanen et al. 2007). The visual appeal of graphics and pictures derives not only from artistic and visual effects but also from the degree to which content appears natural and realistic (Benlian and Hess 2007; Isler and Karnad 2008; Kim et al. 2010). The better the artistic and visual effects are, the greater the value and flow of the digital content. Musicality means the quality of sound effects to match the story background. Therefore, we propose:

Hypothesis 3a: The visual of design is associated with perceived usefulness of e-book by consumers.

Hypothesis 3b: The musicality of design is associated with perceived usefulness of e-book by consumers.

2.2.4 The Content of Digital Characteristics

The content of e-book is used in many different types of fields from entertainment, fun and commercial business. Digital website is available on the website to read business to business e-books, computing & internet e-books and software, or e-books for fun and entertainment (Benlian and Hess 2007). Users tend to concentrate on the entertainment, fun and business value of digital content as their economic status increases (Hughes and Buchanan 2001; Kim et al. 2010; Muramatsu and Ackerman 1998). Therefore, entertainment, fun and business may the important reading purposes in the acceptance process of e-book for users. These arguments lead to the following hypotheses.

Hypothesis 4a: The entertainment of content is associated with perceived usefulness of e-book by consumers.

Hypothesis 4b: The fun of content is associated with perceived usefulness of e-book by consumers.

Hypothesis 4c: The business of content is associated with perceived usefulness of e-book by consumers.

3 Research Methodology

3.1 Operationalization of Constructs and Instrument Design

On considering the specific of different issues of online shopping, hence this study narrows down the scope of each variable and renames it in order to fit the current research needs. The items of scales were collected and merged from the existing measures or on similar scales of literatures. This study changes the scales into more appropriate sentences. Conciseness and clarity of structure in e-book with six items is modified from Jonsson and Gustavsson (2008), Ni et al. (2007) and Finn and Inman (2004). Visual and musically of design in e-book with six items is modified from Isler and Karnad (2008), Seppanen et al. (2007), Sullivan, Ware and Plumlee (2006), and Wang, Li and Shi (2006) in this study. This study, formulates the measurement of entertainment, fun and business of content in e-book with three nine items modified from Koufaris (2002), Muramatsu and Ackerman (1998) and Novak et al. (2000). Perceived usefulness of e-book with three items is all modified from Davis (1989) and Radhakrishnan et al. (2008). Finally, the operational performances with three items are modified from the research of Davis (1989) and Venkatesh et al. (2003). The structured questionnaire was based on academic- and practitioner-oriented literature and interviews. Following the suggestions of Churchill (1979), existing scales were adopted, modified and extended. The constructs in this study are measured using seven-point Likert scales drawn and modified from the existing literatures.

3.2 Sample and Data Collection

The investigation survey was designed originally from the literature review and revised through expert interviews so that the content validity could be assured. The survey population came from on-line customers that who focused their responses from using experience of e-book. Survey data was randomly sampled and collected through e-mail and paper-mail, and after eliminating the same samples, we issued random samples of the survey to 1000 customers. There were 336 effective surveys

(33.6%) collected. The 336 surveys received covered different corporate features, which were used for analysis of the e-book usage with digital content for on-line learning environment. The statistical software tool SPSS for Windows 12.0 and partial least squares software (PLS) were used for corresponding data analysis, and was also applied to determine variable validity and reliability.

Structural equation modeling procedures implemented in a partial least squares graph were used to perform a simultaneous evaluation of both the quality of measurement (the measurement model) and to construct interrelationships (the structural model). The PLS method is primarily intended for causal–predictive analysis and to explain complex relationships by following a component-based strategy (Stewart and Gosain 2006), as is the case with this research. Hence, structural equation modeling procedures implemented in PLS Graph 3.0 (Chin 1998) was used to perform a simultaneous evaluation of both the quality of measurement and construct interrelationships. Moreover, PLS Graph can model latent constructs even under conditions of non-normality and using small- to medium-sized samples (Stewart and Gosain 2006).

4 Research Results

4.1 Sample Demographics

We chose on-line customers who focused their responses from using experience of e-book. Besides, the managerial and academic implications are both higher. 336 on-line customers (154 women and 182 men) outpatients receiving the experimental website through a university-based training center participated in this study. The sample was with a mean age of 23.4 years (SD = 7.3, range = 19-45). All participants gave informed consent.

4.2 Measurement Result

Reliability analysis assured the consistency and stability of the measuring result. Reliability analysis takes the functionality to measure the degrees of data confidence within each factor and its related attributes (Nunnally 1978). If the confidence cronbach α value is greater than 0.7 then confidence in the investigated data is assumed to be sufficient. In general, the cronbach α value for each factor in this research was greater than 0.8 (see table 1). The average variance extracted values were all above the recommended threshold of .50 (Churchill 1979) and the square root of those values were all greater than the construct correlations (Fornell and Larcker 1981). The convergent and discriminate validity tests were both satisfied.

Hypotheses were tested within the structural model shown in Figure 2. All of the nine constructs links were positive and significant in supporting the eight proposed hypotheses. Of the twenty-seven loadings, twenty were above 0.8, indicating that each measure was accounting for 50% or more of the variance of the underlying latent variable.

Table 1. Scale Properties

Construct	Mean	S.D	Alpha	Inter-construct Correlations				
				Structure	Design	Content	Perceived Usefulness	Actual e-book use
Structure	5.44	1.41	0.823	0.76				
Design	5.01	1.44	0.851	0.355*	0.73			
Content	4.98	1.33	0.834	0.276*	0.287*	0.82		
Perceived Usefulness	5.89	1.23	0.896	0.434*	0.401*	0.338*	0.88	
Actual e-book use	5.20	1.38	0.883	0.402*	0.347*	0.376*	0.445*	0.84

a. Diagonal elements (in italics) represent square root of AVE for that construct.

The path coefficients for the research constructs are expressed in a standardized form. Six of the eight path coefficients were above 0.3, with the lowest path coefficient being 0.19, indicating that they are meaningful and significant (Chin 1998). The significance levels of paths in the research model were determined using the partial least squares jackknife re-sampling procedures (Sambamurthy and Chin 2004). Finally, the results suggest a satisfactory fit of the model to the data. As for R-square values, the factors of construct, design and content in the e-book explain 35% of the variance in perceived usefulness of e-book. And perceived usefulness of e-book reaches 42% of the variance in adoption of e-book. The paths are all significant at $p < .01$.

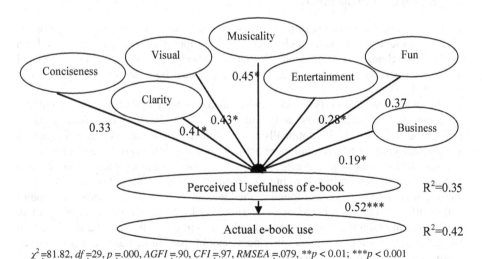

$\chi^2 = 81.82$, $df = 29$, $p = .000$, $AGFI = .90$, $CFI = .97$, $RMSEA = .079$, **$p < 0.01$; ***$p < 0.001$

Fig. 2. Results of path coefficients and hypothesis testing

5 Conclusion

In the year 2008, Kindle of Amazon created a new business model by successfully integrated content provider, hardware or software supplier, telecommunication companies and internet service providers, caused a global revolution in the digital publish industry and further attracted company investments and academics researches and development. The e-book industry is currently in the product introduction stage, to allow it entering the product growth Stage, it is necessary to further analyze consumer habits, content preferences, industry business model and the development of information technology. Therefore, this research based on the consumer's perspective view, TAM and digital characteristics to explore the key factors that influence the intention of using e-book. Using questionnaire as a study tool and recovered 336 valid questionnaires, then PLS empirical study to analyze the influential factors.

This research provides implications useful for the development of an acceptance model of e-book: successful digital website adoption development entails high investment in and conciseness and clarity of the structure, visual and musicality of the design, entertainment, fun and entertainment value of the content for on-line learning environment. Digital content of e-book providers can also build a long-term relationship with users by paying greater attention to the emotional quality of elements such as visual and musicality. Thus, in order to gain a competitive edge in the digital content industry, providers should establish strategies for logical and systematic development of digital content, thereby improving the quality and value of digital content of e-book for on-line learning environment(Finn and Inman 2004; Dillon 2001).

5.1 Managerial Implications

In recent history, we have seen technology represent an increasing percentage of the core competencies of many industries. Though the publisher industry has been resistant to change, it is undeniable that an e-book with a plausible and viable electronic distribution scheme is better than a physical book for several applications (Tsai et al. 2008).

The characteristics of e-book acceptance, such as structure, design, and content, are important for the value users assign to adoption it. Users of e-book acceptance process become absorbed emotionally by the contents by means of elements such as visual, musicality and entertainment value. Sound effects and background music appropriate for a situation can affect on-line customers' responses and allow them to experience flow for on-line learning environment (Dillon 2001). The results of this study confirmed what we have found in the literature that on-line customers are major and significant users of e-books and e-learning books. When the contents bring about more fun, entertainment, and business needs, the sense of value can be increased. In the foreseeable future, e-book providers should pay more attention to product innovation and channel models. The results of this study are served as references to those e-book related providers.

5.2 Research Implications

Additional research on consumer acceptance behavior is needed as digital content based on e-book become practical. Determining the accuracy of predictions regarding the e-book of a new product based on TAM theory is important in assessing the potential for applying the methodology to other evolving communications sectors.

The e-book behavior of customers and the impacts of this behavior have been investigated in considerable depth. Clearly the main lesson learnt was that if e-books are of good quality, needs recommend and reference them in their e-reading lists and if they are made widely and easily accessible they will be used in large numbers.

References

1. Barclay, D.W., Thompson, R., Higgins, C.: The partial least squares (PLS) approach to causal modeling: personal computer adoption and use an illustration. Technology Studies (1995)
2. Benlian, A., Hess, T.: A contingency model for the allocation of media content in publishing companies. Information & Management 44, 492–502 (2007)
3. Calcaterra, J.A., Bennett, K.B.: The placement of digital values in configural displays. Displays 24, 85–96 (2003)
4. Chin, W.W.: Issues and opinion on structure equation modeling. MIS Quarterly 22(1), vii-xvi (1998)
5. Churchill, G.: A paradigm for developing better measures of marketing constructs. Journal of Marketing Research 16(1), 64–73 (1979)
6. Davis, F.D.: Perceived usefulness, perceived ease of use, and user acceptance of information technology. MIS Quarterly 13(3), 319–339 (1989)
7. Dillon, D.: E-books: The University of Texas experience. Library Hi-Tech. 19(2), 113–124 (2001)
8. Finn, S., Inman, J.G.: Digital unity and digital divide: Surveying alumni to study effects of a campus laptop initiative. Journal of Research on Technology in Education 36(3), 297–317 (2004)
9. Fong, A., Hui, S.: An end-to-end solution for Internet lecture delivery. Campus-Wide Information Systems 19(2), 45–51 (2002)
10. Fornell, C., Larcker, D.: Evaluating structural equation models with unobservable variables and measurement error. Journal of Marketing Research 18(1), 39–50 (1981)
11. Hughes, C.A., Buchanan, N.L.: Use of electronic monographs in the humanities and social sciences. Library Hi Tech. 19(4), 368–375 (2001)
12. Isler, V., Karnad, N.: The role of information in the Cop-Robber Game. Theoretical Computer Science 399, 179–190 (2008)
13. Jonsson, P., Gustavsson, M.: The impact of supply chain relationships and automatic data communication and registration on forecast information quality. International Journal of Physical Distribution & Logistics Management 38(4), 280–295 (2008)
14. Kim, C., Oh, E., Shin, N.: An empirical investigation of digital content characteristics, value, and flow. Journal of Computer Information Systems 50(4), 79–87 (2010)
15. Koufaris, M.: Applying the technology acceptance model and flow theory to online consumer behavior. Information Systems Research 13, 23–33 (2002)

16. Langston, M.: The California State University e-book pilot project: implications for cooperative collection development. Library Collections, Acquisitions and Technical Services 27(1), 19–32 (2003)
17. Muramatsu, J., Ackerman, M.S.: Computing social activity, and entertainment: A field study of a GameMUD. Computer Supported Cooperative Computing 7, 87–122 (1998)
18. Ni, Q., Yarlagadda, P.K.D.V., Lu, W.F.: A configuration-based flexible reporting method for enterprise information systems. Computers in Industry 58, 416–427 (2007)
19. Novak, T.P., Hoffman, D.L., Yung, Y.: Measuring then customer experience in online environment: A structural modeling approach. Marketing Science 19(1), 22–42 (2000)
20. Nunnally, J.C.: Psychometric theory. McGraw-Hill, New York (1978)
21. Piccoli, G., Ahmad, R., Ives, B.: Web-based virtual learning environments: A research framework and a preliminary assessment of effectiveness in basic IT skills training. MIS Quarterly 25(4), 401–426 (2001)
22. Radhakrishnan, A., Zu, K., Grover, V.: A process-oriented perspective on differential business value creation by information technology: An empirical investigation. OMEGA: The International Journal of Management Science 36, 1105–1125 (2008)
23. Roskos, K., Brueck, J., Widman, S.: Developing analytic tools for e-book design in early literacy learning. Journal of Interactive Online Learning 8(3), 218–240 (2009)
24. Sambamurthy, V., Chin, W.W.: The effects of group attitudes toward alternative GDSS designs on the decision-making performance of computer supported groups. Decision Sciences 25(2), 215–241 (1994)
25. Sellitto, C., Burgess, S., Hawking, P.: Information quality attributes associated with RFID-derived benefits in the retail supply chain. International Journal of Retail & Distribution Management 35(1), 69–87 (2007)
26. Seppanen, M., Brattico, E., Tervaniemi, M.: Practice Strategies of Musicians Modulate Neural Processing and the Learning of Sound-Patterns. Neurobiology of Learning and Memory 87, 236–247 (2007)
27. Stewart, K.J., Gosain, S.: The impact of ideology on effectiveness in open source software development teams. Minformation Systems Quarterly 30(2), 291–314 (2006)
28. Sullivan, B., Ware, C., Plumlee, M.: Linking audio and visual information while navigation in a virtual reality kiosk display. Journal of Educational Multimedia and Hypermedia 15(2), 217–241 (2006)
29. Tian, S.: The World Wide Web: a vehicle to develop interactive learning and teaching applications. Internet Research: Electronic Networking. Applications and Policy 11(1), 74–83 (2001)
30. Taylor, J.C.: Teaching and learning online: The workers, the lurkers and the shirkers. Journal of Chinese Distance Education 8, 31–37 (2002)
31. Tsai, H., Lee, H., Yu, H.: Developing the Digital Content Industry in Taiwan. Review of Policy Research 25(2), 169–188 (2008)
32. Venkatesh, V., Morris, M.G., Davis, G.B., Davis, F.D.: User acceptance of information technology: Toward a unified view. MIS Quarterly 27(3), 425–478 (2003)
33. Waller, A.O., Jones, G., Whitley, T., Edwards, J., Kaleshi, D., Munro, A., MacFarlane, B., Wood, A.: Securing the delivery of digital content over the Internet. Electronics & Communication Engineering Journal, 239–248 (2002)
34. Wang, C., Li, J., Shi, S.: The design and implementation of a digital music library. International Journal on Digital Libraries 6(1), 82–97 (2006)

35. Welty, C., Guarino, N.: Supporting ontological analysis of taxonomic relationships. Data & Knowledge Engineering 39, 51–74 (2001)
36. Xie, H., Xie, J., Yu, B.: Conciseness and efficiency of file encryption. Computer Engineering and Design 26(12), 3378–3382 (2005)
37. Zhu, K., Kraemer, K.L.: Post-adoption variations in usage and value of E-business by organizations: Cross- country evidence from the retail industry. Information Systems Research 16(1), 61–84 (2005)

Human Computer Interface for Handicapped People Using Virtual Keyboard by Head Motion Detection

Ondrej Krejcar

University of Hradec Kralove, FIM, Department of Information Technologies,
Rokitanskeho 62, Hradec Kralove, 500 03, Czech Republic
Ondrej.Krejcar@ASJournal.eu

Abstract. Nowadays, computers and modern technologies are available to almost all people. However, we should not forget about disabled or handicapped customers. Those people should have access to a PC as well as to the Internet. This project implements the boundary for people with movement disability. Although, the basic assumption is that they can at least move their heads and eyelids. These parts of the body are monitored by web cameras and with their help the user moves around the virtual keyboard shown on the screen. In this way the user is able to utilize their own PC. Moreover, through winking an eye he/she can virtually press a button and write the required symbol. The whole software for detection system was established in a development environment of the Visual Studio 2010 from Microsoft company. The main advantage of our solution is a cheaper price in compare to any existing solutions. By this fact the solution is possible to use even in third countries like in Africa continent.

Keywords: HCI, face detection, users' interface, handicapped people, Open CV libraries.

1 Introduction

As mentioned in the abstract, the developed software is aimed at operating a PC using movements of the head and eyelids. In order to detect head motion using web camera, it is necessary to implement face detection of a particular user and to determine the direction of its movement. The face detection is basic and important technology from the range of scanning systems which are used as a boundary between a PC and a person [1]. A large number of authors have already dealt with such problems. They have published many books and developed considerable numbers of algorithms and methods for both object detection in production processes and face detection. In recent years, face detection algorithms gave rise to extremely precise methods for face localization in images [2].

The main issue of each face detection system is a camera which contains a CCD (Charge-Coupled Device) sensor. With ample speed this sensor generates images of space in front of the camera. These images are then saved by the camera onto a hard disk in video file format. It is mostly the format of AVI (Audio Video Interleave) [3].

R. Katarzyniak et al. (Eds.): Semantic Methods, SCI 381, pp. 289–300.
springerlink.com   © Springer-Verlag Berlin Heidelberg 2011

The detection system then searches according to the chosen algorithm for a face in the images of the video, eventually in its segments and identifies their position based on the number of pixels from the left margin and the number of pixels from the upper margin. The programmed software's objective is to contrast the offset of detected parts in comparison with former images. According to the offset of parts, it enables the movement on a virtual keyboard. The motion of eyelids is then analyzed as the pressing of a certain button on the keyboard. A person usually blinks with both eyes at once, which is the reason why the program cannot respond to this movement.

Therefore it is clear that two components have the most significant impact on the system's quality. These are the apparatuses for face scanning of a user and the algorithm for face detection from a scanned image. It is very important to know the value of resolution in a web camera. If using a camera with small resolution, the user's motion does not have to be detected with required certainty. On the other hand, if using a high-quality camera with high resolution and speed of image scanning, it might take longer for older computers to implement detection algorithm and face location. Perhaps the most complicated issue of detection system concerns choosing the correct algorithm for face detection. First of all, it is necessary to consider the uncountable number of variations in human appearance. People differ from one another not only through the shapes and sizes of individual body parts, but also through skin pigmentation, eye color and hair and eyebrow lengths. A woman has an entirely differently shaped face than an unshaved man etc. Another disturbing impact might be caused by a movement behind a user, i.e. a sudden change of light intensity in a monitored scene, or the entry of another person to the scanned zone. All of these and many other obstacles are supposed to be eliminated by this system in order to accomplish the highest levels of quality and reliability. On the other hand, there are also some advantages such as the homogeneity of faces in comparison with other categories of detected objects. All of these share a certain canonical structure, including two eyes, a nose and a mouth. These are symmetrically placed with regard to longitudinal axis of oval contour (although not all of the components are traceable from all visual angles) [4].

It is suitable to use face static components for detection of head movement. These components represent the motion of a whole face (i.e. nose, birthmarks).

2 Problem Definition

In recent times, modern technologies and computers became essential parts of our lives. They are present in our homes, schools, offices etc. Access to the internet and working on computers became part of a routine for ordinary people. However, how do you access the internet or send emails to friends living on different continents, when you cannot move your hands and fingers? Fortunately, a large number of experts started to be concerned about this issue and developed many alternative boundaries for the operating of computers even without a mouse and a keyboard.

The first very interesting system was proposed and designed in the Czech Republic. It is called the *i4control*. The system *i4control* is the new type of computer

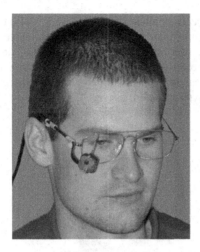

Fig. 1. i4control User Interface

periphery which enables disabled immobile users to operate personal computers by eye motion (eventually by head movement). This system substitutes an ordinary mouse and offers to its user a simple way to communicate with installed applications by eye movement [Fig. 1] [7].

Another product which uses a similar system is called LifeTool IntegraMouse. Basically it is a mouse controlled by lips' motion and the pressing of a button is accomplished by sipping or blowing. This system is ideal for users with various physical problems such as whole body paralysis, after amputation, muscular dystrophy or disseminated sclerosis [Fig. 2] [8].

Fig. 2. IntegraMouse User Interface.

The system *My Voice* is another very interesting option. This program enables users to operate a computer and install programs exclusively by vocal commands. Using these commands, users are able to carry out the same actions as would be otherwise realized by mouse and keyboard. The program is capable of immediately recognizing vocal commands from random person and therefore, it is not necessary for the user to narrate anything before usage.

Fig. 3. SmartNav User Interface.

Finally, there is a product SmartNav from NaturalPoint company. SmartNav uses infrared (IR) camera for monitoring user's head motions [Fig. 3] [10]. The system is based on scanning the user's head using infrared LED. Special reflex facets are placed on the user's head (glasses) and then the radiation is reflected back to a camera which analyses its motion.

All described solutions are very expensive [Table 1]. This fact was one of motivation to develop new solution as a low cost.

Table 1. Existing user interface solutions for handicapped users

Solution name	Price [USD]	Principle
i4control	2200	Eyes movement
IntegraMouse	3400	Lips motion
My Voice	650	Voice recognition
SmartNav	400-500	Head motion
Our Solution	free	Head and eyes motion

As mentioned in the introduction, in this project a virtual camera that scans user's head motion was used in order to enable the user to move over a virtual keyboard. However, this solution method gives rise to certain problems mainly during head detection and its individual parts. Another person might approach the visual field of the camera which might lead to a fault in a detection algorithm. Moreover, it is necessary to consider the distance in which the user's face will be scanned. In order for the system to be cheap and available to everyone, it is also necessary that the system is not composed of parts whose purchase price is equal to hundreds of USD. This is an example of an HD camera. Therefore it is necessary to use ordinary web cameras that are found in most of the notebooks.

These cameras usually have a resolution of around 1.3 megapixels. This means that the scanned image has proportions of 1280x1024 pixels. The image is often programmatically adjusted into a resolution of 640x480 pixels in the case of video scanning in order to fasten the transfer of information on the internet. In this case it is necessary to consider another restricting factor – the distance of a user from camera. An ordinary person is capable to move his/her head in the range of 20 cm. If the person is in distance of 60 cm from the camera, the head movements are going to be scanned in a segment of 150x150 pixels. These values are just approximate, because there are no two web cameras from various producers that would be the same. Furthermore, every one of them has a different defined angle in which its lens is able to scan an image and carry it on CCD sensor. The recognition of head movements in a range of 140x140 pixels is very impractical for mouse movement on a working surface with a resolution of 1280x1024 pixels. A mouse pointer would be likely to move by jumps over the screen which would in turn be caused by slight unwanted head movements of a user.

The mouse movement would be then possible to realize using virtual touchpad with the function which is offered by every notebook. An especially interesting solution might be designed by using accelerometer to scan head motion. However, that is not the objective of this project.

The alternative ways of computer operating which are described in the first part of this chapter are very sophisticating. However, every one of these solutions brings certain disadvantages. Firstly, for using the i4control system a user needs to wear glasses with a camera that is placed a couple of centimeters away from the user's eyes. This might be a very disturbing factor for some people. Moreover, a data cable is needed for connection of the glasses to the computer. This cable might also get attached to some object causing the glasses to fall. A similar situation may happen with *IntegraMouse*. Another system, *My Voice*, can be difficult for usage outside of the home because of disturbing factors such as television or radio in an office or other public place. The last system, *SmartNav*, which is similar to the solution proposed in this project, might be slow for the user because of the necessity of special head movement in order to press a button. Therefore a great emphasis in this project was put on the utilization of all systems' described advantages and their interconnection. This leads to highest possible level of comfort for the user.

3   New Solution

This chapter might be imaginarily divided into two parts. In the first part a necessary theory about programmed detection of face and other parts is given, including information about basic functions of developed programs. The latter part of the chapter introduces the structure of the program itself with description of its individual parts and linkage.

3.1   Face Detection

A detection of a face and both eyes is used for keyboard operating. The face movement represents the movement across the keyboard. Moreover, a reasonably long wink using both eyes gives the system order for pressing a button. Open CV library from Intel company is used in this project in order to launch the camera and detect individual parts. Open CV is a free and open multiplatform library designed for manipulation with image. It is aimed mainly at computer vision and real-time image processing [11]. The CV is an abbreviation which stands for Computer Vision. However, it only works in C/C++ setting which constitutes a disadvantage. Therefore, this project also uses Emgu CV system [12] which is a cross platform for .NET that works as a cover for Open CV libraries indented for image processing.

Moreover, it enables CV libraries to use a function for languages compatible with .NET such as C#, VB, VC++, IronPython etc. [12]. Emgu CV then allows in just few commands image scan by web camera and detection of individual objects on created image. In order to obtain the highest possible speed of the whole application the scanned image is converted into black and white color. The detection itself is then undertaken by DetectHaarCascade command which is based on a high-speed detection algorithm called Viola & Jones (it is the first boundary for detection of objects that carries out real-time detection; it was created in 2001 by Paul Viola and Michael Jones [13]) and another algorithm called Adaboost. Adaboost constitutes an abbreviation for Adaptive Boosting. It is an algorithm of machine learning established by Yoav Freundem a Robert Schapirem. Moreover, it is a meta-algorithm which might be used in conjunction with other learning algorithms in order to improve their performance [14].

The very first argument of a command DetectHaarCascade is so-called Haarcascade (cascade haar style) which is a file of .XML type saved in the same folder as the used application. The detected object is parameterized in this file.

The detection algorithm returns a face variable of VAR data type. Then it is necessary to search only for the wanted face using a command foreach and transfer it to a variable of rectangle type – in this way a rectangle which is represented by coordinates of upper right corner and the size of its sides is acquired. The testing of detection algorithm uncovered that any insignificant zooming in or out of the user's face from the camera would cause unstable changes of the rectangle's size (or rather square, because both width and height are consistent). For this reason, the upper left corner coordinates of the square are inapplicable and the middle of this square was used for the movement across a keyboard. The middle coordinates do not considerably change regarding to different sizes of human faces. These coordinates might be calculated using the following formula:

$$X_{Middle} = X_{Corner} + \frac{W}{2} \; ;$$

where W is the width of a square.

According to a similar formula it is even possible to calculate coordinates Y. In this way those coordinates which are necessary for movement across the virtual keyboard are obtained. It was discovered during another testing of the detection algorithm that the best ratio between distance and sensitivity of a control is within the 211x211 or 295x295 pixels of detected square. (The size of square changes by jumps.) The distance of user's face from the camera in these circumstances was maximally 65cm. If this distance becomes greater, it would not be possible to operate the keyboard just by slight turn of a head, because the face detection itself would fail due to the face being turned too much. Moreover, it would lead to a failure of detection of eyes. In the case when the detected face had size of 211x211 pixels, user's head could be turned by the range of 140x140 pixels.

3.2 Application Structure

The proposed virtual keyboard [Fig. 4] has 7 rows and maximally 14 buttons in one single line. Therefore, the boundary was invented in a way to move to the next button if there is a change by 10 pixels upward or in width. In this way slight head motions and failures of detection algorithm are eliminated.

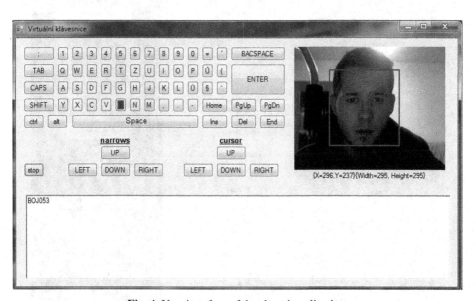

Fig. 4. User interface of developed application

The pressing of a button might be represented in two ways - eye winking or special mouth movement (i.e. opening of a mouth). The first way was chosen for this project, because mouth movement was discovered to cause frequent failures of the system.

The program analyzed a beard as a part of a mouth of an experimental subject. In order to remove this problem it would be necessary to modify or create absolutely new cascade haar style.

Moreover, the detection would have to be done in a colorful image which would significantly increase the time necessary for accomplishment of detected algorithm and reduce the speed of developed application. It was much more convenient to use the scanning of eye motion and, in order to press a button, to close both eyes for a reasonable period of time.

This solution is also well-founded because during the turning of the head, the eyes, especially in limit value, might be covered by the nose. The system is then capable of detecting only one of the eyes. However, at least one of the eyes should be almost always recognizable by algorithm on the acquired image from a web camera. If the algorithm is unable to find either of the eyes for 0.5 – 1.5 s and then detects at least one of them, it will evaluate it as the pressing of a button. In this way it is possible to eliminate shorter autonomous eye winking and also the cases in which eyes would not be detected for longer periods of time (i.e. if a user moves too far from the camera).

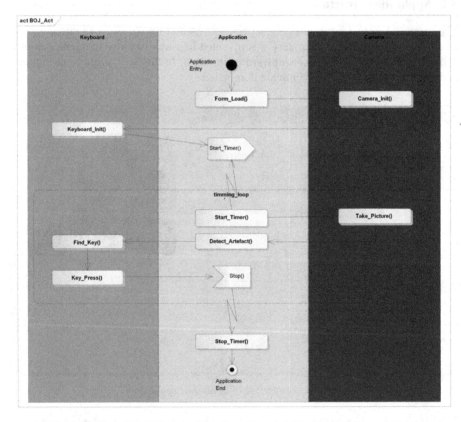

Fig. 5. UML activity diagram of developed user interface for handicapped users

4 Implementation

This chapter is concerned about the practical implementation of the new solution.

The application was developed using the C# language of .NET Framework in Visual Studio 2010 environment from Microsoft Company. The whole graphic page of the application was invented in GUI which is a library of graphical user's boundary. Apart from the usual elements as Button, Picture box or Label, another two components were used to run in the background. Those were Timer and FolderBrowserDialog.

FolderBrowserDialog is a graphic component which serves as an instrument for browsing folders in a computer and after choosing one of the folders it reveals the route towards it. This function was used in the initialization part of the program. The whole initialization runs in Form1_Load method. It is necessary within the program because the application works with two external XML files, in which the two above mentioned cascade styles for detection of face and eyes are saved. Those files are called haarcascade_eye.XML and haarcascade_frontalface_alt2.XML.

After the first initiation of the application, the program firstly determines whether the ini_data.txt file is located in the application's directory. If this file is missing, the application creates it. Consequently, it initiates a system message for a user in which it passes on the information about necessary initialization of the program and the user is asked to insert the path towards files *.XML with cascade styles. The component FolderBrowserDialog is used for inserting this path.

After the next running of the application the program investigates again the presence of initializing file ini_data.txt. If this file is found its content is uploaded and if there are any files .XML with cascade styles detected in the content, a form with virtual keyboard is produced. If the initializing file ini_data.txt is detected, but it does not contain valid data, it is deleted and the user is informed using MessageBox that the application will be closed and it is necessary to reopen it (after a new start-up the whole situation will be running in the same way as during the first initiation of the program).

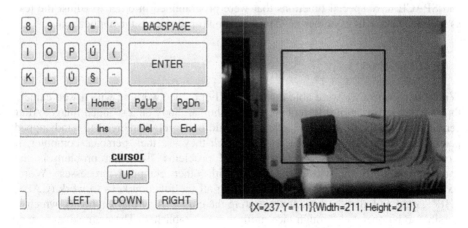

Fig. 6. Part of the application user interface. User face did not recognized – the black rectangle is showed.

When the main window of the whole application is showed, user need to sit down in front of the web camera and position his face into the black square [Fig. 6]. When positioning user face into the middle of the image, the black square should become red and begin to move together with the detected face.

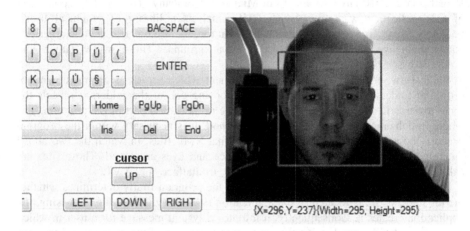

Fig. 7. Part of the application user interface. User face is detected and recognized – the red rectangle is showed around green rectangles at eyes positions

After successful detection of user face, the position and size of user face and also user eyes is being scanned. A red colored active key moves around the keyboard together with the face movement [Fig. 4]. The pressing of this key is done by closing of user eyes for the period of 0.5 – 1.5s. This time depends mainly on the speed of a used computer (processing of an image from the web camera, implementation of detecting algorithm) and also on the speed of the used web camera. The required letter is written into the text field under the keyboard. The buttons ENTER, BACKSPACE and SPACE have special functions that were programmed in order to adjust the text as it is standard with any other keyboard. The button stop ends the application.

5   Conclusions

Within this project we were able to successfully practice working with a web camera through the detection of different parts of the human body on a scanned image. After certain changes the developed application could be largely beneficial to handicapped people and it could simplify the way in which they use their personal computer. It would be necessary to solve two essential problems. The first problem is the communication of the virtual keyboard with other computer processes (Word, Explorer, etc.). The second problem is getting all the function keys to work (CAPS, SHIFT, etc.) and also the keys for operating the cursor. A virtual keyboard, which is possible to be operated through face motion, was established. The main advantage of this project is that the user does not have to buy an expensive hardware (if he has a camera integrated in a notebook). The only thing necessary is to install Emgu.cv software (which is free) and initiate the invented application.

Acknowledgement. This work was supported by „SMEW – Smart Environments at Workplaces", Grant Agency of the Czech Republic, GACR P403/10/1310, We also acknowledge support from student Robin Bojko.

References

1. Zhu, H., Koga, T.: Face detection based on AdaBoost algorithm with Local AutoCorrelation image. IEICE Electronics Express 7(15), 1125–1131 (2010), doi:10.1587/elex.7.1125
2. Sznitman, R., Jedynak, B.: Active Testing for Face Detection and Localization. IEEE Transactions on Pattern Analysis and Machine Intelligence 32(10), 1914–1920 (2010), doi:10.1109/TPAMI.2010.106
3. http://cs.wikipedia.org,
   http://cs.wikipedia.org/wiki/AVI (accessed 8.10.2010)
4. Hershler, O., Golan, T., Bentin, S., Hochstein, S.: The wide window of face detection. Journal of Vision 10(10), Article number 21 (2010), doi:10.1167/10.10.21
5. http://www.inb.uni-luebeck.de,
   http://www.inb.uni-luebeck.de/publikationen/pdfs/
   BoHaRiMaBa09.pdf (accessed 8.10.2010)
6. Peng, Z., Ai, H., Hong, W., Liang, L., Xu, G.: Multi-cue-based face and facial feature detection on video segments. Journal of Computer Science and Technology 18(2), 241–246 (2003) ISSN: 1000-9000
7. http://www.i4control.cz/,
   http://www.i4control.cz/ (accessed 19.10.2010)
8. http://www.lifetool.at/,
   http://www.lifetool.at/show_content.php?sid=218
9. http://www.fugasoft.cz,
   http://www.fugasoft.cz/index.php?cont=myvoice#o-programu
10. http://www.naturalpoint.com,
    http://www.naturalpoint.com/smartnav/products/about.html
11. http://cs.wikipedia.org,
    http://cs.wikipedia.org/wiki/OpenCV (accessed 5.11.2010)
12. http://www.emgu.com,
    http://www.emgu.com/wiki/index.php/Main_Page
13. http://en.wikipedia.org,
    http://en.wikipedia.org/wiki/
    Viola-Jones_object_detection_framework
14. http://en.wikipedia.org, http://en.wikipedia.org/wiki/AdaBoost (accessed 5.11.2010)
15. Kostuch, A., Gierłowski, K., Wozniak, J.: Performance analysis of multicast video streaming in IEEE 802.11 b/g/n testbed environment. In: Wozniak, J., Konorski, J., Katulski, R., Pach, A.R. (eds.) WMNC 2009. IFIP Advances in Information and Communication Technology, vol. 308, pp. 92–105. Springer, Heidelberg (2009)

16. Brad, R.: Satellite Image Enhancement by Controlled Statistical Differentiation. In: Innovations and Advances Techniques in Systems, Computing Sciences and Software Engineering, International Conference on Systems, Computing Science and Software Engineering, ELECTR NETWORK, December 3-12, pp. 32–36 (2007)

17. Mikulecky, P.: Remarks on Ubiquitous Intelligent Supportive Spaces. In: 15th American Conference on Applied Mathematics/International Conference on Computational and Information Science, pp. 523–528. Univ. Houston, Houston (2009)

18. Tucnik, P.: Optimization of Automated Trading System's Interaction with Market Environment. In: Forbrig, P., Günther, H. (eds.) BIR 2010. Lecture Notes in Business Information Processing, vol. 64, pp. 55–61. Springer, Heidelberg (2010)

19. Horak, J., Unucka, J., Stromsky, J., Marsik, V., Orlik, A.: TRANSCAT DSS architecture and modelling services. Journal: Control and Cybernetics 35, 47–71 (2006)

20. Horak, J., Orlik, A., Stromsky, J.: Web services for distributed and interoperable hydro-information systems. Journal: Hydrology and Earth System Sciences 12, 635–644 (2008)

21. Labza, Z., Penhaker, M., Augustynek, M., Korpas, D.: Verification of Set Up Dual-Chamber Pacemaker Electrical Parameters. In: 2010 The 2nd International Conference on Telecom Technology and Applications, ICTTA 2010, Bali Island, Indonesia, March 19-21, vol. 2, pp. 168–172. IEEE Conference Publishing Services, NJ (2010) ISBN 978-0-7695-3982-9, doi:10.1109/ICCEA.2010.187

22. Brida, P., Machaj, J., Benikovsky, J., Duha, J.: An Experimental Evaluation of AGA Algorithm for RSS Positioning in GSM Networks. Elektronika IR Elektrotechnika 104(8), 113–118 (2010) ISSN 1392-1215

23. Chilamkurti, N., Zeadally, S., Jamalipour, S., Das, S.K.: Enabling Wireless Technologies for Green Pervasive Computing. EURASIP Journal on Wireless Communications and Networking 2009, Article ID 230912, 2 pages (2009)

24. Chilamkurti, N., Zeadally, S., Mentiplay, F.: Green Networking for Major Components of Information Communication Technology Systems. EURASIP Journal on Wireless Communications and Networking 2009, Article ID 656785, 7 pages (2009)

25. Dondio, P., Longo, L., Barrett, S.: A translation mechanism for recommendations. In: IFIPTM 2008/Joint iTrust and PST Conference on Privacy, Trust Management and Security, Trondheim, Norway, June 18-20. IFIP Proc., vol. 268, pp. 87–102 (2008)

26. Liou, C., Cheng, W.: Manifold construction by local neighborhood preservation. In: Ishikawa, M., Doya, K., Miyamoto, H., Yamakawa, T. (eds.) ICONIP 2007, Part II. LNCS, vol. 4985, pp. 683–692. Springer, Heidelberg (2008)

27. Liou, C., Cheng, W.: Resolving hidden representations. In: Ishikawa, M., Doya, K., Miyamoto, H., Yamakawa, T. (eds.) ICONIP 2007, Part II. LNCS, vol. 4985, pp. 254–263. Springer, Heidelberg (2008)

Automated Understanding of a Semi-natural Language for the Purpose of Web Pages Testing

Marek Zachara[1] and Dariusz Pałka[2]

[1] AGH University of Science and Technology, Krakow, Poland
mzachara@agh.edu.pl
[2] Pedagogical University of Krakow, Poland
dpalka@ap.krakow.pl

Abstract. The chapter addresses the issues related to software requirement specification (SRS). SRS is usually a part of an agreement between
a software producer and a customer - and is also subject to misunderstanding between these parties. Since formal methods have not been
widely accepted for this process, we propose to use a semi-natural language for this purpose, which is easily understandable for a customer and
yet provides means for automatic modeling (and testing) of the software.
The chapter is focused specifically on one potential area of utilization -
a web page design and testing, although other areas of use are also possible. Practical issues related to implementation of a flexible parser and
web page modeling are described, based a real application developed by
the authors, with a purpose of description-based web page testing.

Keywords: software specification, natural language, web page testing.

1 Introduction

This article addresses a certain aspect of defining the requirements and verifying
a new software, introducing use of a semi-natural language for the purpose of
testing the software against the set requirements.

Over the years, increased capabilities of computer systems made them a desired component of everyday life. Both businesses and individuals alike rely on
software applications - even if they are not aware of this fact. At the same time,
we can observe that the users of the software are rarely satisfied with what they
can get "off the shelf" - a typical product that the producer deemed to be best
for an average customer. Knowing the flexibility of the software systems, and
the relative easiness of their modification, users demand customizations for their
needs, or even custom made applications.

1.1 Customer - Supplier Relations

In times when computers were scarce and their presence limited mostly to academic and other highly specialized laboratories, the end user of the software was
usually also its designer and implementer. From the design point of view this

R. Katarzyniak et al. (Eds.): Semantic Methods, SCI 381, pp. 301–309.
springerlink.com   © Springer-Verlag Berlin Heidelberg 2011

was a preferable situation, since the authors of the software knew exactly what they wanted the software to do.

Nowadays however, software design and implementation is a domain reserved for professionals (i.e. programmers, architects, analyst and so on), who rarely if not never are the users of the produced software. Hence, the need for both parties to agree on a way to convey the customer's needs and requirements to the people producing it. A typical (although very simplified) process of custom software development is presented in Fig. 1.

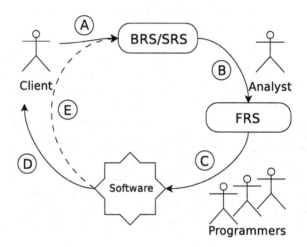

Fig. 1. The client puts down their requirements (*A*) in a form of *Basic Requirement Specification* (*BRS*) or *Software Requirement Specification* (*SRS*), which differ only in the level of details. These requirements are then processed into more detailed technical specification (*B*) usually referred to as *Functional Requirement Specification* (*FRS*). On the basis of the FRS, the programming team creates the code (*C*) which is then handed over to the client (*D*). In case the software does not completely meet customer requirements, additional specification is prepared (*E*) and the cycle repeats.

The major drawbacks associated with this scenario are:

- The BRS/SRS is never complete. Some of the authors had been working in a software development businesses for years, and yet never seen an SRS for anything but a trivial task that completely described customer's needs right from the start.
- The time to develop the software (B) and (C) based on the SRS is long. Thus potential caveats and ambiguities are not known until well into the development process, which result in either low quality code, arguments with the customer, not meeting customer needs - or all of them.

The software industry of course tries to address these issues e.g. by introducing agile methodologies like SCRUM [4] that shorten the B/C time or formal modeling languages like UML [1] that help specify the requirements in more formal way. The formal approach has not gained however a wide audience yet - there are no large scale polls published, but the authors estimate around 20-30% of companies uses UML at all to larger or smaller extend. This is due to the fact that clients does not know (nor want to know) the formal modeling language because they usually do not understand the needs of the programmers, at the same time the programmers do not understand customer's needs. Consecutive iterations of development can lower the gap between client's expectations and reality, but the time it takes to complete each cycle generates costs and lowers the overall satisfaction of the both sides.

1.2   A 'Natural' Functional Description Language

There is a simple and elegant, although still mostly a theoretical solution to the above mentioned issues. If the software could be automatically generated based on the provided SRS, that would eliminate the B/C time issue, allowing the client to check the result and adjust the SRS in real-time. Hopes for such solution tend to rise with new inventions (e.g. UML), yet none of the technologies so far has completely met the expectations. In authors' opinion, one of the reasons is the fact that such technologies and high-level languages still require specific skill from the user, which an average client doesn't have.

An ideal solution would be to process any description in a natural language and verify it against the actual functionality of the software application. However, the natural language understanding faces a lot of problems [3] which does not seem to be overcame in a near future, this making it unsuitable for the task right now. Natural language processing is a huge research subject on its own, with still unresolved issues of precise disambiguation, cultural/knowledge context etc. Yet, the common knowledge of the natural language can be utilized while avoiding the primary issues related to it. If the functional specification could be written in a language that is close enough to 'natural' (e.g. English) to not require the client to learn it, that would provide a noticeable benefit to the process, allowing the client to review and accept it with a high level of confidence. If, at the same time the language would allow for automated parsing and processing of its statements, that could dramatically speed up the development process.

Because of the level of complexity of the subject, the following analysis focuses on a limited functionality related to testing of web pages design against the set requirements, put down in such semi-natural language. Yet the concepts presented can be extended to other functional areas.

2   Graph-Based Modeling of a Web Page

A web page consists of various objects (e.g. text, images, inputs, etc.), each with certain properties (e.g. color, size) and location. This is a direct result of HTML specification [8]. Therefore a web page can be presented in a form of graph, with objects being the graph nodes and the graph edges representing relation between objects, as presented in Fig. 2 (only a small set of possible objects is presented). In this figure the two input fields are declared to be inside of the *Form box* with one input field above the other. Some of the objects also have certain properties declared (e.g. the form box is blue).

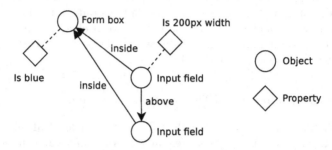

Fig. 2. Sample web page's objects and their relation represented in graph form

To translate a description in a natural language into a similar graph form, several steps need to be performed:

- The textual description is parsed. This phase utilizes knowledge about grammar structures of a natural language (or at least a reasonable subset of these) and a dictionary of words categorized by their type (i.e. object names, available properties and their values). By limiting the natural language to the extend that eliminates certain ambiguities, and by limiting the allowed dictionary to the specific tasks requirements (i.e. web page description), it was possible to parse such description with a parser generated with ANTLR [6].
- Object to object relations and object properties are extracted. Having each element of each phrase labeled (this is done either by matching them with lists for each designated class or by their position within the phrase) it is possible to identify objects (generally nouns), object properties (mostly adjectives) and object relations.
- Objects are identified and matched across the sentences. A unified graph is constructed.

Usually from a single phrase the parser extracts a single node with its properties (e.g. colour, size) or two nodes and their relative relation. Thus after parsing the textual description a large set of such small 'clusters' is available. By matching the objects across sentences (using the method described below), the nodes are joined forming a one graph that supposedly holds all the information about all the elements described.

This process is presented in Fig. 3.

The Meaning/Class of the Words is covered by supplying categorized dictionaries. Limited scope of application (web design) make it possible to classify all (or most) of the words that might be used for SRS/FRS. This in turn allows for more precise matching of various language constructs (e.g. elliptical sentences).

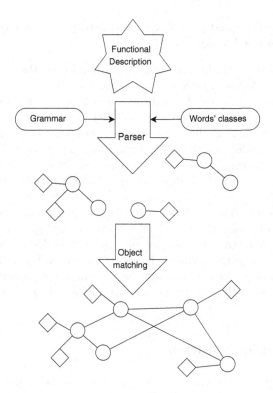

Fig. 3. Translating a textual description of a web page into a graph form

The Positional Context is resolved by utilizing articles (i.e. *a, the*), and by matching usage of the same object names across neighbor sentences. The last assumption is backed by the results of the experiment [2] which confirmed that people mostly describe their requirements linearly (one element at a time and moving from one element to another). If the description needs to refer to an object defined much earlier in the scope, a direct object naming can also be used.

3   Model Verification

As mentioned in the introduction, the textual description has been utilized for verification of an actual web page. In order to do so, two separate graphs are created:

– One graph (A) is constructed on the basis of the specification, as described above.
– The other graph (B) is constructed by analyzing the supplied web page - by rendering the page and traversing its DOM, looking for elements with a physical representation. This way object, their properties and relative

locations are extracted - and a graph can be constructed. The method used for this task is described in detail in [7].

The comparison of the two graphs is done in two stages: first, the object/node properties are matched, followed later by constrains resolution.

During the *first stage*, each node of the graph A is compared against each node of the graph B, taking into account nodes' types (e.g. input, button, etc.) and properties (e.g. color, size). This allows for creation of one-to-many assignments between the graph (A) nodes and the objects identified in the page (denoted by graph B nodes). At this stage, having at least one node in graph A that does not match any node in the other graph means that the page does not meet the provided specification.

During the *second stage*, the algorithm attempts to achieve a one-to-one relation between the graphs A and B nodes. In the reference implementation this is done by a brute-force matching of possible combination of nodes. Although this is not a computationally effective solution, it does work reasonably well since the number of nodes in graph A does not exceed a few dozens in most cases. Besides, comparing two graphs is really a graph isomorphism issue, which is generally of NP class [5].

4   Reference Implementation

As a proof of concept, a reference implementation has been prepared. Its schematic diagram has been presented in Fig. 4. In order to parse the supplied description, an ANTLR parser generator was used. Based on the supplied rule sets two parsers were generated.

The first parser processes the description into an Abstract Syntax Tree (AST). This is a commonly used process in text parsing. At this stage, the tree reflects the structure of the textual description and not the actual web page being described. The parser leaves however in the tree nodes additional information related to recognized object properties (e.g. name, type, etc.) that is later used for identifying the nodes/objects.

The second parser (tree parser) traverses the AST, retrieving the information from the nodes and builds the final graph. During this process, while executing semantic actions it uses the tree structure to determine contextual information (e.g. to identify what 'this' or 'the button' refer to). It also uses the information previously stored in the AST nodes to identify the same objects across separate branches. This allow for aggregation of properties described in separate phrases that relate to the same object.

Finally, the properties that are identified are classified as either

- attributes - which are the properties that are related to the abject alone,
- constrains - that describe relative relationship between two object/nodes

The final product of the specification analysis is an XML document. A fragment from such document is presented below:

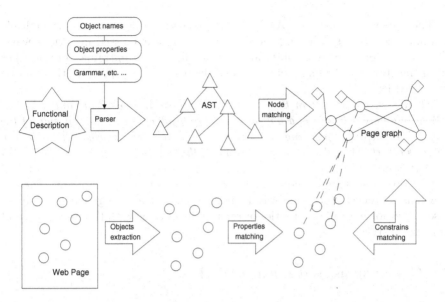

Fig. 4. Key steps performed for verifying an existing web page against a written specification

```
<object id="7" type="button">
    <attribute name="color" value="gray"/>
    <constrain name="left side of">
        <participant id="1" type="page"/>
    </constrain>
    <constrain name="has">
        <participant id="8" type="text"/>
    </constrain>
</object>
```

Independently of the specification, the page supplied for testing is analyzed. For this purpose an open-source Qt WebKit [9] engine is used. It allows for loading and rendering of the provided web page, providing access to the constructed DOM. While traversing the DOM, all the elements of a non-zero physical size are identified. As a result of web page analysis, a list of objects with their properties (retrieved from the DOM), physical location and size is constructed.

The XML from the specification analysis and the list of objects from the web page traversing is then fed into a comparator that utilizes the described above two-stage matching, with brute-force verification of constrains during the second stage. Practical experiments show that the matching time for the test cases were negligible (most of the time was spent on rendering the web page).

The parser generated by ANTLR with provided rules can handle multiple grammar constructs, allowing the description to be expressed in various ways. This allows for more readability and relative use by an untrained person right from the start. An example of functional description (a part of it) that can be processed is:

There is a red header in the page. It is centered and has italic and bold text "Welcome here". There is a button below the header and to the left side of the page. The button is gray and has a text "Press me". To the right of the button there is a text "Why don't you want to press the button?". The header is written in font 24px.

The object matcher handles *it, the* or reuse of object name as reference to previous sentences, the same way humans do in natural language. The parser also allows for multiple properties or relation definitions with the use of *and* and similar constructs.

5   Conclusions and Future Work

In this chapter we propose use of a semi-natural language for functional description of the software specification. The initial implementation confirms it can already be used for automation of certain phases of the development cycle, i.e. functionality tests of a web page design. Automatic construction of a web page based on such description seems to be achievable and will be a natural next step in this project development. Another obvious area that needs to be addressed is utilizing a similar approach for the purpose of describing a web service functionality - the area which is very often subject to misunderstanding between the client and the developer of the software.

Another area that could benefit from further work is an introduction of more generalizations to the parser. Although ANTLR is a quite a flexible tool, utilizing it for the purpose of even 'semi' natural language processing requires a vast amount of rules (mostly variations) as well as rather artificial injection of word categories as lists of alternatives in the match expressions. Therefore building of a specific parser or a generator of ANTLR rules from more general descriptions is being considered.

Finally, the parser could benefit from an adaptive grammar capability. Since the language constructs tend to differ slightly between groups of people, matching the unknown phrases with existing pre-set expressions in terms of grammar and vocabulary used, would allow the parser to get the user confirmation on meaning being exact to one of the known constructs. This would in turn allow for automatic extension of the set of rules - and at the same time building a personal extension to the grammar, covering the *personal* (or *cultural*) context which is specific for individual persons. This is yet a very preliminary work though.

References

1. Booch, G., Rumbaugh, J., Jacobson, I.: Unified Modeling Language User Guide. Addison Wesley Longman Publishing Co. (1999) ISBN:0-201-57168-4
2. Dutkiewicz, L., Grobler, K., Orzechowski, P., Pałka, D., Piskor-Ignatowicz, C., Zachara, M.: Research on automation of web service building processed. In: Automatyka, pp. 841–851. Wydawnictwa AGH, Krakow (2010)
3. Sætre, R.: Natural Language Understanding (NLU) - Automatic Information Extraction (IE) from Biomedical Texts Masters Thesis, Norwegian University of Science and Technology (2003)
4. Schwaber, K.: SCRUM Development Process. In: Proceedings of the 10th Annual ACM Conference on Object Oriented Programming Systems, Languages, and Applications (OOPSLA), Unites States, pp. 117–134 (1995)
5. Skiena, S.: §5.2 Graph Isomorphism. In: Implementing Discrete Mathematics: Combinatorics and Graph Theory with Mathematica. Addison-Wesley, Reading (1990)
6. Terence, P.: The Definitive Antlr Reference: Building Domain-Specific Languages Pragmatic Bookshelf (2007) ISBN: 0-978-73925-6
7. Zachara, M., Pałka, D., Dutkiewicz, L., Grobler, K.: Automatic detection and separation of web services template. In: Borzemski, L., et al. (eds.) Information Systems Architecture and Technology: New Developments in Web-Age Information Systems. Oficyna Wydawnicza Politechniki, Wrocławskiej (2010) ISBN 978-83-7493-541-8
8. HTML 4.01 Specification W3C Recommendation, http://www.w3.org/TR/1999/REC-html401-19991224/
9. QtWebKit - Online Reference Documentation, http://doc.qt.nokia.com/qtwebkit.html

Emerging Artificial Intelligence Application: Transforming Television into Smart Television

Sasanka Prabhala and Subhashini Ganapathy

User Experience Research Group, Intel Corporation
2111 NE 25th Avenue, Hillsboro, OR 97124 U.S.A.
{sasanka.v.prabhala,subhashini.ganapathy}@intel.com

Abstract. The science of Artificial Intelligence (AI) has been applied in different fields such as financial systems, health-care systems, transport systems, and military systems. However, there has been very little research and application of AI solutions in the field of consumer electronics (CE) especially as related to TV and TV related products even with the arrival of Smart TV and Smart TV products in the marketplace. This paper examines the user research on TV practices, discusses the experiences consumers' desire and want, and the AI tools necessary for the development efforts of true Smart TV and Smart TV products.

Keywords: Artificial Intelligence, Smart Television, Context Rich Interactions, User Experience.

1 Introduction

The science of Artificial Intelligence (AI) has been applied in different fields. For example, AI has been applied to (a) financial systems for stock investment [1] and property management [2], (b) health-care systems for patient monitoring [3] and medical diagnosis [4], (c) transport systems for air traffic control [5] and automatic incident detection [6] and (d) military systems for command and control of unmanned aerial vehicles in search, rescue, and destroy missions [7, 8] as well as in dynamic path planning [9]. Although AI solutions have been successfully applied in these fields, there has been very little research and application of AI solutions in the field of consumer electronics (CE). Specifically, as related to television (TV) and TV related products.

In the past 75 years, research and development in the field of TV has come a long way and evolved from black and white to color TVs, from recorded to live programming, from prime-time only to 24 hour programming, from over-the-air to cable/satellite broadcasting, from analog to high-definition (HD) programming, and from dumb TVs to 3D-TVs. However, there seems to be a lack of research and application of AI solutions in this field.

In recent times, advancements in silicon, networking and digital signal processing technologies as well as the widespread adoption of high speed Internet led to the development of Smart TV and Smart TV products. These include standalone products

R. Katarzyniak et al. (Eds.): Semantic Methods, SCI 381, pp. 311–318.
springerlink.com   © Springer-Verlag Berlin Heidelberg 2011

(such as Chumby*, Kodak Theater HD Player*, Roku*, Vudu*, VuNow*), set-top-boxes (such as Archos TV+*, Moxi*), gaming consoles (such as PS3*, Xbox 360*, Wii*), Blu-ray Disc* players (such as LG* BD390, Sony* BDP-N460), and Internet enabled TVs (such as HP Media Smart*, Panasonic VieraCast*, Samsung Info Link*, Sharp AquosNet*).

As the name indicates Smart TV and Smart TV products enables consumers access to (a) Internet-based content, (b) place shifting and time shifting of content, (c) display of photos stored on PCs or photo sharing sites, and (d) streaming of high definition premium content and movies in addition to delivering broadcast content previously unavailable on the TV screens. Although, these products are called Smart TV products, in essence, these products are not smart enough. For example, these products do not really know and understand consumers in terms of content they like and content they view to remind them or record the content automatically if that channel s not tuned in.

Developing Smart TV and Smart TV products is much more complicated than simply integrating a Wi-Fi* or an Ethernet jack, or attaching a cable modem or a digital media adapter (DMA), or a smart set-top-box (STB). So, the challenge here is how to make the Smart TV and Smart TV products really smart. In other words, what features and capabilities need to be enabled on these products to make them smart that consumers would want in their living room.

This paper outlines the research, design, and development of a User Experience (UX) driven features and capabilities that truly make the Smart TV and Smart TV products smart. Specifically, the paper outlines the user research on TV practices, discusses how to enable experiences consumers' desire and want, and concludes with a list of AI tools that are critical to driving Smart TV experiences.

2   User Research on TV Practices

The Intel User Experience Research Group is an interdisciplinary team that investigates the daily lives of people around the world and their relationship with technology. Through multi-sited user research in homes in a growing number of countries, the team identified core attributes and experiences consumers love about TV. For instance, consumers talk about TV as many things: as an object, as a technology, as a language, as a relationship, as an addiction, and as an experience. Furthermore, consumers value TV because it is always there and always working. Specifically, consumers highly value TV because it is Social, Flexible, and Simple.

First, TV is social because it enables (a) multiple viewers more often than not, (b) different sets of viewers in a typical day, (c) a wide range of viewers (different ages, gender, and interests), and (d) people to spend time together and to interact both in front of it, while co-located and while away.

Second, TV is flexible because it entertains and does much more: it acts as a companion, advisor, time-killer, necessity, educator, social life enabler, baby-sitter, boredom-buster, stress-fighter, lullaby, and low-maintenance friend. Even when the TVs are old, they are still viewed as adaptive or flexible as they take on new roles in the home such as TV for playing games or TV for watching movies.

Third, TV is simple because it's undemanding. From a consumer's standpoint, TV is always there and always works. There is no blue screen of death. Also, TV doesn't demand a user's full attention or continuous attention to provide value; it is easily multi-tasked and makes other activities undertaken while it's on more bearable – eating meals, washing dishes, cleaning the house and other types of work, for example.

In addition the team also focused on understanding (a) consumers core values, beliefs, and cultural practices around TV, (b) consumers issues and frustrations with TV and how they adapt to address the issues and the frustrations, (c) consumers interaction with the TV, and (d) consumers desires and expectations from the TV. The below two examples of William (section 2.1) and Karen (section 2.2) provides insights into the experiences consumers love and want from their TV.

2.1 Case-in-Point: William

William is a 50-year-old operations manager in a factory. His wife is a retired childcare provider, who still works occasionally. They live in Leeds, UK in a home in which they raised their two daughters. William likes interacting with people he already knows. He's very safety and security conscious, protective of his daughters, of his home, and of his privacy, which is reflected in the fact that he was a member of the special constabulary for almost 20 years. William's home has a security camera for the front door, and he has installed alarm systems in each of his daughters' homes who live 10-15 minutes away. He doesn't use the front door, but rather the side door – that way he knows if someone comes to the front, he doesn't know them (and likely won't answer the calling bell). William has long been interested in communication technologies. For William, technologies are not for making contact with unknown, 'new' people, but for staying in contact with those he already knows 'offline'.

In general, William likes to get a "good deal", and to save money. Because he is generally interested in technology and tinkering with technology, he likes to buy old items inexpensively, fix them up and then make a profit when he resells them, generally at boot sales (flea markets, in American English). He hardly ever buys home electronics new – almost always used, and usually from someone he knows – that way he knows and trusts that the product will work. He's also quite content to wait until the price on a CE product that he wants comes down to an amount that he's willing to pay. In addition to only buying used CE products - or very occasionally a new one at a store (which happens only if he knows "a chap" who works there, and can therefore get him a good price) – William is given old TVs and other CE products by family and friends, and has been known to 'take in' technologies from friends and relatives who were otherwise going to throw them out. He can't bear the thought of a working television being thrown out, so he takes them in, just in case his daughters might need one sometime in the future.

William has a total of 11 TVs. He has a small portable TV in his kitchen, which gets 5 channels. In his living room he has two televisions; one hooked up to a cable box and the other hooked up to aerial antenna. William says that watching shows (depending on the time of the day) on 3 different TVs helps him quickly find the shows he wants in case he missed watching them.

For William, television is an important part of his day and an activity that evokes both pleasure and guilt. As he explained, he likes to watch when he comes home from work, and he finds it easier and more enjoyable to watch TV than to do work around the house, or to work on his hobbies.

2.2 Case-in-Point: Karen

Karen is a 28-year old part-time student in suburbs of London, UK. Everyday Karen's alarm goes off at 6:30AM. First thing she reaches for are her two mobile phones (work and personal) to check for urgent work messages that need an immediate response, and Facebook updates that orient her to what is happening in her social world. Next the television is turned on (from bed) while she gets dressed. The TV in the Lounge is switched on as soon as she goes downstairs for the information on GMTV (a morning TV show for news headlines, celebrity gossip, weather, etc) and for the time, which is displayed in larger numbers in one corner of the TV screen. On her train commute into London, she uses here wireless adapter for Internet access to her netbook – she checks and writes email, Facebook, and works on her MBA course-work.

2.3 Making Sense of User Research

In general if consumers were asked what their desires are or what experiences they want from their TV, most often than not, there would be no answer. This is because consumers don't know what they want. However, from the user research it is possible to get a good understanding of what consumers' desires are. Through the multi-sited TV user research and through William and Karen's TV practices, it's clear what experiences consumers' desire and want from their TV.

- Consumers want smart and personalized TV that can understand profile and preferences and can provide personalized content
- Consumers want continuous entertainment that can connect content across products, platforms and people with TV at the center
- Consumers want context rich content interaction that can connect content across multiple application frameworks in an integrated user interface (UI)
- Consumers want advanced forms of interaction with TV that can enable navigation and control of fast responsive 3D UIs

3 Enabling Experiences Consumers Desire and Want

From the user research on TV, it is clear that consumers desire and want new experiences on their TV, but, expect their TV to be social, flexible, and simple. Therefore, it is critical to deliver these new experiences on the TV platform without breaking what consumers love about their TV. In addition, enabling these new experiences on the TV platform would also differentiate products in the marketplace.

TV user research also provided direction for designing UIs, interaction models, content expectations and delivery of content.

Different usage scenarios that correspond to the experiences indicated in section 2.3 were designed as storyboards. These storyboards were put through User Experience Assessment (UXA) method to measure consumers' likes, dislikes, thoughts, perceptions, expectations, desirability, attitude, and values [10, 11]. Typically, UXA study is conducted by recruiting real consumers (from targeted consumer segmentation); interacting with product features or usage concepts or prototype; in a focus group or one-on-one setting; assessing consumer usage and desired experiences. UXA study was critical as it provided data from a consumer viewpoint, on whether the usage scenarios designed delivered the experiences that consumers desire and want. Section 3.1 – 3.4 provides examples of the usage scenarios associated with each of the experiences that consumers desire and want.

3.1 Smart and Personalized TV

Fig. 1 specifies how consumers can access content they like on the TV irrespective of where the content is coming from (broadcast TV, broadband TV, DVR, VOD, etc). In this scenario, the TV provides consumers with content that suits their taste depending on who is watching. So, from the consumers' point of view, they can watch a show on the TV itself even if the show is not a broadcast show.

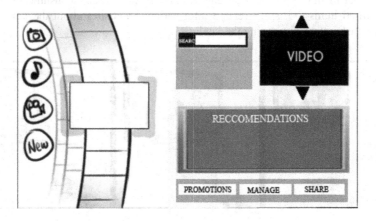

Fig. 1. Access to personalized content from different sources

3.2 Continuous Entertainment

Fig. 2 specifies how mobile products can be used as extensions of TV screen for continuous entertainment. In this scenario, consumers while watching TV content can use a mobile product (a laptop, photo frame, or any mobile Internet product) to browse for electronic program guide (EPG), Internet-based content, and/or personal content. Consumers can also send content from their mobile products to the TV. The TV content is paused while switching content from the mobile products. In addition, the mobile products can also act as a remote control to change content sources if desired.

Fig. 2. Mobile products as extensions of TV screen

3.3 Context Rich Content Interaction

Fig. 3 specifies access to embedded contextual content for context rich content interaction. In this scenario, consumer TV recognizes objects on-screen and links them to other media content (e.g., identifying the Eiffel Tower). Consumers can select the identified items to display related information – websites about the landmark or home videos/pictures containing the landmark.

Fig. 3. Embedded contextual content interactions

3.4 Advanced Forms of Interactions

Fig. 4 specifies shared experiences with friends through avatars and gesture interactions. In this scenario, connected products enable group of friends to actively participate in a virtual play. To join the play, each user is recognized and their customized avatars are displayed on-screen. User actions and gestures are also captured and mirrored on-screen. Users can also join as spectators and are added to the background for greater enjoyment and entertainment.

Fig. 4. Shared experiences with friends through avatars and gesture interactions

4   What's Needed in Driving These Experiences?

The TV user research provided insights into consumers' daily TV practices and their desires for new TV experiences. This led to the creation different usage scenarios which were put through UXA process to understand from consumer viewpoint the experiences that consumers desire and want. However, in order to drive the experiences into reality, it is essential for the TV to have intelligence. For example, to drive context rich content interaction experiences, it is essential for the TV to have intelligence such as context sensing and Inferencing, image processing, object recognition, and seamless connectivity with other trusted products. Table 1 lists the AI tools that are critical in order to drive the Smart TV experiences (smart and personalized TV, continuous entertainment, context rich content interaction, and advanced forms of interactions) that consumers' desire and want.

Table 1. List of AI tools and methods critical to drive Smart TV experiences

- Machine learning
- Context sensing and interfacing
- Computer vision and image processing
- Real and virtual world blending
- Object recognition
- Gesture/motion recognition and processing
- Video analytics
- Low power sensing
- Speech and audio recognition and processing
- Cloud computing
- Digital rights management and seamless connectivity

Although, the list of AI tools in Table 1 are not new, the author argues that applying these tools to drive Smart TV experiences is not that straight forward as it sounds. For example, TiVo* recommends shows to the user based on their (a) profile, (b) rating system - thumbs-up and thumbs-down (whether they liked the show or not), and (c) distributed collaborative filtering method [12]. However, TiVo doesn't take into consideration an important factor which is who else is in the room watching the show. As noted in Section 2, TV is highly valued by consumers because it is Social, Flexible, and Simple. Hence, there is definitely a need for more research in refining the AI tools that take into consideration the different factors of how consumers watch and enjoy TV as well as provide them with the new experiences they desire and want.

References

1. Lee, K.C., Kim, H.S.: A Fuzzy Cognitive Map-Based Bi-Directional Inference Mechanism: An Application to Stock Investment Analysis. Intl. Sys. in Acct. Fin. and Mang. 6, 41–57 (1997)
2. Tay, D.P.H., Ho, D.K.H.: Artificial Intelligence and the Mass Appraisal of Residential Apartments. J. of Prop. Val. and Invt. 10(2), 525–540 (1992)
3. Tatara, E., Cinar, A.: Interpreting ECG Data by Integrating Statistical and Artificial Intelligence Tools. IEEE Mag. on Eng. Med. and Biol. 21(1), 36–41 (2002)
4. Wollkind, S., Valasek, J., Ioerger, T.R.: Automated Conflict Resolution for Air Traffic Management using Co-Operative Multi-Agent Negotiation. In: Proceedings of the American Institute of Aeronautics and Astronautics Conference on Guidance, Navigation, and Control, Providence, RI (2004)
5. Srinivasan, D., Xin, J., Chue, R.L.: Evaluation of Adaptive Neural Network Models for Freeway Incident Detection. IEEE Tran. on Intl. Trans. Sys. 5(1), 1–11 (2004)
6. Prabhala, S., Ganapathy, S.: S. Narayanan, Gallimore, J.J., Hill, R.R.: Model-Based Simulation to Examine Command and Control Issues with Remotely Operated Vehicles. In: El-Sheikh, A.A., Al-Ajeeli, A., Abu-Taieh, E.M. (eds.) Simulation and Modeling: Current Technologies and Applications, pp. 199–219. IGI Publishing, NY (2007)
7. Gallimore, J.J., Prabhala, S.: Creating Collaborative Agents with Personality for Supervisory Control of Multiple UCAVs. In: Human Factors and Medicine Panel Symposium on Human Factors of Uninhabited Military Vehicles as Force Multipliers, Biarritz, France (2006)
8. Prabhala, S., Ganapathy, S., Gallimore, J.J.: S. Narayanan, Hill, R.R.: Interactive Optimization for Unmanned Aerial Vehicle Routing. In: Proceedings of the Summer Computer Simulation Conference, Philadelphia, PA (2005)
9. Beauregard, R., Younkin, A., Corriveau, P., Doherty, R., Salskov, E.: Assessing the Quality of User Experience. Intel Techn. J. 11(1) (2007)
10. Prabhala, S.: User Experience Assessment Program, unpublished
11. Ali, K., Stam, W.A.: TiVo: Making Show Recommendations Using a Distributed Collaborative Filtering Architecture. In: Proceedings of the 10th ACM SIGKDD International Conference on knowledge Discovery and Data Mining, Seattle, WA (2004)

Secure Data Access Control Scheme Using Type-Based Re-encryption in Cloud Environment

Namje Park

Department of Computer Education, Teachers College, Jeju National University,
61 Iljudong-ro, Jeju-si, Jeju-do, 690-781, Korea
namjepark@jejunu.ac.kr

Abstract. Cloud computing service provider cannot be totally trusted due to data security reasons, risk of data security and violation of privacy factors should be considered. Especially, guaranteeing data confidentiality is required. To solve these problems, S.C. Yu etc. proposed scheme which guarantees data confidentiality and fine-grained access control. However, data confidentiality can be violated by collusion attack of revoked user and cloud server. To solve this problem, we guaranteed data confidentiality by storing and dividing data file into header and body. In addition, the method of selective delegation regarding the whole or partial message according to delegator's reliability towards delegate using type-based re-encryption was specified.

1 Introduction

Cloud computing is a promising computing paradigm which recently has drawn extensive attention from both academia and industry. Especially, cloud computing is used in internet business for data storage service. Cloud computing has five kinds of characteristics (Multitenancy, Massive scalability, Elasticity, Pay-as-you-go, Self-provisioning resources). Enterprise utilizes these characteristics to increase revenue [7].

However, in spite of these useful characteristics, cloud computing should consider several issues to use your existing business environment [1]. In particular, medical data that relate to certain person must be protected to ensure data confidentiality. Thus, we focus on confidentiality. We propose secure protocol model for cloud computing environments in this paper. Also, we suggest specific models to process medical data. On the other hand, protocol model in cloud computing environments must be guaranteed to fine-grained access control as well as data confidentiality. For example, access control should be differentiated by user's attribute such as positions (doctor, nurse, etc). In addition, it must be able to revoke decryption right for data access. Thus, user revocation would require re-encryption of data files accessible to the leaving user.

To ensure these requirements, [6] proposed system model using Key Policy-Attribute Based Encryption (KP-ABE) and Proxy Re-Encryption (PRE). Formally, [6] ensure data confidentiality using KP-ABE and data owner delegate computation overload to proxy using PRE. Also, [6] introduce user revocation on system model. But, proposed scheme is vulnerable to collusion attack of revoked user and cloud server.

R. Katarzyniak et al. (Eds.): Semantic Methods, SCI 381, pp. 319–327.
springerlink.com   © Springer-Verlag Berlin Heidelberg 2011

And there is no selective delegation for level of trust. To resolve these challenges, we divide data files stored on the cloud servers into header, body in [6]. Decryption rights such as head (encrypted key using KP-ABE) and body (encrypted message using symmetric encryption) is concentrated to cloud servers in [6]. For this reason, cloud server is the most powerful. This paper begins to point that cloud servers can't be trusted. So we create the trusted authority named privilege manager to manage header. And body is stored on the cloud servers like existing way. Thus, we will split the power that concentrated to cloud servers. Also we introduce data access privilege management model using Type-based Proxy Re-encryption's concept in mobile cloud environments. Finally, we show scenario about privacy preserving data sharing in health cloud environments and analyze security against collusion attack.

2 Related Work

2.1 Bilinear Mapping

Let G and G_T be two cyclic multiplicative groups with the same prime order q. A bilinear pairing is a map e: $G \times G \rightarrow G_T$ with the following properties:

- Bilinearity: $\forall g_1, g_2 \in G, \forall a, b \in Z_q^*$, we have $e\left(g_1^a, g_2^b\right) = e(g_1, g_2)^{ab}$.
- Non-degeneracy: There exist $g_1, g_2 \in G$ such that $e(g_1, g_2) \neq 1$.
- Computability: There exists an effcient algorithm to compute $e(g_1, g_2)$ for $\forall g_1, g_2 \in G$.

2.2 Attribute Based Encryption

Attribute Based Encryption is classified as CP-ABE (Cipher-text Policy-Attribute Based Encryption) [3] and KP-ABE (Key Policy-Attribute Based Encryption) [8]. In this paper, we should take advantage of KP-ABE. Data are encrypted by set of attributes and user secret key are associated with access structure in KP-ABE. In access structure, internal nodes are threshold gates and leaf nodes are associated with attributes which is used to encrypt data. Thus, if encrypted data's attribute satisfy user secret key's access structure, user is able to decrypt a cipher-text. Formally, KP-ABE is used to encryption and decryption for Data Encryption Key (DEK) in proposed scheme.

2.3 Proxy Re-encryption

Proxy Re-Encryption is cryptographic scheme in which proxy is able to convert cipher-text encrypted under Alice's public key into cipher-text that can be decrypted by Bob's secret key. But, proxy is semi-trusted server so that proxy server should not be seen original plaintext in PRE. Formally, user should update header's secret key component using [5] in proposed scheme.

2.4 Existing Study for Cloud Computing

[6] forecasted that existing business environment will move to cloud computing environment, and proposed a system model which can guarantee data confidentiality. The

problem of data confidentiality in computing environment begins from the untrustworthy cloud service provider. Receiving and sending documents related to each field in plaintext would make it convenient, but if they are transmitted through an untrustworthy cloud service provider, data confidentiality becomes violated. Therefore, encrypting data as attribute set and containing a secret key related to the access structure which satisfies attribute set of data, the concept of KP-ABE which can see data can be utilized. That is, by encrypting data as DEK (Data Encryption Key), body is composed, and by encrypting DEK as attribute set, header is composed. Combining header and body as such, data file is composed to transmit to the cloud service provider. When the user group intends to access data, if the possessed secret key satisfies the attribute set of header, DEK can be gained and read the data of the body.

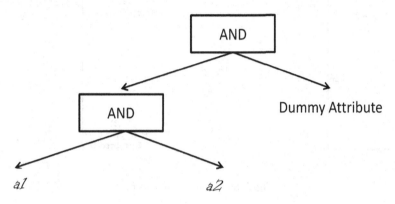

Fig. 1. Access structure in paper [6]

Regarding the Method of [6], if the data owner defines the attribute universe, creates and sets the key related to system, and encrypts data under DEK and encrypts DEK by KP-ABE to send to the cloud service provider, the cloud service provider entirely manages the encrypted data and DEK. That is, the data owner does not have to be on-line. In [6], if the user does not have the same version of attribute while updating the version of attribute using PRE to efficiently revoke the user in the system, operation will be impossible. For the cloud service provider to perform such function, the data possessor should create an access structure including dummy attribute as in Figure 1 so that illegal use of system master key is prevented by delivering access structure without dummy attribute.

However, the Method of [6] can violate data confidentiality through collusion of revoked user and cloud server. Figure 2 is the system model of [6]. Describing the Figure 2, the combination (data file) of encrypted data under DEK (body) encrypted DEK by KP-ABE (header) is sent to the cloud service provider, and when the user group requires data access, the data file (combination of header and body) can be transmitted. In [6], the revoked user disables decryption for body even if receives data file by updating the version of used attribute when encrypting the header. However, if the revoked user and cloud server collude, so if the version of the attribute of the encrypted data for the revoked user is not updated, data access is possible.

The proposed scheme is to prevent the collusion attack by storing and dividing data file into header and body by having trusted authority called privilege manager group. That is, the header is sent to privilege manager group, and the body to the cloud service provider. Through these schemes, data confidentiality is guaranteed by the user group necessarily having to gain admission by privilege manager group.

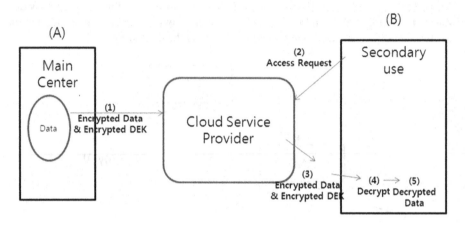

Fig. 2. System model in paper [6]

3 Proposed Scheme

3.1 Main Idea

The proposed scheme, in [6], data file stored in the cloud server is divided into header and body to store to the privilege manager group and the cloud service provider. In the header, DEK used for encrypting data into body is encrypted to KP-ABE. In other words, if the user is not authorized by privilege manager group, the user cannot gain information of DEK within header, and actually, the user intends to gain data within body becomes completely impossible. Thus, data confidentiality is guaranteed by storing and dividing data file stored in cloud server into header and body. Also, by having privilege manager group within cloud, user access is managed, and selectively delegation according to weight (type information) set by the data owner is enabled. To specify such function, Type-based PRE is utilized. According to the level the data owner trusts the user group, the data is divided into the whole part or parts, and type information is granted to each, and is encrypted. Through this, data owner can selectively delegate decryption right for decrypting the whole or part of the data to the user group.

3.2 Scheme Description

Table 1 is description about notation used in our scheme.

Table 1. Notation

Notation	Description
PK, MK	system public key and master key
A_i	public key component for attribute i
a_i	master key component for attribute i
SK	user secret key
sk_i	user secret key component for attribute i
E_i	Cipher text component for attribute i
I	attribute set assigned to a data file
DEK	data encryption key of data m
P	user access structure
L_P	Set of attributes attached to leaf nodes of P
T	Type information set
Att_D	The dummy attribute

Component in our scheme consists of 4 groups (data owner, privilege manager group, cloud service provider (CSP), user group). Our scheme is to add two algorithm level (A Encrypt B, A Decrypt B). And we will present only changed parts in existing scheme. Clearly, the description for unchanged parts shall be excluded.

① System Setup
Data owner enters system attribute set $U=\{1,2,\ldots,N\}$ and security parameter k to set the system level ($PK,MK \leftarrow ASetup(k)$) which generate PK and MK. Utilize proxy re-encryption scheme by adding the part where type information set T is defined for user group.

- Define bilinear group G_1 of prime order p having creator g, bilinear mapping $e: G_1 \times G_1 \rightarrow G_2$, and Hash function $H: \{0,1\}^* \rightarrow Z_q^*$. Define type info set T $(T=(t_1, t_2, \ldots, t_N))$ for user group ($t \in \{0,1\}^*$).
- Creat $Y \in G_2$ ($Y= e(g,g)^y$, $y \in Z_q$) for attribute $A_i \in G_1$ ($A_i = g^{a_i}$), i, $a_i \in Z_q$ $(1 \le i \le N)$.
- Create system public key $PK=(Y, A_1, A_2, \ldots, A_N)$ and master key $MK=(y, a_1, a_2, \ldots, a_N, T)$ and sent (PK , $\delta_{O,PK}$) with the signature on PK to manager group.

② New File Creation
The system level enables the data owner to encode the file before sending it to privilege manager group and CSP.

- Select the ID for data m and $r \in Z_p^*$ randomly. Calculate $DEK = q^{B(r\|t)}$ by entering type info t. $((\tilde{E}, \{E_i\}_{i \in I}) \leftarrow AEncryptH(I, DEK, PK))$
- Define attribute set I for data m and encode $DEK = q^{B(r\|t)}$ into KP-ABE.
- Save the header to privilege manager group and body to CSP in the same way as shown in fig. 3.

	header	body
ID	$I, \dot{E}, \{E_i\}_{i \in I}$	$m \cdot g^{H(r\|t)}$

Fig. 3. Data file's Format

③ New User Grant

The system level where data owner allocates the access structure P of user w when the new user w wants to attend to the system.

- Start with AND Gate only as access structure P is transmitted through privilege manager group. Add dummy attribute Dummy Attribute Att_D. (The leaf node for access structure P should be known to perform user cancel function for privilege manager group. System master key t can be utilized if all access P is sent. Therefore, add Dummy Attribute Att_D to access structure P and send access structure Q except Dummy Attribute Att_D for privilege manager administrator group). Define polynomial $p_i(x)$ for each node i of access structure P. Define $p_r(0) = y$ for access structure P's root node and $p_i(0)$ for leaf node. Create secret key $SK = \{sk_i\}_{i \in L_P}(sk_i = g^{p_i(0)/a_i})$. ($SK \leftarrow AKeyGen(P, MK)$)
- Encode $(P, SK, PK, \delta_{O, (P, SK, PK)})$ with public key of user w and indicate with C. Send $(Q, C, \delta_{O, (Q, C)})$ to privilege manager group.
- privilege manager group checks $\delta_{O, (Q, C)}$ immediately after receiving $(Q, C, \delta_{O, (Q, C)})$. Save Q in user list UL and send C to user w.
- User w decode C with his own secret key and verify if $\delta_{O, (P, SK, PK)}$ is right. Accept P, SK, PK.

④ User Revocation

Withdraw user v's data reading authority. This is the system level where redefined proxy re-encoding key is displayed according to the component update of system master key and open key.

- Select attribute set D from CNF(Conjunctive Normal Form) type short sentence when there is minimum attribute set after designing CNF for access structure ($D \leftarrow AMinimalSet(P)$).
- Data owner shall select $a_i \leftarrow Z_q$ and calculate $A_i' = g^{a_i'}$ and $rk_{i \rightarrow i} \leftarrow a_i'/a_i = R($ $R \leftarrow AUpdateAtt(i, MK))$. Send $Att = (v, D, \{i, A_i', \delta_{O, (i, A_i')}\}_{i \in D})$ to privilege manager group

- Privilege manager group removed user v from system user list UL. Add R to history list AHL_i of attribute $(i, A_i', \delta_{O, (i, A_i')})$.

⑤ *File Access*

This is the system level performed when user u is accessed to the data.

- User u shall creates access requirement REQ to the data and send it to privilege manager group.
- Privilege manager group shall verify user u in UL and search for all proxy re-encoding keys in AHL_i (calculation will not be performed if i is the newst version. The latest MK is assumed as $a_{i,n}$). Stop all subsequent calculation when there is no user u in UL.
- Calculate $E_{i,n} \leftarrow (E_i)^{rk_{i \rightarrow i^n}} = g^{a_i/a_s}$ by calculating $rk_{i \rightarrow i^n} \leftarrow$ $rk_{i \rightarrow i} \cdot rk_{i \rightarrow i^n} \cdots rk_{i,n-v_{i \rightarrow i^n}} = a_{i,n}/a_i$ with searched proxy re-encoding key. ($E_{i,n} \leftarrow A\,UpdateAtt4File(i, E_i, AHL_i)$) CSP will calculate $sk_i' \leftarrow (sk_i)^{rk_{i \rightarrow i^n}}$ $= g^{(p_i/(1)/a_i) \cdot (a_i/a_{i^n})} = g^{(p_i/(1)/a_{i^n})}$ when calculated $(rk_{i \rightarrow i^n})^{-1} \leftarrow rk_{i \rightarrow i} \cdot rk_{i \rightarrow i} \cdot$ $\cdots rk_{i,n-v_{i \rightarrow i^n}} = a_i/a_{i,n}$ and $sk_i' \leftarrow (sk_i)^{rk_{i \rightarrow i^n}} = g^{(p_i/(1)/a_i) \cdot (a_i/a_{i^n})} = g^{(p_i/(1)/a_{i^n})}$ is sent to CSP. ($sk_i' \leftarrow A\,UpdateSK(i, sk_i, AHL_i)$)
- Privilege manager group shall send it to $RBSP_1 = (s^{(a)}, \overline{E}_{i,n}, \{B_{i,n}\}_{i,n \in J})$ while CSP sends $RBSP_2 = (\{i, sk_i', A_i, \delta_{O, (i, A_i)}\}_{i \in I \wedge A_{i,i}}, m \cdot g^{H(r|k)})$ to user u.
- User u shall verify $\delta_{O, (i, A_i)}$ and sk_i', and replace sk_i in SK to sk_i', and calculate $Y^s = e(q, q)^{ys}$ through the calculation of AND Gate and OR Gate.
- Calculate header E to Y^s. Obtain data m by calculating body's $m \cdot g^{H(r|k)}$ with DEK. ($m \leftarrow ADecryptB(m \cdot DEK, DEK)$)

⑥ *File Deletion*

Data owner can delete the data file from CSP. Data owner shall attach the signature with the file ID to be removed and send it privilege manager group to delete the file. Once the signature is approved the group shall request to delete the save data to CSP. CSP is the system level which deletes the data.

4 Security Analysis

4.1 Comparison of Existing Scheme and Proposed Scheme

[6] suggest a model which can guarantee data confidentiality and fine-grained access control. However, as mentioned above, it is vulnerable to the collusion attack of revoked user and cloud service provider. It is because authority related to decrypting such as encrypted DEK, encrypted data (header and body), and etc is concentrated to the cloud service provider. The proposed scheme manages key (header) by having trusted authority called privilege manager group, and the cloud server stores only the encrypted data(body), That is, proposed scheme prevents collusion attack and guarantees data confidentiality.

Additionally, by utilizing Type-based PRE, weight is granted according to the data owner's reliability towards the user to show the whole or part of the data. That is, selectively delegating message is possible. For example, if A institute is more reliable than B institute, the part of the message which can be read can be set to have selective delegation.

4.2 Collusion Attack of Revoked User

The important characteristic of the proposed method is secure against collusion attack. Collusion attack refers to revoked user colluding with cloud server to illegally read data. In [6], there is no direct change in secret key SK which the user possesses even if revocation event occurs. DEK used for encrypting data within body is encrypted as KP-ABE and is composed to the header. Therefore, if update of secret key component is not performed within header which the cloud server possesses and it is transmitted to the revoked user, the header can be decrypted with own secret key SK. That is, by obtaining DEK, data m can be read.

As solution for collusion attack, data file is divided into header and body to be separately stored to privilege manager group and cloud service provider. If the user is not authorized by privilege manager group, it cannot obtain information regarding DEK within header, so decrypting body is completely impossible. That is, proposed scheme is secure against collusion attack of revoked user and cloud server through intervention of privilege manager group.

5 Conclusion

Cloud computing is computing paradigm which provides computing resource as service. Because service user can efficiently invest regarding computing resource in cloud computing, previous service field will be expanded. However, because cloud service provider cannot be totally trusted due to data security reasons, risk of data security and violation of privacy factors should be considered. Especially, guaranteeing data confidentiality is required. To solve these problem [6] proposed scheme which guarantees data confidentiality and fine-grained access control. However, data confidentiality can be violated by collusion attack of revoked user and cloud server.

To solve this problem, we guaranteed data confidentiality by storing and dividing data file into header and body. In addition, the method of selective delegation regarding the whole or partial message according to delegator's reliability towards delegated using type-based PRE was specified. And in chapter 4, health cloud model which processes medical data while protecting privacy was composed and security of the proposed scheme was analyzed.

Meanwhile, to reinforce privacy in cloud computing environment, data owner should be user-centric for user information. To solve these requirements, analysis of PCE (Patient Controlled Encryption)[2] which can control (user-centric) sharing and usage of the patient's own medical record should be conducted as future work.

Acknowledgments. This paper is extended from a conference paper presented at 1st ACIS/JNU Int. Conference on Computers, Networks, Systems, and Industrial Engineering (CNSI 2011), 'Attribute based Proxy Re-Encryption for Data Confidentiality in Cloud Computing Environments'. The author is deeply grateful to the anonymous reviewers for their valuable suggestions and comments on the first version of this paper. The author also thanks Prof. Youjin Song, and Mr. Jeongmin Do of Department of Information Management, Dongguk University for their encouragement and support.

References

1. CSA : Security Guidance for Critical Areas of Focus Cloud Computing, vol. 2.1 (2009)
2. Benaloh, J., Chase, M., Horvitz, E., Lauter, K.: Patient controlled encryption: ensuring privacy of electronic medical records. In: Proceedings of the 2009 ACM Workshop on Cloud Computing Security, pp. 103–114. Association for Computing Machinery, New York (2009)
3. Bethencourt, J., Sahai, A., Waters, B.: Ciphertext-Policy Attribute-Based Encryption. In: Proceedings of the 2007 IEEE Symposium on Security and Privacy, pp. 321–334 (2007)
4. Ibraimi, L., Tang, Q., Hartel, P., Jonker, W.: A Type-and-Identity-based Proxy Re-Encryption Scheme and its Application in Healthcare. In: 5th VLDB Workshop on Secure Data Management, SDM, August 24, pp. 185–198 (2008)
5. Blaze, M., Bleumer, G., Strauss, M.J.: Divertible protocols and atomic proxy cryptography. In: Nyberg, K. (ed.) EUROCRYPT 1998. LNCS, vol. 1403, pp. 127–144. Springer, Heidelberg (1998)
6. Yu, S.C., Wang, C., Ren, K.I., Lou, W.J.: Achieving Secure, Scalable, and Fine-grained Data Access Control in Cloud Computing. In: INFOCOM, 2010 Proceedings IEEE, pp. 321–334 (2010)
7. Mather, T., Kumaraswamy, S., Latif, S.: Cloud Security and Privacy. O'Reilly Media, Sebastopol (2009)
8. Park, N., Kwak, J., Kim, S., Won, D.H., Kim, H.W.: WIPI mobile platform with secure service for mobile RFID network environment. In: Shen, H.T., Li, J., Li, M., Ni, J., Wang, W. (eds.) APWeb Workshops 2006. LNCS, vol. 3842, pp. 741–748. Springer, Heidelberg (2006)
9. Park, N., Kim, H.W., Kim, S., Won, D.H.: Open location-based service using secure middleware infrastructure in web services. In: Gervasi, O., Gavrilova, M.L., Kumar, V., Laganá, A., Lee, H.P., Mun, Y., Taniar, D., Tan, C.J.K. (eds.) ICCSA 2005. LNCS, vol. 3481, pp. 1146–1155. Springer, Heidelberg (2005)
10. Park, N., Kim, H.W., Kim, S., Won, D.H.: Open location-based service using secure middleware infrastructure in web services. In: Gervasi, O., Gavrilova, M.L., Kumar, V., Laganá, A., Lee, H.P., Mun, Y., Taniar, D., Tan, C.J.K. (eds.) ICCSA 2005. LNCS, vol. 3481, pp. 1146–1155. Springer, Heidelberg (2005)
11. Park, N., Kim, S., Won, D.: Privacy Preserving Enhanced Service Mechanism in Mobile RFID Network. In: ASC, Advances in Soft Computing, vol. 43, pp. 151–156. Springer, Heidelberg (2007)
12. Park, N.: Security scheme for managing a large quantity of individual information in RFID environment. In: Zhu, R., Zhang, Y., Liu, B., Liu, C. (eds.) ICICA 2010. Communications in Computer and Information Science, vol. 106, pp. 72–79. Springer, Heidelberg (2010)
13. Park, N., Kim, S., Won, D.H., Kim, H.W.: Security analysis and implementation leveraging globally networked rFIDs. In: Cuenca, P., Orozco-Barbosa, L. (eds.) PWC 2006. LNCS, vol. 4217, pp. 494–505. Springer, Heidelberg (2006)
14. Goyal, V., Pandey, O., Sahai, A., Waters, B.: Attribute-based encryption for fine-grained access control of encrypted data. In: Proc. of CCS 2006. Association for Computing Machinery, New York (2006)

A New Method for Face Identification and Determing Facial Asymmetry

Piotr Milczarski

University of Lodz, Faculty of Physics and Applied Informatics,
Pomorska str. 149/153, 90-236 Lodz, Poland
piotr.milczarski@uni.lodz.pl

Abstract. In the paper the new automatic method for recognizing faces, determining the facial asymmetry and a new frontal facial asymmetry measures are proposed. These new factors and measures are the great help in designing more novel and automatic methods for recognizing not only the faces but also determining the dominance of a proper brain hemisphere. The examples of the implementation of the method are presented and compared with the improved method derived by Milczarski, Kompanets and Kurach [14] and the Anuashvili's [1] method for the estimation of verical axis. The results of the estimation may be used for designing absolutely novel e-learning technologies, for choosing an appropriate type of learning program for bilateral, left-brain, or right-brain thinker, or for synthesis of synergic work collectives. The method may be used for automatic identification of the mentioned above personality types in the future.

Keywords: biometrics, face asymmetry, area asymmetry measure, frontal facial field asymmetry measure, true vertical projectionl axis, horizontal facial axis.

1 Introduction

Facial recognition techniques are currently used in a wide range of applications for access control, security and database indexing etc. Facial asymmetry is a personality feature and has been long known as a factor for face recognition [6], [11], [10], face attractiveness determination [17], [2], and face expression recognition [11], etc. One of the most important features is finding the natural symmetry axis so called true vertical projection axis TVPA.

In the previous papers [14], [9] there were introduced new measures: the area asymmetry measure AAS and two new measures, AAs^+, AAs^-, of appropriate facial regions. The values AAs^+, AAs^- are calculated as the difference of the left and right parts (defined by the vertical axis) of the facial binary images. The general idea of Area Asymmetry Measure has been presented in [6], [8], [9], [14] and followed by other authors in [3], [4].

Now, the author's interest is to develop a method to broaden the use of the facial asymmetry biometric. It is assumed that it may also allow to recognize

R. Katarzyniak et al. (Eds.): Semantic Methods, SCI 381, pp. 329–340.
springerlink.com

automatically the dominating hemisphere from a frontal facial image. This is the second step of the research in this direction, after the paper [14]. The new method and the *Field Asymmetry Measure, FAM,* introduced in this paper are much more robust than the previous one [14]. For example they are invariant on choosing some of the set of initial points of the method and on slight rotation of the face in the x-axis and z-one as well. Both methods are great help for oftalmogeometric facial pattern, beacause they allow to find *True Vertical Projection Axis,* **TVPA**.

The paper is organized as follows. In Section 2, the previous method of deriving the area asymmetry measure and True Vertical Projection Axis *TVPA* is derived. The facial asymmetry measure algorithm is presented also in this section. The method of the area asymmery measure determination and *TVPA* from a frontal facial image and experimental results [14] are described in Section 3. In Section 4 a new author's (*Frontal Facial*) *Area Asymmetry Measure, FAM,* is described, as well as *Right* and *Left Facial Area Asymmetry Measures, RFAM* and LFAM correspondingly. Further, in Section 5 a novel author's method, *FAM-PoliMet,* for determinng facial asymmetry is derived. The results of comparison of the three methods (Anuashvili's [1], Milczarski, Kompanets and Kurach [14]) are presented. The conclusions and future work are outlined at the end. The examples shown in the paper were chosen from [1], becasue they also corresponds to the diagnosed left-brain thinker, right-brain thinker and bilateal thinker.

2   Facial Asymmetry Measure [14]

While identifying a person's face it is crucial to determine some initial points and primitives that constitute the computer's face model. This set of data is obtained from our frontal facial picture and is sufficient to recognize or identify a given human being [8], [15].

In the presented methods it is used the normalized *RGB* color space representation is used to detect a face on an image and gradient intensity information to localize eyes position [9] [14]. The original algorithm of finding the possible iris edge is based on information of region average pixel intensity determines the diameter of iris circle, Dr, and a center point, S. The size of the iris diameter is used to introduce the absolute normalization unit called by the authors $[Muld]$ (for the diameter expressed in pixels) or $[Muld^2]$ (for the area, expressed in $pixel^2$) [8]. The normalization unit has been based on a fact[15] that in 4-5 year of our life, the diameter of non-transparent part of iris of any person becomes constant and is equal to 10±0.56 mm. This procedure makes possible to scale a person's face image to its real dimension.

In the next step of that method the eyelid contours is achieved by means of approximation with three degree polynomials. The coordination points of upper and lower eyelids are detected based on pixel's illumination variation and unique mapping of inner and outer eyes corners. After the facial features extraction, that algorithm of the face asymmetry measure is applied. The algorithm has included automatic selection of facial asymmetry regions based on the proper vertical axis, and the facial size normalization using the *Muld* unit.

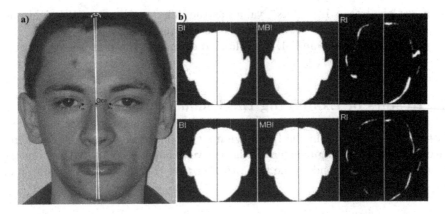

Fig. 1. Graphic illustration of determining the facial asymmetry [14]: a) input image with two candidate axis (0° and 1.5°); b) BI binary images, MBI mirrored binary images, and result images, RI; for 0° axis (upper images) and for 1.5° axis (lower images).

The idea of finding True Vertical Projection Axis using Field Area Asymmetry Measure (see Section 4) is corresponding to the following one. That was described in [7], [8] and [14] also. The facial axis is rotated around the anthropocentric (center between inner eyes' corners) point, Oec, that minimizes the **Area Asymmetry** (AAs) measure of facial silhouette. The value asymmetry measure equals the difference between the **Binary Image** (BI) and its **Mirrored Binary Image** (MBI). The binary image, BI, represents facial silhouette and is obtained from a frontal input image using the Otsu threshold algorithm [16] and morphological operations. The minimal values of AAs define the true vertical projection axis for this method.

In the next step there were derived AAs^+, AAs^- values. The values AAs^+, AAs^- are calculated as the difference of the left and right parts (defined by the vertical axis) of the facial binary images (Fig.1 b) and represent the right-sided and left-sided facial asymmetry, respectively.

The $AAsMeasure$ algorithm to calculate the values of the measures AAs, AAs^+, AAs^- is as follows [14]:
Input data: Oec - point coordinates, $\phi[^o]$ - rotation angle, BI - silhouette's binary image, Dr - right iris diameter.
Output data: AAs, AAs^+ and AAs^- values.

Step 1. Define the vertical axis, $FA(\phi)$, on a binary facial silhouette's image, BI, based on coordinates point, Oec, and the value of angle $\phi[^o]$ (see BI images in Fig.1b)

Step 2. Create a vertically mirrored binary image, MBI, based on the information about BI image and the axis $FA(\phi)$ (see MBI images in Fig.1b)

Step 3. Calculate AAs^+, AAs^- values as the difference between the right and left side of $BI's$ and $MBI's$ images (see RI images in Fig.1b)

Step 4. Use the AAs^+ and AAs^- values to determine the person's asymmetry type.

3   Results of Examining the Facial Asymmetry [14]

In the method to obtain the facial symmetry type a color image of a person's face is taken as an input signal. It is believed that the computed asymmetry types closely correspond to the brain thinker types i.e., bilateral, right-brain or left-brain thinker.

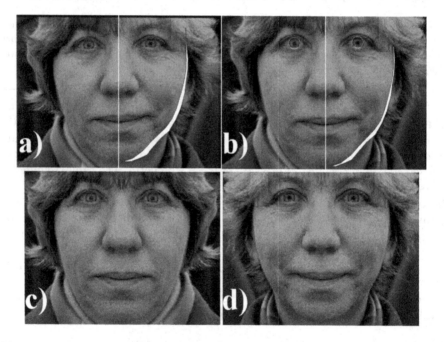

Fig. 2. Person with the left asymmetry and with dominance of the left hemisphere: a) image with axis $0°$ and $AAs=1.04$ $Muld^2$, b) image with axis $-1°$ and $AAs=0.5$ $Muld^2$, c) right-sided, d) left-sided composite.

If the assumption is true, the dominance of the right asymmetry means that the person on the image is right-brained thinker and the left asymmetry dominance determines the left-brain one. When the asymmetry is balanced it means that the person is the bilateral one.

To examine objectively the previous method [14], the obtained results of persons' asymmetry types were compraed with the psychological types described by Anuashvili's [1]. In the first step of the comparison as the base axis was chosen the one calculated from the Anuashvili's composites, i.e. synthesized images

Table 1. Results of asymmetry measure for the face with the left asymmetry

$\phi[^o]$	AAs^-		AAs^+		AAs	
	$[Pixel]$	$[Muld^2]$	$[Pixel]$	$[Muld^2]$	$[Pixel]$	$[Muld^2]$
-2	205	0.31	741	1.14	946	1.45
* -1	**261**	**0.40**	69	0.10	330	0.50
0	676	1.04	0	0.00	676	1.04
1	1682	2.59	0	0.00	1682	2.59
2	4787	7.37	80	0.12	4867	7.49

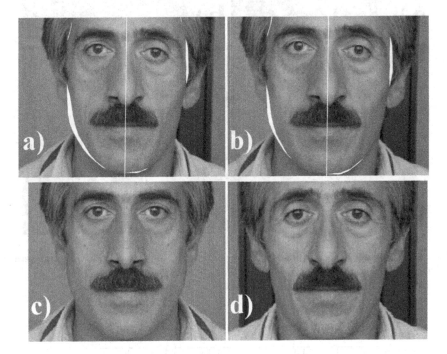

Fig. 3. Person with the right asymmetry and with dominance of the right hemisphere: a) image with axis 0° and AAs=3.79 $Muld^2$, b) image with axis $+1^o$ and AAs=2.49 $Muld^2$, c) right-sided, d) left-sided composite.

Table 2. Results of asymmetry measure for the face with the right asymmetry

$\phi[^o]$	AAs^-		AAs^+		AAs	
	$[Pixel]$	$[Muld^2]$	$[Pixel]$	$[Muld^2]$	$[Pixel]$	$[Muld^2]$
-2	381	0.59	4558	7.02	4939	7.61
-1	416	0.64	3090	4.76	3506	5.40
0	554	0.85	1908	2.94	2462	3.79
* 1	757	1.17	**860**	**1.32**	1617	2.49
2	1498	2.30	232	0.36	1730	2.66

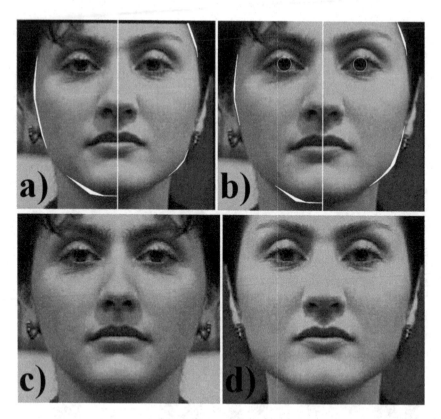

Fig. 4. Person with the symmetrical face and with no brain dominance (bilateral): a) image with axis $0°$ and $AAs=1.15$ $Muld^2$, b) image with axis $+1°$ and $AAs=0.87$ $Muld^2$, c) right-sided, d) left-sided composite.

Table 3. Results of asymmetry measure for the symmetrical face.

	$\phi[°]$	AAs^-		AAs^+		AAs	
		[Pixel]	[Muld²]	[Pixel]	[Muld²]	[Pixel]	[Muld²]
	-2	202	0.31	1835	2.83	2037	3.14
	-1	158	0.24	1197	1.84	1355	2.08
	0	579	0.89	170	0.26	749	1.15
*	**1**	**348**	**0.53**	219	0.34	567	0.87
	2	690	1.06	82	0.12	772	1.18

from left-left and right-right parts of a face (see Figs. 2-4). Construction of the composite was described in [6].

In our opinion the Anuashvili's method [1] of determining axis is based on the *intuitive* technique of choosing the vertical axis (i.e. *visually/manually*). In the previous method, originally described in [8], the vertical axis is the one determined by the automatic adaptation procedure, which is precise.

After the basis axis is determined, it is rotated around the anthropological point, Oec. The axis divides the face into two parts. For each part we build the mirrored images for the left and right sides. Then, there are created two precise facial composites by connecting each part with its mirrored image part.

Figures 2-4 represent the input images of representative persons with previously examined brain type, for cases of left brain thinker, right brain thinker and bilateral thinker, respectively. Each figure contains images with marked (in white) regions of the asymmetry for two axes, the base one 0° (Fig. 2-4a) and the proper one, automatically obtained by the $AAsMeasure$ algorithm (Fig.2-4b). The Fig.2-4c) and d) show the right and left composite for the proper axis.

Tables 1-3 correspond to Figures 2-4, respectively and show the results of the area asymmetry measures AAs for 5 axes. The base axis is rotated left and right according to angles -2°, -1°, $+1^o$, $+2^o$ (the + sign means clockwise axes rotation) around the anthropological point . The values of AAs, AAs^+ and AAs^- are given in $[Pixel]$ and $[Muld^2]$ units. We show the results for determining the vertical axis with the accuracy 1^0 that is enough to illustrate the idea of the proposed asymmetry measures. The proper vertical axis is indicated by the lowest value of the asymmetry measure Aas; symbol (*), in Tables 1-3, denotes the case of minimal AAs (see the *-lines). With the gray colour we show the results for the 0° axis. With bold letters it is highlined bigger from the sided assymetry values where the total asymmetry measure AAs is minimum.

4   Frontal Facial Field Area Asymmetry Measure FAM

As it was mentioned before finding the vertical projection axis $TVPA$ plays crucial role in identification and authentication methods. Some of the methods in which the authors used vertical projection axes were based on an intuition rather than on the objective data [1]. In the previous papers [14] and [8] there were shown objective measures and methods that allowed to find that axis. The problem of introducing satisfactory set of facial elements is very important in many areas: computer games and facial visualization as an avatar [9] human personality characteristics recognition [18] [2] [3] [4] [13] etc.

The algorithm of finding FAM is based on initial points from oftalmogeometrical pattern and a new starting point M that is thought with Oec point as an antropological point through which the $intuitive$ vertical projection axis goes. After edge candidates points estimation, the circle of iris diameter, Dr, and center, S, are extracted. As it was said before it is a significant part of the method because the extracted size of the iris diameter is used as the absolute normalization unit called by the authors $[Muld]$) or $[Muld^2]$ [8]. This procedure allows to resize a personal face image to real dimension. The normalization unit has been based on a Muldashev research [15]. The minimal value of FAM indicates also the truthful vertical projection axis, $TVPA$.

The previous $Area\ Asymmetry\ Measures$ were based on the whole facial silhouette. In this new method with use of $Frontal\ Facial\ Field\ Area\ Asymmetry\ Measure\ FAM$ we do not need to have to use the whole facial silhouette. In the new method we take only into account the eyes and nose/mouth part of the face. With facial rotation by some angles the measures can change dramatically as well as the projection axes . The advantage of such choice is invariance of the method on the initial choice of the philltrum point, M.

5 $FAMPoliMet$ - New Method of Determining the Facial Asymmetry

The advantage of the FAM measure is that we do not need so much information as in AAM one. We need a lot less data about the ezamined face. It is clearly visible, that the method does not need the whole face but only eyes, nose and mouth area, see Figures 5-7. The other advantage is that the face can be slightly rotated around x (width) axis and of course z-axis (the face can be upside-down either). In the previous method deriving the $TVPA$ were based on the fact that the picture is a frontal face one.

The method $FAMPoliMet$ is based on a frontal facial color image. After initial transformations of the image there are derived:

Step 1. the diameters of the eyes' irises;

Step 2. (4 times) 5 points from the frontier of scleraand eyelid (upper and lower for each eye);

Step 3. inner and outer eyecorners as an intersection of the appropriate aproximated upper and lower eye curves: RO - right outer, RI - right inner, LI - left inner, LO - left outer

Step 4. the anthropological point, Oec as a middle point of inner corners;

Step 5. the philtrum point, M;

Step 6. the base axis (so called 0° axis) as the line $OecM$;

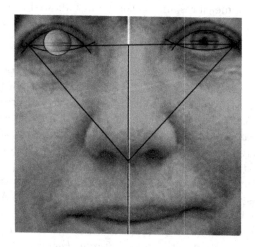

Fig. 5. Field Asymmetry Measure, FAM, for the person with the left asymmetry

Table 4. Results of field asymmetry measure, *FAM*, for the person with the left asymmetry

$\phi[°]$	$\dfrac{LFAM}{[Muld^2]}$	$\dfrac{RFAM}{[Muld^2]}$	$\dfrac{FAM}{[Muld^2]}$
-2	0.356	0.23	0.379
* -1	**0.331**	0.0	0.331
0	0.496	0.167	0.663
1	0.717	0.390	1.107
2	0.924	0.600	1.524

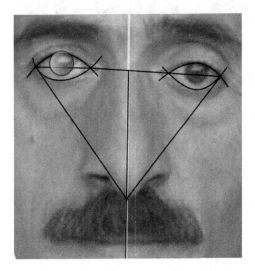

Fig. 6. Field Asymmetry Measure, *FAM*, for the person with the right asymmetry

Table 5. Results of field asymmetry measure, *FAM*, for the person with the right asymmetry

$\phi[°]$	$\dfrac{LFAM}{[Muld^2]}$	$\dfrac{RFAM}{[Muld^2]}$	$\dfrac{FAM}{[Muld^2]}$
-2	0.624	0.929	1.553
-1	0.420	0.723	1.143
0	0.205	0.505	0.710
1	0.0	0.296	0.296
* **1.875**	0.0	**0.294**	0.294

After the basis axis is determined, it is rotated around the anthropological point, *Oec* i.e. it is enough to rotate the inner and outer corners and *M* point. We obtain two composites in shape of poligons. For each part the mirrored images for the left and right side poligons are built. Then, there are created two precise

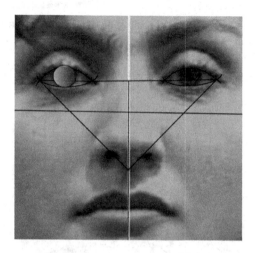

Fig. 7. Field Asymmetry Measure, *FAM*, for the person with the symmetrical face

Table 6. Results of field asymmetry measure, *FAM*, for the person with the symmetrical face

$\phi[^o]$	$LFAM$ $[Muld^2]$	$RFAM$ $[Muld^2]$	FAM $[Muld^2]$
* -1.375	**0.007**	**0.033**	0.040
-1	0.082	0.109	0.191
0	0.205	0.505	0.646
1	0.519	0.550	1.069
2	0.714	0.747	1.461

poligon facial composites by connecting each part with its mirrored image part. At last $RFAM$, $LFAM$, and $FAM = RFAM + LFAM$ values are counted.

We are given almost the same axis up to the initial method acuracy set by the user. To show the *FAM* more clearly in the tables (Tabs 4-6) and figures (Figs 5-7, correspondingly) below there are only values in $[Muld]$ and $[Muld^2]$ units respectively. In the grey rows there are shown the results for the initial 0° axis as before (see Section 3). In the * row there are results for the axis *TVPA* were the *FAM* takes the minimal value with the corresponding values of right (*RFAM*) and left (*LFAM*) *FAM*. The angles given in the rows show the angles of rotation around Oec point of the initial axis. There is no point in showing the *FAM* values in *pixel* or $pixel^2$ measures.

6   Conclusions

While using ophtalmogeometrical pattern it is crucial to have the vertical projection axis, derived in an objective way. Hence the new method, *FAMPoliMet*, fulfills that conditions. After comparison of the three methods it is easily visible that the previous method based on the area asymmetry AAs mesures and the new method, $FAMPoliMet$, give the same result with the method acuracy, which is 1° degree, starting using the same initial points. In Anuashvili's case it is supposed that the projection axis is chosen deliberately. For the two last examples the exact angles for TVPA axes were given with the 0.125^o degree acuracy, which can be bigger of course. In the case of symmetrical face both measures are relatively close.

The new results of the computational solutions of the person's asymmetry type have been considered. Just like before (for AAs) all of the method's procedures are automatic, and include biometrics solutions such as face detection, eye localization, iris extraction, inner eye corners finding and the based on poligons messures determination. For the proposed method the new measures (FAM, $LFAM$ and $RFAM$) of the facial asymmetry are introduced.

The asymmetry measure FAM is used in the special adaptation procedure for precise determination of the proper facial vertical axis. The proper axis as well as the values $RFAM$ and $LFAM$ are the sufficient determinants of the facial asymmetry type. Moreover, those precise values are crucial to get information from a frontal facial image about correlation between the facial asymmetry type and supposed inclinations towards brain hemisphere dominance determination. What is important, they are invariant to facial expression and rotation.

The both methods results confirm that the facial asymmetry is important and informative. Furthermore, in the new method we can objectively compute (from a frontal facial image) a precise appropriate person's characteristics based on the introduced asymmetry measures. Those characteristics can be used as biometric in person's identification/authentication as well as other applications concerning psychological types determination. In author's opinion, just like with AAs measure further research may result in a pioneer chance to combine computer methods with non-traditional computing psychological methods [5], [13], which can be used in e-learning and biometric safety systems. There is designed fuzzy logic system-adviser for deriving person's asymetry type with appropriative experts "if-then" rules (for example, like [12]). This is used in author's current research.

References

1. Anuashvili, A.N.: Fundamentals of Objective Psychology, 5th edn. Institute of Control, Psychology and Psychotherapy, Warsaw-Moscow (2008)
2. Chen, A.C., German, C., Zaidel, D.W.: Brain Asymmetry and Facial Attractiveness. Neuropsychologia 35(4), 471–476 (1997)
3. Kamenskaya, E., Kukharev, G.: Recognition of psychological characteristics from face. Methods of Applied Informatics, Pol. Acad. Sci. 1, 59–73 (2008)
4. Kamenskaya, E., Kukharev, G.: Some aspects of automated psychological characteristics recognition from the facial image. Methods of Applied Informatics, Pol. Acad. Sci. 2, 29–37 (2008)
5. Keirsey, D., Bates, M.: Please Understand Me: Character and Temperament Types. Prometheus Nemesis. Del Mar, CA (1984)

6. Kompanets, L.: Biometrics of Asymmetrical Face. In: Zhang, D., Jain, A.K. (eds.) ICBA 2004. LNCS, vol. 3072, pp. 67–73. Springer, Heidelberg (2004)
7. Kompanets, L.: Facial Composites, Ophthalmic Geometry Pattern, and Based on Stated Phenomena the Test of Person/Personality Virtuality/Aliveness. In: 7th Intern. Conf. on Intelligent Systems Design and Applications, pp. 831–836. IEEE Computer Society Press, Los Alamitos (2007)
8. Kompanets, L., Kurach, D.: On Facial Frontal Vertical Axes Projections and Area Facial Asymmetry Measure. Intern. J. of Computing, Multimedia and Intelligent Techniques 3(1), 61–88 (2007)
9. Kurach, D., Milczarski, P.: Test Of Aliveness/Virtuality Based on Ophthalmo-geometry and Facial Asymmetry Characteristics. Polish Journal of Environmental Studies 17(4C), 497–502 (2008)
10. Liu, Y., Schmidt, K., Cohn, J., Mitra, S.: Facial asymmetry quantification for expression invariant human identification. Computer Vision and Image Understanding 91, 138–159 (2003)
11. Liu, Y., Weaver, R.L., Schmidt, K., Serban, N., Cohn, J.: Facial Asymmetry: A New Biometric. The Robotic Institute of Carnegie Melon Univ. (2001)
12. Martin, M.A., Mendel, J.M.: Flirtation, a Very Fuzzy Prospect: a Flirtation Advisor. (1995)
13. Myers-Briggs, I., Myers, P.: Gifts Differing: Understanding Personality Type. Davies-Black Publishing (1995)
14. Milczarski, P., Kompanets, L., Kurach, D.: An Approach to Brain Thinker Type Recognition Based on Facial Asymmetry. In: Rutkowski, L., Scherer, R., Tadeusiewicz, R., Zadeh, L.A., Zurada, J.M., et al. (eds.) ICAISC 2010. LNCS (LNAI), vol. 6113, pp. 643–650. Springer, Heidelberg (2010)
15. Muldashev, E.R.: Whom Did We Descend From? OLMA-Press, Moscow (2002)
16. Otsu, N.: A threshold selection method from gray-level histograms. IEEE Trans. Systems, Man and Cybernetics 9, 62–66 (1979)
17. Reis, V.A., Zaidel, D.W.: Functional Asymmetry in the Human Face: Perception of Health in the Left and Right Sides of the Face. Laterality 6(3), 225–231 (2001)
18. Rutkowska, D.: An Expert System for Human Personality Characteristics Recognition. In: Rutkowski, L., Scherer, R., Tadeusiewicz, R., Zadeh, L.A., Zurada, J.M. (eds.) ICAISC 2010. LNCS, vol. 6113, pp. 665–672. Springer, Heidelberg (2010)

3W Scaffolding in Curriculum of Database Management and Application – Applying the Human-Centered Computing Systems

Min-Huei Lin[1] and Ching-Fan Chen[2]

[1] Aletheia University, Tamsui Dist., New Taipei City, Taiwan (R.O.C.)
lmh@email.au.edu.tw
[2] Tamkang University, Tamsui Dist., New Taipei City, Taiwan (R.O.C.)
cfchen@mail.tku.edu.tw

Abstract. When teachers teach and strive to achieve successful students learning outcomes at our university, most of them have experienced difficulty in communicating with students. It's due to the large number of students, various communication styles of students and the different contextual definition between teachers' and students'. There are only a few students can learn well, early and timely, lots of students learn helplessness or give up halfway under successive frustration instead. So this study tries to use the 3W approach, 'When, Who, What', to investigate how to find who needs what assistance when they need under the scheduled progress, then teachers have chance to design various scaffolding to support various students' clusters. It aims to achieve effective teaching and learning. We use the information technology to collect the learning data of students (textual data), and extract value from these data by the recognition and concepts of them through the human-centered computing system, the result is that clusters of students with different knowledge structures appear. While teachers observe and interpreter the features of various clusters, they can design appropriate and necessary scaffolding for every individual cluster to assist the students learning.

1 Introduction

Instruction is a systematic action, its goal is to induce students learning, besides let students learn well, it also should let students develop their potential abilities and construct the abilities of learning actively continuously. Under the high enroll rate of university entrance examination in Taiwan, promoting the learning effects of under-achieved students is the urgent responsibility, challenge and goal for teachers.

Innovation diffusion theory of Rogers is usually used to explain the acceptance of innovative products by consumers on the markets, but when innovators broadcast the new functions of innovative products through advertisement media, only 13.5% early adopters purchase them(Rogers 2003). And Moore (Moore 1999) indicated that there exists a large chasm between early adopters and early majority in 'inside the tornado', failing to cross the chasm, the spreads of innovation finally failed. Students are the main core of education, when teachers teach in some courses it may happen that the

R. Katarzyniak et al. (Eds.): Semantic Methods, SCI 381, pp. 341–351.
springerlink.com　　　© Springer-Verlag Berlin Heidelberg 2011

diffusion of content knowledge cannot succeed because of the difficult communication between students and teachers usually.

Data modeling and database development is usually a required and core curriculum in department of MIS. Its content knowledge contains entity-relationship model, relational data model, SQL, normalization etc [3, 5, 8]. The researcher has taught this course since 13 years ago, many students learned the database analysis and design hard but not well. When they attended the project analysis and implementation course, they eventually sensed that the insufficient ability for developing database will influence the progress and quality of developing and implementing information systems.

In this study, the researcher investigates the teaching and learning database modeling and developing course in department of MIS in Aletheia University. Teachers play the role of innovator and plan scheduled progress in courses, they make use of the taxonomy of Bloom's educational objectives in cognitive domain to precisely specify learning objectives, activities and assessments. Every new or unknown course unit is an innovative product for students, and students are divided into leading adopters and underachievement majority according to the time they need to achieve knowledge cognition level. And we try to use the 3W approach, 'When, Who, What', to collect students learning data through data mining and text mining technology and analyze data by using human-centered computing systems (Hong 2009), our goal is to extract the features of students knowledge clusters and discover when is the proper time, who needs assistance and what scaffolding they need, to support students learning effectively.

2 Literature Review

2.1 Diffusion of Innovation

Rogers defined the diffusion process as the process of a new idea spread to end users or users in 1995, his experiment of diffusion of innovation model divided individuals into five kinds of groups according the time of accepting innovation, innovators, early adopters, early majority, late majority, and laggards (Fig.1). He also notified us unceasingly that only about 20% of early adopters may accept innovation whenever new products emerge, because they have some characteristics for accepting innovation quickly and early, use and develop the useful and creative value of innovation. And Moore indicated that there exists a large chasm between early adopters and early majority in 'inside the tornado', if not identifying the early adopters and their creative value, the innovation diffusion will be terminated.

In this study, students are divided into leading adopters and underachievement majority according to the time they need to achieve knowledge cognition level, and try to extract the features of clusters of students and let the chasm appear. When teachers discover the chasm, they have chance to build a platform to support teachers and students crossing chasm. When the underachieved majority may achieve more, the diffusion of innovation may happen.

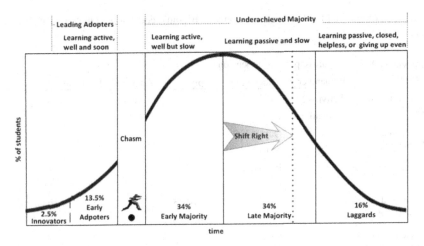

Fig. 1. Clusters of students and chasm in Learning and Teaching

2.2 Scaffolding Instruction

Scaffolding instruction theory is originated from the constructivism, it emphasized that one's knowledge is constructed by individual, and the one's constructive process is through the interaction with others in social systems. The meaning of scaffolding extends to instruction, it stands that adults or teachers provide temporary scaffolding or support form to assist learners to develop their learning ability. Such assistance is the same as the scaffolding built when buildings established, reinforced or beautified (Wood et al. 1976).

The key point of implementing scaffolding instruction is whenever the adults provide assistance they can really discover the zone of proximal development of learners, they just can provide appropriate assistance at the right moment.

2.3 Revised Taxonomy of Bloom's Educational Objectives in Cognitive Domain

Bloom's Taxonomy has, in the past, provided a foundation for developing learning objectives designed for learners to acquire knowledge. The taxonomy system is not only an assessment tool but also a common specification used for teachers when they designate learning objectives. In the revised taxonomy in cognitive domain, educational objectives are divides into knowledge dimension and cognitive process dimension, the former helps teachers distinguish what to teach, the latter aims to promote the retention and transfer of knowledge learned by students.

In the knowledge dimension, there are four classifications of knowledge, factual knowledge, conceptual knowledge, procedural knowledge and metacognitive knowledge. The cognitive process dimension contains remember, understand, apply, analyze, evaluate and create [1, 2]. The two-way taxonomy table is listed in Table 1, teachers usually fill learning objectives, learning activities, assessments etc in the taxonomy table.

Table 1. Taxonomy table of revised taxonomy of Bloom's educational Objectives in cognitive domain

Knowledge Dimension	Cognitive process dimension					
	Remember	Understand	Apply	Analyze	Evaluate	Create
Factual Knowledge	Activity 2 Evaluation 1	-	-	Activity 1	-	-
Conceptual Knowledge	-	Object 1~2 Activity 2~3 Evaluation 1 Evaluation 2	-	-	-	-
Procedural Knowledge	-	-	Activity 3~4 Evaluation 2	-	-	Evaluation 3
Metacognitive Knowledge	-	-	-	-	-	-

3 Methodology

In this study, we applied innovation diffusion model and the human-centered computing system, and proposed the U-framework for crossing the chasm in teaching and learning. The experimental subject is the sophomores of the department of MIS in Aletheia University in the Database management and application course. And we try to use the 3W approach, 'When, Who, What', to investigate how to find who needs what assistance when they need under the scheduled progress in this course by analyzing data collected from students.

3.1 Framework of Using Scaffolding to Cross Chasm of Teaching and Learning

The framework is depicted as Fig.2, there are three roles defined in the innovation diffusion social system, i.e. teachers, leading adopters and underachieved majority. At first, teachers play innovators, they design materials, assessments, activities and instructional scheduled progress, classify the course knowledge and related cognitive process, and then start teaching. Whenever a new unit is being conducted, the new unit is an innovation to students. Students still practice their daily life, attend class, ask questions, discuss, consult references, teach peers, do assignments, and undergo tests. Then the teaching and learning is proceeding, teachers continuously gather the students learning data, and use data mining technology to extract the various students clusters of various knowledge cognition from these learning data. Students are divided into leading adopters and underachievement majority according to the time they need to achieve knowledge cognition level, the leading adopters' actual development level is almost consistent with teachers defined, and the underachieved majority's actual development level is apparently incomplete.

Teachers can recognize the features of every cluster and design appropriate scaffolding for every cluster, and students can acquire needed learning support. After the scaffolding activities finished, teachers observe the transformation trend of students' actual development level and potential development level.

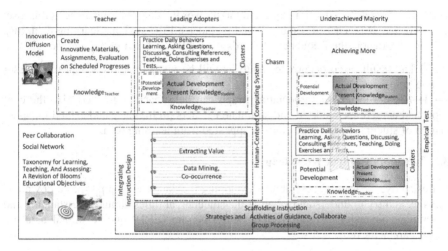

Fig. 2. U-framework for crossing chasm

3.2 Human-Centered Computing Systems – Extract the Features of Clusters of Students

In this study, we consider the process that underachieved majority accept the content knowledge from the view of scaffolding instruction. So teachers collect the students learning data and perform data analysis at first to extract the actual development level of students, and generate the associative network of students' knowledge cognition. The interactive steps are as follows:

Step 1: Data preprocess
1-1H[1]) Teachers define the key words and concept terms according to content knowledge and cognition.
1-1C[2]) Teachers specify a time period and course unit, and then retrieve the required students learning data from the database on Moodle.
1-2H) Teachers recognize the students learning data by their domain knowledge, and then tag words, eliminate meaningless words, and append the concept label with words.

Step 2: Words co-occurrence analysis
2-1C) The associative value of two words can be computed as formula 1. It is based on their co-occurrence in the same sentence.

$$assoc(W_i, W_j) = \sum_{s \in D} \min\left(\left|W_i\right|_s, \left|W_j\right|_s\right) \tag{1}$$

---

[1] 'H' means performing by human.
[2] 'C' means doing by computer.

where W_i and W_j is the ith word and the jth word; s denotes a sentence and is a set of words; D is a set of sentences and includes all the students learning data. $|W_i|_s$ and $|W_j|_s$ denote the frequency of words W_i and W_j occurred in the sentence s.

2-2C) The result of co-occurrence analysis will be visualized to co-occurrence associative graph.

2-1H) The co-occurrence associative graph can help teachers to recognize the concepts and categories inside it, and stimulate teachers preliminary understand the association of students knowledge clusters appeared from learning data.

3.3   Learning Content Knowledge

In this study, we use the case about developing the database of mp3 music download website, and try to understand whether the students' abilities about understanding and executing conversion of entity-relationship diagram (ERD) and relational database schema are achieved.

Content Knowledge. Classification on the revised Bloom's taxonomy in cognitive domain   Teachers classify the content knowledge based on the revised Bloom's taxonomy in cognitive domain, and list these dimensions in Table 2.

Table 2. Content Knowledge Classification

Knowledge Dimension	Content
Factual Knowledge	entity, relationship, simple attribute, single-value attribute, key attribute, multi-valued attribute, compounded multi-valued attribute, table, primary key, foreign key
Conceptual Knowledge	entity-relationship diagram, cardinality, relational database diagram
	ERD can be transformed to table: 1.regular entity type is converted to table, 2.multi-valued attribute is converted to table, 3.many-to-many relationship type is converted to table, 4.one-to-many and one-to-one relationship type are converted to field.
Procedural Knowledge	The method and steps about convert ERD to table: 1.regular entity type: add the simple and single-valued attributes, and choose one key attribute to act as primary key, the other key attributes are served as unique key; 2.multi-valued attribute: add the primary key of original entity's table(one set of foreign key) and the multi-valued attribute, both fields forms primary key; 3.many-to-many relationship type: add two participated entities' primary keys(two sets of foreign keys), both two is primary key; one-to-many and one-to-one relationship type (others omitted).
Metacognitive Knowledge	-

Learning Activities and Learning objectives. We provide the ERD (drawn via ER Assistant) and relational database diagram (MS SQL Server 2008) depicted in Fig.3 to students, and the objectives and activities are (listed in Table 3) designed that students should express the relation between the ERD and database diagram.

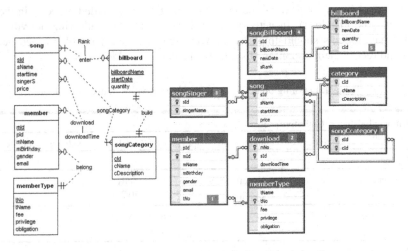

Fig. 3. Entity-relationship diagram and relational database diagram

Table 3. Learning objectives and learning activities

Objectives		Activities	
objective1	Understand the concept of conversion between ERD and relational database schema	activity1	Teachers explain the activities
			Teachers show the ERD and correspondent database diagram
			Teachers tell students that they need to use the knowledge about entity-relationship model and relational data model to describe how the assigned items in database diagram are correspondent to which part of ERD.
objective2	Execute the converting ERD to relational database schema	activity2	Differentiate essential information
			Teachers inquire students the learning tasks, (1) What do I need to do? (2)What should I need to know? to let students focus on differentiating essential information of entity-relationship model and relational data model by comparing two pictures.
		activity 3	How about tNo of member table
		activity 4	How about download table
		activity 5	How about songSinger table
		activity 6	How about songBillboard table
		activity 7	How about cId of billboard table
		activity 8	How about songCategory table

And the learning objectives and learning activities are recorded properly in the taxonomy table (Table 4). In activity 1, teachers explain the following activities and tasks that students should implement. After activity2, teacher will proceed activity 3~8 and record their responses of students in database on Moodle.

Table 4. Learning objectives and learning activities in taxonomy table

Knowledge dimension	Cognitive process dimension					
	remember	understand	apply	analyze	evaluate	create
Factual	-	-	-	-	-	-
Conceptual	activity 3~8	objective1 activity 3~8	-	activity2	-	-
Procedural	-	-	objective2 activity 3, 7	-	-	-
Metacognitive	-	-	-	-	-	-

4 Data Analysis and Findings

After the instructional implementation was completed, we retrieved related students' data from Moodle and started using the human-centered computing system to analyze these data. We tried to visualize data into associative graph to find the clusters within leading adopters and the clusters within underachieved majority. These clusters of students were linked to various content knowledge and misconceptions, so we could understand what a cluster learned and underachieved. In the following works, we'll try to design scaffolding for various clusters to assist their learning.

4.1 Data Resource

The experimental data were collected from 20 students in the practice course of database management and application on 10 March, 2011, and the learning objectives and activities have been described in '3.3 Learning Content Knowledge'.

4.2 Human-Centered Computing Phase: Extract the Knowledge Clusters of Students

Based on the framework of integrating human-intelligence, data mining and text mining technology, teachers read students learning data detailed, and tokenize words (Fig.4), and define three conceptual labels according to content knowledge in Table2 and Table 4, they are [one-to-many relationship to field], [many-to-many relationship to table] and [multi-valued attribute to table], the three conceptual labels are belonged to the category-{ERtoTable_C}. After reading students data, teachers define five wrong conceptual labels according to the wrong responses answered by students in addition, they are [entity to foreign key], [fail to convert one-to-many relationship], [fail to convert many-to-many relationship], [fail to convert multi-valued attribute]

Student1 memberType(E)_member(T)tNo(F) download(R)_download(T) singerS(MVA)_songSinger(T)
enter(R)_songBillboard(T) category(E)_billboard(T)cld(F) songCategory(R)_songCategory(T)
Teacher belong(R)_member(T)tNo(F) download(R)_download(T) singerS(MVA)_songSinger(T)
enter(R)_songBillboard(T) build(R)_billboard(T)cld(F) songCategory(R)_songCategory(T)

Fig. 4. Tokenization - key words for students' learning data and teachers' data[3]

[ERtoTable_W] Student1 memberType(E)_member(T)tNo(F) [entity to foreign key]
category(E)_billboard(T)cld(F) [entity to foreign key]
[ERtoTable_C] Student1 download(R)_download(T) [many-to-many relationship to table]
singerS(MVA)_songSinger(T) [multi-valued attribute to table] enter(R)_songBillboard(T) [many-to-
many relationship to table] songCategory(R)_songCategory(T) [many-to-many relationship to
table]
[ERtoTable_C] Teacher belong(R)_member(T)tNo(F) [one-to-many relationship to field]
download(R)_download(T) [many-to-many relationship to table] singerS(MVA)_songSinger(T)
[multi-valued attribute to table] enter(R)_songBillboard(T) [many-to-many relationship to table]
build(R)_billboard(T)cld(F) [one-to-many relationship to field] songCategory(R)_songCategory(T)
[many-to-many relationship to table]

Fig. 5. Append the category and conceptual labels with words

and [meaningless conversion], the five conceptual labels are belonged to the category-{ERtoTable_W}. And then, words with similar meanings with one concept are belonged to the concept, and append the concept label with the words (Fig.5).

And then, through computing the frequency and co-occurrence of terms, the associative graph will be generated (Fig.6). In the associative graph, we separate two divisions, top and bottom, according to the achievement and underachievement of learning. In the bottom of the associative graph, the content knowledge conceptual graph is defined by teachers, the header of these blocks is yellow and their body is blue. In the top of the graph, the partial content knowledge conceptual graph is presented by students' data, the header of these blocks is pink and their body is white, white color means empty and underachievement.

From the associative graph, we can see the status of connection between Student1 and conceptual blocks, and find that there is some difference between the Student1's knowledge cognition and teachers'. The student has understood the concept about one-to-many relationship converted to a set of fields, but she made mistake when execute such a conversion, she answered 'entity to foreign key'. When teachers observe the feature of the students, they can design proper scaffolding to support the student.

In Fig.7, there contained twenty students' learning data and a teachers' knowledge structure in the associative graph, and there existed six students clusters labeled with G1~G6. The students in the same cluster were of similar features of knowledge cognition, teachers could observe the underachieved content knowledge of every cluster. For example, clusters G1, G2 had no connection with teacher's knowledge, it meant that these students failed to apply knowledge and failed to understand knowledge, and

---

[3] (E) means entity type,(R) means relationship type, (MVA) means multi-valued attribute, (T) means table, (F) means field.

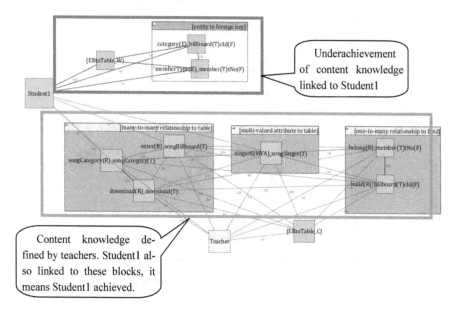

Fig. 6. One student's knowledge cognition and teachers' knowledge cognition associative graph

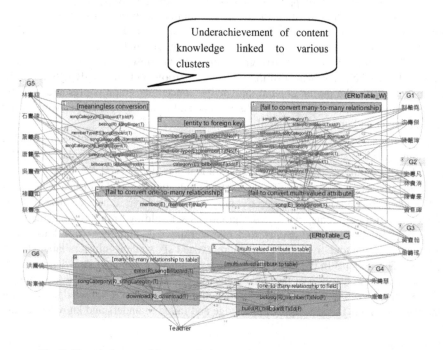

Fig. 7. Knowledge cognition associative graph – students clusters' vs. teacher's

cluster G3 needed to enhance understanding and applying some procedural knowledge, and G5 had lots of links with meaningless conversion, it meant that teachers should require to pay much more attention and concern to them, although they had understood some procedural knowledge, but from their wrong responses, we found that they were seriously lack of the knowledge about entity-relationship model and relational data model, so they just presented several meaningless conversion. Both G6 and G4 were belonged to the leading adopters, the actual development level of G6 was achieved the level defined by teachers, but G4 still needed to enhance applying procedural knowledge.

5 Suggestions

Teachers plan scheduled progress in courses, make use of the taxonomy of Bloom's educational objectives in cognitive domain to precisely master learning objectives, activities and assessments. During students participate the activities, teachers collect the learning data from students, define the conceptual labels according to content knowledge cognition in courses, and the conceptual labels may be appeared from students learning data, and then teachers code students learning data with conceptual labels predefined. When the knowledge cognition associative graph is created, teachers can indeed observe and interpreter the features of various clusters of students, and these features can be provided for teachers to refer and design appropriate scaffolding for proper students just when they need assistance.

References

1. Yeh, L.C., Lin, S.P.: The study of the Revised Taxonomy of Educational Objectives in Cognitive Domain. Journal of Education Research 105, 94–106 (2003)
2. Lee, K.C.: Taxonomy of Educational Objectives in Cognitive, Affective, and Psychomotor Domains: Applications in Assessment. High Education Press (2009)
3. Chen, P.S.: The Entity Relationship Model, Towards a Unified View of Data. ACM Transactions on Database Systems 1(1), 9–36 (1976)
4. Hong, C.-F.: Qualitative Chance Discovery – Extracting competitive advantages. Information Sciences 179, 1570–1583 (2009)
5. Halpin, T.: Bloesch A: Data Modeling in UML and ORM: A Comparison. Journal of Database Management 14(4), 4–14 (1999)
6. Moore, G.: A Inside the Tornado: Marketing Strategies from Silicon Valley's Cutting Edge. Harper Business, New York (1999)
7. Rogers, E.M.: Diffusion of Innovations. Free Press, New York (2003)
8. Connolly, T.M., Begg, C.E.: A Constructivist-Based Approach to Teaching Database Analysis and Design. Journal of Information Systems Education 7(1), 43–53 (2006)
9. Wood, C., Bruner, J.S., Ross, G.: The role of tutoring in problem solving. Journal of Child Psychology and Psychiatry 17, 89–100 (1976)

Geoparsing of Czech RSS News and Evaluation of Its Spatial Distribution

Jiří Horák[1], Pavel Belaj[1], Igor Ivan[1], Peter Nemec[2], Jiří Ardielli[1], and Jan Růžička[1]

[1] VSB Technical University of Ostrava, Institute of Geoinformatics, 17. listopadu 15,
70833 Ostrava-Poruba, Czech Republic
{jiri.horak,pavel.belaj,igor.ivan,jiri.ardielli,
jan.ruzicka}@vsb.cz
[2] Software602 a. s., Hornokrčská 15, 140 00 Praha 4, Czech Republic
pnemec@602.cz

Abstract. Geoparsing assigns geographic identifiers to textual words and phrases in documents. The specific problem is how to apply geoparsing in languages where changes of word termination occur. An appropriate method requires a flexible solution reflecting different strategies and priorities. Sixteen Czech RSS news channels were evaluated according to ten criteria. Three selected RSS channels were monitored for more than two years. The applied geoparsing included successive steps of different filters' application and utilized the generation of different grammatical cases for recognized entities. Various problems with geographical names are classified and documented. The quality assessment shows satisfactory results namely for identification of names in domiciles (94%). The pessimistic strategy is applied to analyze a geographical balance of news distribution. The results show significant differences between distribution of news in monitored channels and document a high concentration of cultural and national news in several locations.

Keywords: RSS, Geoparsing, Geocoding, News, Czech TV.

1 Introduction

In 2003 it was estimated that approximately 80% of all information is stored in textual documents, meaning in a form of unstructured data. Extraction and processing of useful information head towards transformation into structural data and classification or indexing text according to selected criteria, usually on the base of user's queries. A substantial support in these processes is found in Knowledge organization systems such as thesauruses, classification schemes, subject heading systems, and taxonomies in the frame of semantic web. New solutions require deeper linguistic analyses of text. It is necessary to deal with different forms of words, with spelling, to utilize contextual information, build a thematic thesaurus to employ relationships between terms (descriptors).

R. Katarzyniak et al. (Eds.): Semantic Methods, SCI 381, pp. 353–367.
springerlink.com   © Springer-Verlag Berlin Heidelberg 2011

Automated processing of natural languages is one of the most demanding tasks of artificial intelligence. Usually keyword analysis, syntactic-semantic analysis and recognition of named entities are utilized in the process of understanding natural language. Named entities are phrases that contain the names of persons, organizations, locations, times and quantities (Erik, 2002). Automatic recognition of named entities is a subject of wide research. Besides English the research enables processing of various other languages like German, French, Spanish, Swedish, Greek and Italian. From Slavonic languages we can find Polish (Piskorski 2004), Romanian (Cucerzan, Yarowsky 1999), Russian (Popov et al. 2004),) and Bulgarian (Da Silva et al. 2004). Processing is usually oriented to systematically label (tag) recognized named entities in the text (Chowdhury, 2003).

Geographical names represent one of the named entities applied in automated processing of natural languages. They use spatial relationships to improve results of recognition.

Principles of geocoding have been well known in geographical information systems for more than 20 years (Aronoff 1989). The original idea is to obtain geographical location of data by matching individual parts of postal addresses (for data) and interpolate the location from the range of house numbers. Geoparsing (or geotagging) deals with unstructured texts, and must overcome the uncertainty connected with writing geographical information into the text.

Geoparsing is still more disseminated in English speaking environments. Usually countries or selected cities are located. Geoparsing which recognizes more detail locations is still infrequent (Lee, Lee 2005).

The easiest way of monitoring news media is to explore its RSS channels because most media provide such internet channels. The news from different providers is usually organized according to sections, such as economy, national news, world news, sports, culture, regional news etc. The objective of the study is to identify a suitable (flexible and efficient) method of geoparsing Czech (or other languages where changes of word termination occur) geographical named entities from RSS media channels and to evaluate the geographical and temporal balance of news, because any deviation from a regular distribution may create bias of the media image of particular localities.

2   Geocoding, Geoparsing and RSS

Geocoding identifies occurrences of geographical entities in structured location references such as postal addresses, and assign appropriate geographical coordinates to them. The process usually compares sections of the given address with data in a reference layer (like set of municipalities, set of streets). Currently geocoding services are provided by many agencies such as Microsoft, Google, and Yahoo. Services obtain addresses for geocoding and return answers containing geographical coordinates, country, town, ZIP, street, number, etc. The answer may contain also ancillary information like accuracy or warnings.

Geoparsing assigns geographic identifiers to textual words and phrases in documents with unstructured content.

Geocoding utilizes only structured information, while geoparsing is based on the processing of unstructured text, where geographical information is usually disseminated in different places, combining quantitative values (like 10km from ..) and qualitative values (name of locations, administrative units).

Besides classic textual documents, other forms of media may also be geoparsed (i.e. audio content).

Basically geoparsing consists of 2 main steps:

- Entity extraction (separate character strings matching the search text),
- Geotagging (selection of appropriate geographical identifiers for identified phrases, solving different ambiguities).

The results of geoparsing may be joined with an original text to create new documents suitable for geographical application (Caldwell, 2009).

Geoparsing is utilized for various purposes.

Beaman and Barry (2003) describe the application of these methods to improve streamlining and automate acquisition of biogeographic data. The multi-step successive process includes pre-processing text for language, locale or project specific anomalies and phrase analysis. Text parsing and pattern matching involve detecting feature types (i.e. National Park, Island), place names, and their inter-relationships, calculation of geographic offsets (i.e. 2.5 km WNW of something) and recording.

Geoparsing of disease alerts (Keller et al. 2008) using the gazetteer approach and utilization of neural networks is applied to improve the georeferencing capability of the HealhMap server (www.healthmap.org).

Geoparsing of Czech texts is still quite rare because of problems with different words' terminations. A web prototype for geoparsing of the Liberec Region (http://geoparser.kraj-lbc.cz) was presented by Košková and Kafka (2009).

RSS (Really Simple Syndication) is a family of Web feed formats used to publish frequently updated works – such as blog entries, news headlines, audio, and video – in a standardized format. An RSS document (refers also as "feed", "web feed", or "channel") includes full or summarized text, plus metadata such as publishing dates and authorship. A web feed (or news feed) is a data format used for providing users with frequently updated content http://en.wikipedia.org/wiki/RSS_(file_format).

A RSS document usually contains headlines and text of news and publishes them on a unique URI address. By its internal form a RSS document is a subset of XML (RSS Specifications, 2011).

GeoRSS represents a geographical extension of RSS. A GeoRSS document is created by adding spatial information to RSS. The main advantages of GeoRSS can be seen in an effective conceptual solution of geoparsing and probably more accurate localization. Usually it is applied for special events like floods, wildfires or earthquakes, where an accurate geographical location is crucial.

3 Evaluation and Selection of RSS Channels

The objective of the study is to evaluate a geographical distribution of Czech news media. First, the extended list of various available RSS channels providing media news was created. Next, an evaluation and a selection of RSS channels were required using appropriate qualitative criteria. It is hardly to find any recommendation concerning relevant evaluation of RSS channels' quality; usually a maximum number of available channels (i.e. all collected RSS feeds (Sia, Cho 2007)) or selection based on popularity are applied (i.e. popularity in Gmail's "web clips" and Bloglines' „most popular feeds," (Jun, Ahamad 2006)). We suggest an own set of criteria.

All channels were evaluated using the following criteria: validity of RSS, change of address or structure, number of news items in the channel, average number of news items per day, content, domicile, special setting, GeoRSS, quality of formatting and metadata. Every criterion was ranked on a scale of 0 to 4 according to the level of satisfaction.

Explanations of criteria follow:

• Validity of RSS

We applied „Feed Validation Service" provided by W3C (http://validator.w3.org/feed/). A valid RSS channel is evaluated with 4 (for small recommendations) or 3 (for large recommendations). Invalid channels with small number of errors are evaluated with 2 or 1. Occurrences of important errors limit the evaluation to 0.

• Change of address or structure

Unfortunately the RSS specification still does not provide tools how to announce coming changes of the address. In the test period we recognized the change of address in the case of one channel (denik_kultura). The change of address was accompanied by the change of structure (seven new sections). In several other cases we recognized the changes of structure. I.e. channels of CT24 have extended their contents with multimedia elements.

• Number of news items in the channel

The criterion reflects the required situation with the optimal number of news items (not too large but also not too small). Large amounts of news items cause a long processing time, occurrence of out-of-date news or server overloading. Too small of a number of news items may demand frequent refreshing of content and increases the risk of data loss (with inadequate frequency of reading by consumers). Analysis of 245 news channels (Jun, Ahamad 2006) indicates the average counts for the news channels range from 10 to 126 and 65% channels have a fixed entry count. We recommend an optimal range of 20 to 50 news items available in the channel. Lower or higher numbers are penalized by lower evaluation of this criterion.

• Average number of news items per day

Higher average number of news items represents a better channel providing more information. The number of news items oscillates, thus the average number calculated for a long time testing interval is recommended. The evaluation should take into account the natural differences in number of news items according to the thematic

dedication of the channel. Usually channels oriented to travelling or culture provides significantly less news than regional or national channels.
• Content
The criterion evaluates the structure and length of the element <description>. This element contains a part (usually the first part) of the news. It is often called „perex". The evaluation covers a way of perex publishing, length of the perex and the delivery of the whole content of the news (some channels like CT24 provide the full content of news in the element <content:encoded>).
• Domicile
A domicile usually represents a place of residence; in our case it means a basic location of the news. The presence of domicile facilitates and refines geoparsing of the news. Channels may also differ in quality of the domicile indication.
• Special settings
The criterion reflects various necessary modifications which have to be done before testing. I.e. CT24 channels require sending security strings. Such modifications complicate utilization and thus they are penalized in the evaluation.
• geoRSS
The best situation for the georeferencing of news is when geoparsing is superfluous due to a presence of direct geotaggs in the text. The text may contain tags <georss:point> for point localization of object/events or <georss: line> for line segments (i.e. roads with traffic jam). The criterion evaluates a presence of such geotags in the text and also their reliability (in some cases news contains geotags only for part of localities situated in the text).
• Quality of formatting
The criterion evaluates a quality of formatting rules inside the text. Also occurrences of various special characters (like quotation marks, apostrophes, ampersands, hyphens) which complicate text processing are taken into account.
• metadata
The criterion addresses a presence and quality of metadata. Especially elements like <ttl> (minimal interval for update) or <lastBuildDate> (date and time of last updating) are valuable for processing.

A total evaluation uses multicriteria evaluation. The weight assigned to every criterion provides information about the significance of the criterion in the total evaluation. To find out the appropriate value of weights in the system of 10 criteria, analytic hierarchy process (AHP) was applied. This technique was described by Saaty (1994) and Saaty and Vargas (2001). A final suitability of each RSS channel is expressed by weighted linear combination (WLC). WLC combines criteria scores and criteria weights. The usability of RSS channels (tab. 1) varies between 2.005 („novinky-cestovani" - travelling) and 3.645 (CT24 national news).

According results we decided to select 3 channels for further processing: CT24_domaci (CT24 national news), ct24_regionalni (regional) and ct24_kultura (cultural).

Table 1. Criteria scores and total suitability for channels (Nemec 2010).

RSS channel	Validity of RSS	Change of address or structure	No. of news items in the channel	Average number of news per day	Content	Domicile	Special settings	GeoRSS	Quality of formatting	Metadata	Suitability
cn_cestovani	1	4	4	1	3	3	4	0	2	2	2.245
cn_domov	1	4	4	4	3	3	4	0	2	2	2.703
ct24_cestovani	4	3	4	0	4	4	0	2	3	2	3.035
ct24_domaci	4	3	4	4	4	4	0	2	3	2	3.645
ct24_doprava	4	3	4	2	4	4	0	2	3	2	3.340
ct24_kultura	4	3	4	1	4	4	0	0	3	2	3.102
ct24_regionalni	4	3	4	3	4	4	0	2	3	2	3.493
ct24_sport	3	3	4	3	4	4	0	2	3	2	3.278
denik_kultura	2	1	4	4	3	3	4	0	4	2	2.815
denik_domov	2	4	4	4	3	3	4	0	4	2	2.979
mf_zpravodaj	3	4	3	4	1	2	4	0	3	4	2.604
novinky_cestovani	4	4	1	0	3	0	4	0	4	2	2.005
novinky_domaci	4	4	1	4	3	0	4	0	4	2	2.616
novinky_krimi	4	4	1	4	3	0	4	0	4	2	2.616
novinky_kultura	4	4	1	4	3	0	4	0	4	2	2.616
novinky_sport	4	4	4	4	3	0	4	0	4	2	2.869

4   Establishing a List of Geographical Entities

The applied system of geoparsing can be seen as a member of hand-crafted grammar-based systems with spatial extension.

Before geoparsing it is necessary to select an appropriate list of geographical entities, and the named entities to be searched in news media. According to the objective of the study, the official list of municipalities in the Czech Republic is selected (Registry of territorial units from the Czech Statistical Office). The level of municipalities seems to be adequate for both an initial estimation of news volume and anticipated frequency of individual location.

The initial list of municipal names has to be pre-processed before geoparsing in the following steps:

• First, duplicate names are deleted. The modified list contains only unique geographical names. During geoparsing it is possible to assess probabilities of linkage between the found geographical name and municipalities with duplicate names and to assign the news to the more appropriate municipality.

- Next, parentheses are eliminated.
- The next step is to delete (or substitute) roman numerals from the municipal names (11 municipalities).
- Optionally, hyphens are substituted by spaces (57 municipalities).

The final number of municipalities dropped from 6249 to 5328 (by approx. 15%).

Next, the process has to solve grammatical cases and different word forms.

The Czech language nouns are declined (7 grammatical cases) which causes changes of word termination. To find out geographical entities in natural language it is necessary to identify grammatical variants in the news. One of possible solutions is to extend the existing list of municipal names with variants given by all potential grammatical cases.

A PHP script generates all variants based on the lemat (root of the word) and cases termination (i.e.: Ostrava, Ostravy, Ostravě, Ostravou). These grammatical variants are added to the list of municipal names. All variants are equal in geoparsing. Nevertheless, these extensions cause also data redundancy and increase the risk of conflicts.

Unfortunately, the current script is unable to deal with multi-word names (21% of all municipal names in the Czech Republic). Variants of termination for such cases have to be prepared manually.

5   Problems of Georeferencing and a Strategy Selection

Problems in georeferencing may be classified into the following groups:

- "Overlays of municipal names" represents mistakes when one municipal name is a part of another municipal name (one name "covers" or "overlays" the second one; i.e. "Hradec" is a part of "Hradec Králové", "Opatovice" is included in „Opatovice nad Labem") including matching a municipal name in some grammatical case termination (i.e. "Karlov" and "Karlovy Vary" when "Karlovy" is one of possible Karlov's case terminations). Possible solutions include:

    - The overlays should be eliminated by subsequent operations when multi-word names of municipalities are identified, then they are removed and afterwards single words are searched.
    - Assign a lower probability for matching to one-word names

- "Overlays of municipal names with other geographical (non-municipal) names." Examples include wrongly matched river names ("Dyje" is both a river name and a municipal name), municipal parts ("Vítkov" is a part of Prague but also a name of small municipality in Silesia), street names ("Výsluní" is a street name in Pilsen city as well as a municipal name in the Chomutov district), area names (ski area „Lipno" has the same name as a municipality „Lipno nad Vltavou"). It is impossible to find universal solutions for these problems. Basically, a list of other geographical names (i.e. "geonames", "point of interests" and similar products may be used) may help to discover potential conflicts. Also contextual information may significantly decreases the risk of mismatching. Finally, it is useful to deal with a different occurrence probability for searched strings.

• "Similarity to natural language" refers to issues with names of municipalities identical to common words of natural language. The Czech word "Příčina" (means a "reason") is the same as the municipal name "Příčina". The way of solution may be similar to previous type of mistakes.

• "Joint municipal names" represents issues with artificial combination of municipal names created by authors to shorten the text. For example "Mariánské a Františkovy Lázně" is created by combination of „Mariánské Lázně" (Marianske Spa) and „Františkovy Lázně" (Frantiskovy Spa). The usual geoparsing identifies only the last municipal name. A possible solution may utilize a list of such joint names with decoding and substitution by individual names.

• "Inaccurate municipal names " means using municipal names which do not match the official names. Nevertheless they are used in common language, thus we cannot consider them as errors. "Věstonice" is not an official name; there are two municipalities instead of one: „Horní Věstonice" (upper Vestonice) and „Dolní Věstonice" (lower Vestonice).

• The last group of problems represents various spelling errors.

It is clear that there are many reasons why the searched geographical entity is wrongly assigned to the news. It is necessary to select an appropriate strategy for geoparsing to eliminate undesirable impacts. Due to uncertainty it is possible to find a wide spectrum of various strategies between two extreme strategies which can be called the "optimistic" and the "pessimistic" strategy:

• The optimistic strategy: Users wish to find a maximal number of occurrences even though some of them may be not appropriate (mistakes in georeferencing). The risk of missing news assigned to a given municipality is minimized while the risk of false positive occurrence is maximal. Mistakes represent errors of commission. Such a strategy should be useful for a mayor interested in all possible news about his/her municipality (and does not mind reading some superfluous news).

• The pessimistic strategy: Users wish to find only exact occurrences. Thus only a perfect matching is applied. The risk of a false positive occurrence is minimized while the risk of missing news assigned to a given municipality is maximal. Mistakes represent errors of omission. The strategy may be appropriate for analytical purposes when we are interested i.e. in correlation between a size of population and the number of news items (and deviation from this relationship).

The final selected strategy may be situated between these extreme strategies and may also include other approaches such as:

• Different probabilities of conflict entities – higher probability is given to larger municipality, to more frequent names etc.

• Different priorities to different group of names (entities) – i.e. spa locations may have a higher priority.

6 Recording and Geoparsing of Media News

RSS channels of CT24 publish news according to a structural element called item. Its structure is explained bellow.
The <item> element contains following 5 child elements:

- <title> the headline of the story,
- <link> the hyperlink to the item,
- <guid> a unique identifier for the item,
- <pubDate> the day of issue this item,
- <description> - the abbreviated version of the original story. This element is divided into two separated parts:
  - Domicile – specifies the location of news,
  - Content - the full story is included using unstructured text. The geographical references may also be included.

The domicile can be written in different variants and this fact complicates an automatic separation of the domicile from the content.

The RSS channels of CT24 are refreshed approximately every hour by publishers. Old news is deleted and fresh news is added. A continuous analysis of RSS channels and the recording of all new items to a database table is supported by the second PHP script. All news are segmented into detail elements and stored into database MS SQL server 2008.

Geoparsing works with reference data and the table of recorded news. It assigns locations from the reference data (in this case it is a list of municipal names) to all news. This task is completed by a third PHP script. Any news description is segmented into words, which are compared to every case termination generated from the reference data table.

As it was documented above, geoparsing is negatively influenced by an ambiguity of municipal names, which leads to assign more candidates for one name. The last PHP script searches for any additional information in the content to improve the success rate. The procedure utilizes supplemental information like the name of a district (i.e. Karlovarsko). If no supplemental information is available, the script takes into account other municipalities recognized and recorded for the news (item). It calculates Euclidean distances and selects the candidate closest to other municipalities (Nemec, Horák, 2009).

After geoparsing, the located geographical entities were transformed into geometry data types (currently point and polygon data types) into a spatial database. The spatial database system is a standard database system with the spatial data types that allow effective use of spatial indexing and spatial operations (Güting, 1994). The spatial database will enable us to effectively manipulate with spatial data, analyze spatial relationships, use spatial operators and create simple maps. Finally spatial operators (especially topological relationships) and features will be applied to refine the location of media news.

7 Quality Assessment

Geoparsing and auxiliary operations (namely pre-processing) create a sequential process where each step brings a certain simplification and a possible loss of some names which can cause ambiguous localization. The original number of municipalities is reduced by 15%. Even though the processing of remaining municipal names is not errorless and still a possibility to confuse places exists.

The result of automated processing is compared with an output of manual processing of a randomly selected sample. The rate of successful localization is 95% records geolocated by domicile, which is quite similar to processing of English texts which can be geoparsed with a success rate to 94% (Košková and Kafka 2009). The recorded news contains in 36 cases also a locality in the text. The successful localization from the text was recognized in 25 news' cases which represents 69%. It is possible to conclude that the geocoding based on domicile is highly successful; unfortunately the geocoding in the text is not yet fully satisfactory.

8 Spatial Distribution of News

The three RSS channels of CT24 were recorded and processed for the period from 1 January 2007 to 20 May 2009. The results of georeferencing number of regional news are presented in two statistical maps. The first map (Fig. 1) shows the number of CT24 regional news by municipalities where geoparsing was applied to all content of news. The second map (Fig. 2) depicts the same collection of news but the municipal name was searched only in the domicile. When both figures are compared, there is evident identical or a very similar number of news items for large cities in the Czech Republic (mainly in the case of all 14 centers of NUTS3 Regions). Many smaller municipalities disappeared in the second map. This fact indicates a low probability of recognition of names of small municipalities in the domicile and so the full text of news must be processed.

The most occurring names of cities are also the most populated cities and often the centres of NUTS3 Regions – Prague, Karlovy Vary, Pardubice, Brno etc. The mutual close relationship between the number of regional news items and the size of population was proved (Pearson Correlation 0.866). It is clear that not only Pardubice but mainly Karlovy Vary show an outstanding number of news items according to their population size. A different situation is in Ostrava (third biggest city), Olomouc and Zlín. These Moravian cities (also regional centers) have smaller number of regional news than other regional centers (also the less populated) in the Bohemia (west part of the Czech Republic). This fact could indicate some heterogeneity in spatial distribution (imbalance between west and east part). The strong position of Karlovy Vary is caused by its cultural (discussed later) and tourism (balneology and mineral springs) importance and by the strong minority of the Russian population which is often discussed in the news.

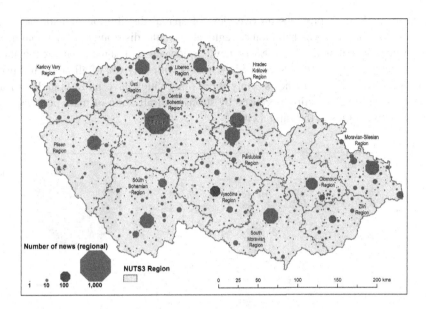

Fig. 1. Number of news items from CT24 regional RSS channel (1.1.2007-20.5.2009, geoparsing of full text).

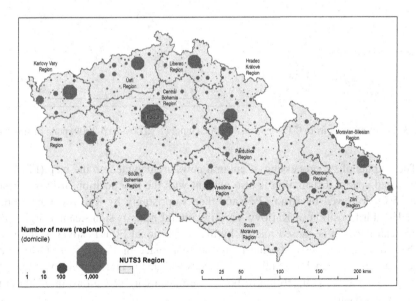

Fig. 2. Count of news from CT24 regional RSS channel (1.1.2007-20.5.2009, geoparsing in domicile).

The map (Fig. 3) provides an overview of spatial distribution of national news (CT24). Contrary to the distribution of regional news the distribution of national news is significantly unbalanced – 70% of news is located in Prague. The second most often occurring city is Brno (second biggest city) with only 6% of all news, followed by Ostrava with 2% of all news. Only three other cities have more than 1% of all national news. This situation is caused by strong centralization of main political, and many other activities in the capital city.

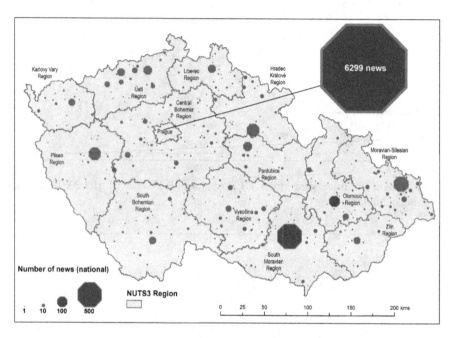

Fig. 3. Count of news from CT24 national RSS channel (1.1.2007-20.5.2009, pars. of full text).

The outputs of geoprocessing of cultural news are depicted on the map (Fig.4). An unbalanced spatial distribution is also evident with the key position of Prague (more than 75%), followed by Brno (2%) and then surprisingly by Karlovy Vary (almost 2%). The third position of Karlovy Vary in cultural news is significantly higher than in national news. It is caused mainly by the International Film Festival which is one of the oldest of film festivals (the first year took place in 1946). Only Ostrava and Pilsen have more than 1% of all cultural news and this is due to their large size, but also by their mutual competition to be the European City of Culture in 2015 (in the end won by Pilsen).

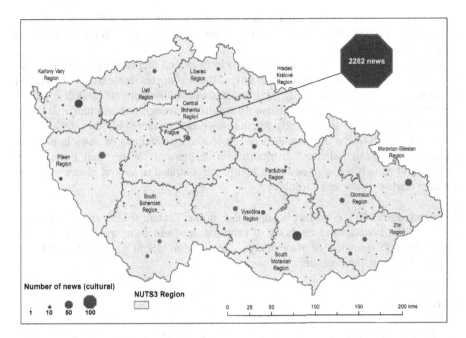

Fig. 4. Count of news from CT24 cultural RSS channel (1.1.2007-20.5.2009, pars. of full text).

9 Conclusions

Geoparsing represents a main method for identification of geographical named entities in unstructured textual documents. The combination of geoparsing with other suitable techniques (mainly pre-processing) was applied to recognize geographical location of media news and evaluate issues in performing such tasks. Media news is explored from RSS news channels.

Sixteen Czech RSS news channels were analyzed and evaluated. The list of criteria includes validity of RSS, change of address or structure, number of news items in the channel, average number of news items per day, content, domicile, special setting, GeoRSS, quality of formatting and metadata. Each channel obtains value 0-4 for every criterion. Weights for every criterion were assigned using Saaty AHP process. A final evaluation based on multi-linear weighted combination helped to select three representative channels – CT24 regional news, CT24 national news and CT24 cultural news. These channels were scanned between 1.1.2007 and 20.5.2009 using a PHP script. News items were segmented and stored to the spatial database.

Geoparsing of news tries to identify names of Czech municipalities. Pre-processing of the municipal register includes elimination of duplicate names, removing of special characters like parentheses, hyphens and roman numerals. Next, the second PHP script generates all variants of municipal names using cases terminations.

The analyses of various matching shows typical problems of georeferencing such geographical names - overlays of municipal names, overlays of municipal names with other geographical (non-municipal) names, similarities to natural language, joint municipal names, inaccurate municipal names and spelling errors. Possible solutions for individual types of problems help to eliminate their impacts. Basic strategies of geoparsing are explained. The process of matching words (phrases) in news and prepared list of names (with variants) is provided by a third PHP script. It also utilizes contextual information, occurrence of other names in the text and Euclidean distances to candidate names. The quality assessment shows satisfactory results for recognition of municipal names in domicile (94%) and significantly worse results for the content of news (69%).

The results show the differences between spatial distributions of news in different channels. The most balanced distribution is recognized for regional news, while cultural and namely national news are highly concentrated in several places – Prague, but also Karlovy Vary and Pardubice.

Acknowledgment. The project is supported by the Technical University of Ostrava SP2011/131 "Localization methods of criminal acts and their spatial aspects" and SP/2010188 "Evaluation of data analysis options in MS SQL SERVER 2008 by Applied Business Intelligence".

References

1. Aronoff, S.: Geographic Information Systems: A Management Perspective. WDL Publicatios, Ottawa (1989)
2. Beaman, R.S., Conn, B.J.: Automated geoparsing and georeferencing of Malesian collection locality data. Telopea. 10(1), 43–52 (2003)
3. Caldwell, D.: Geoparsing Maps the Future of Text Documents,
   http://www.directionsmag.com/article.php?article_id=3268
4. Chowdhury, G.G.: Natural language processing. Annual Review of Information Science and Technology 37(1), 51–89 (2003)
5. Cucerzan, S., Yarowsky, D.: Language Independent Named Entity Recognition Combining Morphological and Contextual Evidence. In: Proc. Joint Sigdat Conference on Empirical Methods in Natural Language Processing and Very Large Corpora., pp. 90–99. The Association for Computational Linguistics, Stroudsburg (1999)
6. Da Silva, J.F., Kozareva, Z., Lopes, G.P.: Cluster Analysis and Classification of Named Entities. In: Proc. Conference on Language Resources and Evaluation, pp. 321–324. LREC, Lisbon (2004)
7. Erik, F.T.K.S.: Introduction to the CoNLL-2002 Shared Task: Language-Independent Named Entity Recognition. In: Proc. of CoNLL 2002, Taipei, Taiwan, pp. 155–158 (2002)
8. Güting, R.H.: An Introduction to Spatial Database Systems. VLDB Journal 3(4), 357–399 (1994)
9. Jun, S., Ahamad, M.: FeedEx: Collaborative exchange of news feeds. In: Proc. of the 15th International Conference on World Wide Web, pp. 113–122. ACM, New York (2006)
10. Keller, M., Brownstein, J. S., Freifeld, C. C.: Expanding a Gazetteer-Based Approach for Geo-Parsing Disease Alerts (2008),
    http://prior-knowledge-language-ws.wdfiles.com/
    local-files/start/keller_slides.pdf

11. Košková, I., Kafka, Š.: Geoparser – automatické vyhledávání geografických lokalizací v textu. In: Proceedings of Geoinformační Infrastruktury Pro Praxi., p. 100. MSD, Brno (2009)
12. Lee, S., Lee, G.G.: Heuristic Methods for Reducing Errors of Geographic Named Entities Learned by Bootstrapping. In: Dale, R., Wong, K.-F., Su, J., Kwong, O.Y. (eds.) IJCNLP 2005. LNCS (LNAI), vol. 3651, pp. 658–669. Springer, Heidelberg (2005)
13. Nemec, P.: Evaluation of trends and structures of news using geoparsing. Diploma Thesis, p. 83. VŠB-TU Ostrava (2010)
14. Nemec, P., Horák, J.: The Geographical Balance of Regional News of Czech TV CT24. In: Proc. of international Symposium GIS Ostrava 2009, p. 10. TANGER, Ostrava (2009)
15. Piskorski, J.: Extraction of Polish Named-Entities. In: Proc. Conference on Language Resources and Evaluation, pp. 313–316. LREC, Lisbonne (2004)
16. Popov, B., Kirilov, A., Maynard, D., Manov, D.: Creation of reusable components and language resources for Named Entity Recognition in Russian. In: Proc. Conference on Language Resources and Evaluation, pp. 309–312. LREC, Lisbonne (2004)
17. RSS Specifications, http://www.rss-specifications.com
18. Saaty, L.T.: Fundamentals of decision making and priority theory with analytic hierarchy process. RWS publications, Pittsburgh (1994)
19. Saaty, L.T., Vargas, L.G.: Models, methods, concepts, and applications of the analytic hierarchy process. Kluwer Academic, Boston (2001)
20. Sia, K.C., Cho, J.: Efficient monitoring algorithm for fast news alerts. IEEE Transactions on Knowledge and Data Engineering 19(7), 950–961 (2007)

Author Index

Adamski, Michał 3
Ardielli, Jiří 353

Belaj, Pavel 353
Bobek, Szymon 265
Burdka, Łukasz 3

Chen, Ching-Fan 341
Chiu, Tzu-Fu 183, 199
Chiu, Yu-Ting 183
Chu, Shuchuan 93

Ganapathy, Subhashini 311
Gecow, Andrzej 159

Ha, Quang-Thuy 23
Holban, Ştefan 103
Hong, Chao-Fu 183, 199, 215
Horák, Jiří 353
Hsu, Chia-Ling 171

Ivan, Igor 353

Jaksik, Roman 135

Kaczor, Krzysztof 57
Katarzyniak, Radosław P. 3, 147
Kluska-Nawarecka, Stanisława 13
Kluza, Krzysztof 57
Korytkowski, Przemysław 35
Krawczyk, Bartosz 115
Krejcar, Ondrej 67, 289
Kusztelak, Grzegorz 127
Kusztina, Emma 47

Lee, Jer-Haur 199
Li, Yan-Ru 231, 277
Lihu, Andrei 103
Lin, Min-Huei 341
Lin, Mu-Hua 199, 215
Lin, Woo-Tsong 215
Lin, Yuh-Chang 199, 215
Luu, Cong-To 23
Łysik, Łukasz 57

Maleki, Hamed 81
Marczyk, Michał 135
Meissner, Adam 253
Milczarski, Piotr 329

Nalepa, Grzegorz Jacek 57, 265
Nemec, Peter 353
Nguyen, Ngoc Thanh 147
Nowostawski, Mariusz 159

Pałka, Dariusz 301
Pan, Jengshyang 93
Park, Namje 319
Pham, Huyen-Trang 23
Polańska, Joanna 135
Polański, Andrzej 135
Prabhala, Sasanka 311

Regulski, Krzysztof 13
Różewski, Przemysław 47
Ruhl, Dawn Michele 241
Růžička, Jan 353

Sarfaraz, Amir Homayoun   81
Skorupa, Grzegorz   3
Sobecki, Janusz   147
Stańdo, Jacek   127
Sun, Chaoli   93

Tadeusiewicz, Ryszard   47
Tsai, Wei-Chen   277

Vu, Tien-Thanh   23

Wang, Ai-Ling   241
Wilk-Kołodziejczyk, Dorota   13
Wiśniewski, Tomasz   35

Yang, Hsiao-Fang   215

Zachara, Marek   301
Zaikin, Oleg   47
Zeng, Jianchao   93
Zhang, Yunqiang   93